AF322760

Animal Nutrition with Transgenic Plants

CABI Biotechnology Series

Biotechnology, in particular the use of transgenic organisms, has a wide range of applications in areas such as agriculture, forestry, food and health. Evidence has shown that work in biotechnology could have a major impact on the development of plants and animals that are able to resist stresses and diseases, thereby increasing food security. As well as the potential to produce pharmaceuticals in plants and provide foods that are nutritionally enhanced, genetically modified organisms can also be used in cleaning up pollution and contamination. However, the application of biotechnology has raised concerns about biosafety, and it is vital to ensure that genetically modified organisms do not pose any risks to health or the environment. To understand the full potential of biotechnology and the issues that relate to it, scientists need access to information that not only provides an overview of and background to the field but also keeps them up to date with the latest research findings.

This series, which extends the scope of CABI's successful Biotechnology in Agriculture Series, addresses all topics relating to biotechnology including transgenic organisms, molecular analysis techniques, molecular pharming, *in vitro* culture, public opinion, economics, development and biosafety. Aimed at researchers, upper-level students and policy makers, titles in the series provide international coverage of topics related to biotechnology, and include both current data and discussions regarding future research perspectives.

Titles Available

1. Animal Nutrition with Transgenic Plants
   Edited by G. Flachowsky

Animal Nutrition with Transgenic Plants

Edited by

Gerhard Flachowsky

_Senior Visiting Scientist, Institute of Animal Nutrition,
Friedrich-Loeffler-Institute (FLI), Federal Research Institute for
Animal Health, Braunschweig, Germany_

CABI is a trading name of CAB International

CABI
Nosworthy Way
Wallingford
Oxfordshire OX10 8DE
UK

CABI
38 Chauncey Street
Suite 1002
Boston, MA 02111
USA

Tel: +44 (0)1491 832111
Fax: +44 (0)1491 833508
E-mail: info@cabi.org
Website: www.cabi.org

T: +1 800 552 3083 (toll free)
T: +1 (0)617 395 4051
E-mail: cabi-nao@cabi.org

A catalogue record for this book is available from the British Library, London, UK.

Library of Congress Cataloging-in-Publication Data

Animal nutrition with transgenic plants / edited by Gerhard Flachowsky.
    pages ; cm -- (CABI biotechnology series ; 1)
  Includes bibliographical references and index.
  ISBN 978-1-78064-176-8 (alk. paper)
  1. Food animals--Nutrition. 2. Food animals--Feeding and feeds--Risk assessment. 3. Transgenic plants--Risk assessment. I. Flachowsky, G., editor of compilation. II. C.A.B. International, issuing body. III. Series: CABI biotechnology series ; 1.
  [DNLM: 1. Animal Nutritional Physiological Phenomena. 2. Plants, Genetically Modified. 3. Animal Feed. 4. Animals, Domestic. 5. Food Supply. 6. Food, Genetically Modified. SF 95]

  SF95.A636 2014
  632'.8--dc23

                          2013024394

ISBN-13: 978 1 78064 176 8

Commissioning editor: David Hemming
Editorial assistant: Alexandra Lainsbury
Production editor: Lauren Povey

Typeset by Columns Design XML Ltd, Reading, UK.
Printed and bound in the UK by Bertforts Information Press Ltd.

Contents

**3   Guidance Documents for Nutritional and Safety Assessment of Feeds
from GM Plants**   **30**
*Yi Liu, Anna Christodoulidou, Antonio Fernandez Dumont, Andrea Germini and
Claudia Paoletti*

**4   Compositional Analysis for Nutritional and Safety Assessment of
Feeds from GM Plants**   **50**
Gijs Kleter and Esther Kok

Contributors

Berheim, Aude, INSERM U 1001 (Robustness and Evolvability of Life) Université Paris Descartes Sorbonne Paris Cité, Faculté de medicine, 24 rue de Faubourg Saint-Jacques, 75014 Paris, France. E-mail: aude.bernheim@gmail.com

Christodoulidou, Anna, European Food Safety Authority (EFSA), GMO Unit, Via Carlo Magno 1A, IT-43126 Parma, Italy. E-mail: anna.christodoulidou@efsa.europa.eu

De Loose, Marc, Institute for Agricultural and Fisheries Research (ILVO), Burgemeester Van Gansberghelaan 115 bus 2, 9820 Merelbeke, Belgium, and Department of Plant Biotechnology and Bioinformatics, Ghent University. E-mail: marc.deloose@ilvo.vlaanderen.be

Depicker, Ann, Department of Plant Biotechnology and Bioinformatics, Ghent University, and Department of Plant Systems Biology, VIB, Technologiepark 927 Ghent 9052, Belgium. E-mail: anpic@psb.vib-ugent.be

Dillen, Koen, European Commission – Joint Research Centre – Institute for Prospective Technological Studies, Edificio Expo, c/Inca Garcilaso 3, 41092 Seville, Spain. E-mail: koen.dillen@ec.europa.eu

Dumont, Antonio Fernandez, European Food Safety Authority (EFSA), GMO Unit, Via Carlo Magno 1A, IT-43126 Parma, Italy. E-mail: antonio.fernandezdumont@efsa.europa.eu

Einspanier, Ralf, Freie Universität Berlin, Institute of Veterinary Biochemistry, Oertsweg 19b, 14163 Berlin, Germany. E-mail: ralf.einspanier@fu-berlin.de

Flachowsky, Gerhard, Senior Visiting Scientist, Institute of Animal Nutrition, Friedrich-Loeffler-Institute (FLI), Federal Research Institute for Animal Health, Bundesallee 50, 38116 Braunschweig, Germany. E-mail: gerhard.flachowsky@t-online.de

Germini, Andrea, European Food Safety Authority (EFSA), GMO Unit, Via Carlo Magno 1A, IT-43126 Parma, Italy. E-mail: andrea.germini@efsa.europa.eu

Kleter, Gijs, RIKILT Wageningen UR, Akkermaalsbos 2, NL-6708WB Wageningen, Netherlands. E-mail: gijs.kleter@wur.nl

Kok, Esther, RIKILT Wageningen UR, Akkermaalsbos 2, NL-6708WB Wageningen, Netherlands. E-mail: esther.kok@wur.nl

Kuntz, Marcel, Cnrs/CEA/Inra/Université Joseph Fourier, Grenoble, Laboratory Physiologie cellulaire végétale, 17 rue des Martyrs, F-38054 Grenoble cedex 9, France. E-mail: kuntz@ujf-grenoble.fr

Liu, Ranran, Institute of Animal Sciences, Chinese Academy of Agricultural Sciences, P.R. China. E-mail: liuranran112@126.com

Liu, Yi, European Food Safety Authority (EFSA), GMO Unit, Via Carlo Magno 1A, IT-43126 Parma, Italy. E-mail: yi.liu@efsa.europa.eu

Paoletti, Claudia, European Food Safety Authority (EFSA), GMO Unit, Via Carlo Magno 1A, IT-43126 Parma, Italy. E-mail: claudia.paoletti@efsa.europa.eu

Paris, Alain, INRA, met@risk, AgroParisTech, 16 rue Claude Bernard, F-75231 Paris cedex 05, France. E-mail: aparis@paris.inra.fr

Pascal, Gérard, INRA, Le Breuil, 63220 Saint Alyre d'Arlanc, France. E-mail: gerard.pascal@paris.inra.fr

Qaim, Matin, Georg-August-University of Goettingen, Department of Agricultural Economics and Rural Development, Platz der Goettinger Sieben 5, 37073 Goettingen, Germany. E-mail: mqaim@uni-goettingen.de

Ricroch, Agnes E., Université Paris-Sud, CNRS, AgroParisTech, UMR 8079, Labortoire d'Ecologie, Systematique et Evolution, Bât, 360, F-91405 Orsay cedex, France. E-mail: agnes.ricroch@u-psud.fr

Rodríguez-Cerezo, Emilio, European Commission – Joint Research Centre – Institute for Prospective Technological Studies, Edificio Expo, c/Inca Garcilaso 3, 41092 Seville, Spain. E-mail: emilio.rodriguez-cerezo@ec.europa.eu

Scholderer, Joachim, Aarhus University, School of Business and Social Sciences, Department of Business Administration, Bartholins Allé 10, DK-8000 Aarhus C, Denmark. E-mail: sch@asb.dk

Snell, Chelsea, School of Biosciences, University of Nottingham, Sutton Bonington Campus, Loughborough, Leicestershire, LE12 5RD, UK. E-mail: stycres@nottingham.ac.uk

Tillie, Pascal, European Commission – Joint Research Centre – Institute for Prospective Technological Studies, Edificio Expo, c/Inca Garcilaso 3, 41092 Seville, Spain. E-mail: pascal.tillie@ec.europa.eu

Verbeke, Wim, Ghent University, Faculty of Bio-Science Engineering, Department of Agricultural Economics, Coupure Links 653, B-9000 Ghent, Belgium. E-mail: wim.verbeke@ugent.be

Wen, Jie, Institute of Animal Sciences, Chinese Academy of Agricultural Sciences, P.R. China. E-mail: wenj@iascaas.net.cn

Wright, Atte von, University of Eastern Finland, Institute of Public Health and Clinical Nutrition, Finland. E-mail: atte.vonwright@uef.fi

Preface

In 1962, Rachel Carson published *Silent Spring* and she described the situation of chemical plant protection in the USA and the 'rest of the world', with dramatic consequences for insects, fish, birds, mammals and mankind. On page 289, she concluded that:

> High hopes now attend tests with *Bacillus thuringiensis* – originally discovered in Germany in 1911 in the province of Thuringia, where it was found to cause a fatal septicaemia in the larvae of the flour moth. This bacterium actually kills by poisoning rather than by disease. Within its vegetative rods there are formed, along with spores, peculiar crystals composed of a protein substance highly toxic to certain insects, especially to the larvae of the mothlike lepidopteras. Shortly after eating foliage coated with this toxin the larva suffers paralysis, stops feeding, and soon dies. For practical purposes, the fact that feeding is interrupted promptly is of course an enormous advantage, for crop damage stops almost as soon as the pathogen is applied...
>
> (Carson, 1962)

About 30 years later, genetically engineered maize plants were able to express this protein and protect themselves against the European corn borer and further lepidopteras.

Apart from protection against insects and various chemicals such as herbicides and insecticides, genetically engineered plants are also able to use naturally limited resources such as water, minerals, fuel, etc., more efficiently and to change their composition and their nutritive value in the desired directions.

In the middle of the last century, Norman Borlaug developed high-yielding varieties of wheat and other plant species and contributed to overcoming the hunger and starvation of large regions in Asia and South America. In 1970, the Nobel Committee awarded Dr Borlaug the Nobel Peace Prize for the combination of his scientific and humanitarian achievements, resulting in the so-called (first) 'Green Revolution'. Norman Borlaug stated in his Nobel lecture, 'If you desire peace, cultivate justice, but at the same time cultivate the fields to produce more bread; otherwise there will be no peace.'

During the past few years, the global population has increased to 7 billion and in 2050 will probably reach more than 9 billion people. How to feed a growing population is a very old question, but now, it is present with a higher explosive effect.

Plant breeding should be considered as the starting point of the whole food chain. Therefore, high and stable yields of plants with low external inputs of non-renewable resources, low emissions of gases with greenhouse potential during cultivation, high resistance against biotic and abiotic stressors (including adaptation to potential climate change), a low concentration of undesirable substances in the plants and an increase of the nutritive value determining components of plants are real challenges for plant breeders in the future.

It is possible to realize these objectives by traditional plant breeding, but genetic engineering may be faster and can contribute substantially to achieving these goals. This objective of plant breeding is of global significance and need for all those involved in this topic.

Public research also should contribute towards solving the problem of feeding the growing population, and public–private as well as public–public partnerships should be formed with the mission of reaching the set goals in the coming decades.

The objective of this book is to summarize the present stage of knowledge in the cultivation of genetically modified (GM) plants for animal feed, and feeding of such materials to food-producing animals, on the basis of scientific studies published in peer-reviewed journals. The authors of the book have covered free of ideology, policy, public mainstream and commercial interests, with the following topics in their chapters:

- Plant breeding as the starting point of the food chain and challenges for plant breeding with the focus on feed (Chapter 1)
- Fundamentals of plant biotechnology (Chapter 2)
- International guidance documents, especially for safety assessment of feed from GM plants (Chapter 3)
- Compositional analysis of GM feed (Chapter 4)
- Types of animal feeding studies and results of such studies (Chapters 5–8)
- Fate of DNA and newly expressed proteins (Chapter 9)
- Influence of GM feed on composition/quality of food of animal origin (Chapter 10)
- Feed additives from GM microorganisms (Chapter 11)
- Future developments/trends (Chapter 12)
- GM cultivation and socio-economic aspects (Chapters 13–14)
- Public acceptance (Chapter 15)

As the editor, I would like to thank all authors for their enthusiasm and willingness to contribute to the book despite their high workload and the time pressures in their research institutions.

Furthermore, I thank Dina Führmann, Sigrid Herweg, Prof Dr M. Grün, colleagues from the Institute of Animal Nutrition Braunschweig, and my wife Elisabeth for their discussion, scientific advice and much technical support.

In addition, I would like to thank, for their helpful comments and observations, those who reviewed the Table of Contents and the current first edition of the book.

Finally, I would like to thank CAB International, especially Alexandra Lainsbury, for her understanding and help in some critical phases, and Lauren Povey, as well as David Hemming, who convinced me to edit the book. My thanks also go to Chris McEnnerney for copy-editing the manuscript.

Readers are invited to notify the authors of any errors, omissions or irrelevant material they may come across in order that they can be omitted from the next edition.

Dedication

I appreciate the opportunity to have worked with the chapter authors, all experts, on this book. I have learned from them and dedicate this book to them.

Gerhard Flachowsky

1 Introduction and Background – Challenges and Limitations of GM Plants for Animal Nutrition

Gerhard Flachowsky*

Institute of Animal Nutrition, Friedrich-Loeffler-Institute (FLI), Federal Research Institute for Animal Health, Braunschweig, Germany

1.1 Global Food Situation

The world population is still growing and demanding more and better food as well as other products for an improved standard of living. At the end of October 2011, the 7 billionth person was born. Sustainability in feed and food production is a key challenge for agriculture, as has been summarized recently in many papers (Fedoroff *et al.*, 2010; Godfray *et al.*, 2010; Pardue, 2010; Foley *et al.*, 2011; FAO, 2012a; Giovannucci *et al.*, 2012; HLPE, 2012; Flachowsky *et al.*, 2013) and books or proceedings (Zollitsch *et al.*, 2007; Wenk *et al.*, 2009; Behl *et al.*, 2010; Casabona *et al.*, 2010; Welzer and Wiegandt, 2011; Potthast and Meisch, 2012; Viljoen and Wiskerke, 2012; Wals and Corcoran, 2012). In the future, there will be strong competition for arable land and further non-renewable resources such as fossil carbon sources, water (Renault and Wallender, 2000; Hoekstra and Champaign, 2007; Cominelli and Tonelli, 2010; Schlink *et al.*, 2010; Deikman *et al.*, 2012) and some minerals (such as phosphorus; Hall and Hall, 1984; Scholz and Wellmer, 2013), as well as between feed/food, fuel, fibre, areas for settlement and natural protected areas. According to the FAO (2009a,b), the human population will increase globally from currently about 7 billion to more than 9 billion in 2050, but about 70% more meat and milk will be required (Alexandratos and

Bruinsma, 2012; HLPE, 2013). Cereal production has increased from 0.88 billion t (1961) to 2.35 billion t (2007) and is expected to rise to over 4 billion t by 2050 (FAO, 2006).

As vegans demonstrate, there is no essential need for food of animal origin, but the consumption of meat, fish, milk and eggs may contribute significantly to meeting human requirements for amino acids (Young *et al.*, 1989; WHO, 2007; D'Mello, 2011; Pillai and Kurpad, 2011) and some important trace nutrients (such as Ca, P, Zn, Fe, I, Se, Vitamins A, D, E, B_{12}, etc.), especially for children and juveniles, as well as for pregnant and lactating women (Wennemer *et al.*, 2006). Human nutritionists (Waterlow, 1999; Jackson, 2007) recommend that about one-third of the daily protein requirements (0.66–1 g per kg of body weight Rand *et al.*, 2003; Jackson, 2007; WHO, 2007) should originate from protein of animal origin. This means that about 20 g of a daily intake of about 60 g should be based on protein of animal origin, which is lower than the present average consumption throughout the world (without fish: 23.9 g per day; Table 1.1).

The conditions for the production of food of animal origin are also being questioned more and more, especially in the developed countries, as exemplified in Fig. 1.1. Immediately after the Second World War, people were hungry and required all types of

**E-mail: gerhard.flachowsky@t-online.de*

Table 1.1. Intake of milk, meat and eggs as well as protein of animal origin per inhabitant per year and portion (%) of total protein intake (minimum and maximum values, global averages and German values for comparison; kg per inhabitant per year; data from 2005). (From FAO, 2009a.)

Food	Minimum	Average	Maximum	Germany
Milk	1.3 (DR Congo)	82.1	367.7 (Sweden)	248.7
Meat[a]	3.1 (Bangladesh)	41.2	142.5 (Luxembourg)	83.3
Eggs	0.1 (DR Congo)	9.0	20.2 (PR China)	11.8
Edible protein of animal origin (g per human per day)	1.7 (Burundi)	23.9	69.0 (USA)	52.8
Portion of animal protein in per cent of total protein intake per person	4.0 (Burundi)	27.9	59.5 (USA)	53.7

Note: [a]Probably empty body weight (meat plus bones; see Flachowsky and Kamphues, 2012).

food. *Food security* was much more important than *food safety* or aspects of food processing or animal health and welfare. This situation has changed during the past years and food safety is paramount in Western countries today. But nevertheless, the question, 'I am hungry, is there anything to eat?' (see Table 1.1 and Fig. 1.1), is still relevant to many people (about 1 billion; WHO, 2007) and many countries. This is one of the reasons for producing more and better foods of plant and animal origin all over the world. In his Nobel Prize acceptance speech, Norman Borlaug summarized his philosophy in the following statement: 'If you desire peace, cultivate justice, but at the same time cultivate the fields to produce more bread; otherwise there will be no peace' (Borlaug, 1970). Recently, Aerts (2012) formulated the challenge for the future of agriculture as 'more (food) for more (people), with less (inputs and emissions)'.

There is, however, a high variation in the availability and consumption of food of animal origin between persons and countries (between 1.7 and about 70 g of protein of animal origin per person per day; see Table 1.1). If people in the 'developed' countries continue their high consumption and people's intake in the developing countries is to increase, a dramatic rise in the production of food of animal origin on a global scale is necessary. Other reasons for people's consumption of foods of animal origin are the high bioavailability of various nutrients and their considerable enjoyment value of the products. Such food is also considered as an indicator of the standard of living in many regions of the world. Further reasons for the higher demand for food of animal origin in some countries are the increased income of the population (Keyzer *et al.*, 2005) and the imitation of the so-called 'Western lifestyle' (of nutrition). In the next 20 years, up to 3 billion more 'middle-class consumers' ('middle class' is defined as having daily per capita spending of US$10–100 in purchasing parity terms; Kharas, 2012) are expected to have purchasing power (presently about 1.8 billion). In anticipation of these changes, sufficient animal feed should be considered as the starting point for food of animal origin (Zoiopoulos and Drosinos, 2010; Flachowsky *et al.*, 2013). Higher amounts of food of animal origin require higher plant yields and/or a larger area for feed production and more animals and/or higher animal yields, as well as a more efficient conversion of feed into food of animal origin (Powell *et al.*, 2013; Windisch *et al.*, 2013) for various levels of yields or performance, as demonstrated in Table 1.2.

In addition feed/food production causes emissions with a certain greenhouse gas potential, such as carbon dioxide (CO_2) from fossil fuel, methane (CH_4; greenhouse

Fig. 1.1. Past, present and future situation for consumers and policies, as well as the challenges for agricultural research after the Second World War (Flachowsky, 2002a).

Table 1.2. Model calculation on the influence of human intake of protein of animal origin (except fish), yields of plants and performance of animals, as well as the relation between protein from meat and milk, on the need for arable area (adapted by Flachowsky and Bergmann, 1995; Flachowsky, 2002b and Flachowsky *et al.*, 2008, based on other plant yield levels and animal performance).

	Protein of animal origin (g per inhabitant per day)							
	10		20		40		60	
Relation between meat[c] and milk (% of protein)	Intensity level							
	A[a]	B[b]	A	B	A	B	A	B
	Arable land required for feed production (m^2 per inhabitant per year)							
70:30	345	95	690	190	1380	380	2080	570
50:50	290	85	580	170	1160	340	1740	510
30:70	235	75	470	150	940	300	1410	450

Notes: [a]Plant yield level A: 3 t DM (dry matter) cereals; 10 t DM roughage per ha per year. Performance of animals A (per animal per day): 15 kg milk; weight gain: beef: 600 g; pork: 400 g; poultry: 30 g. [b]Plant yield level B: 8 t DM cereals; 20 t DM roughage per ha per year. Performance of animals B: 30 kg milk; weight gain: beef: 1200 g; pork: 800 g; poultry: 60 g. [c]Relation between beef, pork and poultry meat = 20:50:30.

gas factor (GHF) about 23; IPCC, 2006) from enteric fermentation, especially in ruminants, and from excrement management, as well as nitrogen compounds (NH_3; N_2O: GHF about 300; IPCC, 2006) from the protein metabolism in the animals (DEFRA, 2006; Flachowsky and Hachenberg, 2009; FAO, 2010; Godfray *et al.*, 2010; Grünberg *et al.*, 2010; Leip *et al.*, 2010; Flachowsky *et al.*, 2011; Table 1.3).

Apart from the low input of limited resources along the food chain, a low output of greenhouse gases (CO_2, CH_4 and N_2O) and minerals such as phosphorus (Table 1.3) and some trace elements during feed/ food production are very important aims of sustainable agriculture. Presently, about 15% of total global emissions comes from crop and livestock production (HLPE, 2012).

Table 1.3. Effects of animal species, categories and performance on some emissions (per kg edible protein). (From Flachowsky, 2002b; Flachowsky *et al.*, 2012.)

Protein source (body weight)	Performance per day	Nitrogen excretion (per cent of intake)	Methane emission (g per day)[c]	Emissions in kg per kg edible protein			
				P	N	CH_4	CO_{2eq}[d]
Dairy cow	10 kg milk	75	310	0.10	0.65	1.0	30
(650 kg)	20 kg milk	70	380	0.06	0.44	0.6	16
	40 kg milk	65	520	0.04	0.24	0.4	12
Dairy goat	2 kg milk	75	50	0.08	0.5	0.8	20
(60 kg)	5 kg milk	65	60	0.04	0.2	0.4	10
Beef cattle	500 g[a]	90	170	0.30	2.3	3.5	110
(350 kg)	1000 g[a]	84	175	0.18	1.3	1.7	55
	1500 g[a]	80	180	0.14	1.0	1.2	35
Growing/fattening pig	500 g[a]	85	5	0.20	1.0	0.12	16
(80 kg)	700 g[a]	80	5	0.12	0.7	0.08	12
	900 g[a]	75	5	0.09	0.55	0.05	10
Broilers	40 g[a]	70	Traces	0.04	0.35	0.01	4
(1.5 kg)	60 g[a]	60		0.03	0.25	0.01	3
Laying hen	50%[b]	80	Traces	0.12	0.6	0.03	7
(1.8 kg)	70%[b]	65		0.07	0.4	0.02	5
	90%[b]	55		0.05	0.3	0.02	3

Notes: [a]Daily weight gain; [b]laying performance; [c]CH_4 emission depending on composition of diet; [d]equivalent to carbon footprints (sum of greenhouse gas emission of CO_2; CH_4 (× 23) and N_2O (× 300; IPCC, 2006) for edible protein of animal origin).

1.2 Plant Breeding as the Starting Point of the Food Chain

Plant breeding and cultivation are the key elements and starting points for feed and food security in the next years (see Flachowsky, 2008; SCAR, 2008; The Royal Society, 2009; Flachowsky *et al.*, 2013). The most important objectives for plant breeders can be summarized as follows:

- High and stable yields with low external inputs of non-renewable resources (low-input varieties) such as water, minerals, fossil fuel, plant protection substances, etc. (Table 1.4).
- Maximal use of natural unlimited resources such as sunlight, nitrogen and carbon dioxide from the air (Table 1.4).
- Higher resistance against biotic and abiotic stressors (such as drought and increased salinity), including healthy plants and adaptation to potential climate changes.
- Optimization of the genetic potential of plants for a highly efficient photosynthesis.

- Lower concentrations of toxic substances such as secondary plant ingredients, mycotoxins from toxin-producing fungi, toxins from anthropogenic activities or of geogenic origin.
- Lower concentrations of substances that influence the use or bioavailability of nutrients such as lignin, phytate, enzyme inhibitors, tannins, etc.
- Higher concentrations of the components determining nutritive value such as nutrient precursors, nutrients, enzymes, pro- and prebiotics, essential oils, etc.

From the global perspective of feed and food security, plants with low inputs of non-renewable resources and high and stable yields should have the highest priority in breeding. In addition, resistance to insect infestation (Shade *et al.*, 1994; Lee *et al.*, 2013) and low losses in the field during harvest and storage are also important aspects of feed/food security. Furthermore, undesirable substances often cannot be removed from feedstuffs or can be removed only with great effort (Flachowsky, 2006; Morandini, 2010; Verstraete, 2011;

Table 1.4. Potential to produce phytogenic biomass and its availability per inhabitant when considering the increase in population. (From The Royal Society, 2009; Flachowsky, 2010.)

Plant nutrients in the air (N_2, CO_2)	↑↔
Solar energy	↔
Agricultural area	↓
Water	↓
Fossil energy	↓
Mineral plant nutrients	↓
Variation of genetic pool	↑

Note: ↑ = increase; ↓ = decrease; ↔ = no important influence.

Fink-Gremmels, 2012). Therefore, a decrease of undesirable substances in plants is also an important objective of plant breeding. From the perspective of human nutrition, an increase of essential nutrients (e.g. amino acids, fatty acids, trace elements, vitamins, etc.) could be very favourable in meeting the requirements for essential nutrients (see Chapter 7). But this aspect is not so important for animal nutrition in some parts of the world such as Europe because of the availability of the large amount of feed additives on the market. Furthermore, potential aspects of climate change (HLPE, 2012; IPCC, 2012; Schwerin *et al.*, 2012) should be considered by plant breeders, and 'new' plants should be adapted to such changes (Reynolds, 2010; Newman *et al.*, 2011). It is possible to fulfil the objectives of plant breeding mentioned above with conventional breeding (Flachowsky, 2012), but in the future, methods of 'green' biotechnology may be more flexible, more potent and faster (Tester and Langridge, 2010; Whitford *et al.*, 2010). 'New' plants, newly expressed proteins in plants and/or changed composition of plants are real challenges for animal and human nutritionists for safety and nutritional assessment of such products (see Fig. 1.3).

Increasing feed/food demands requires higher plant yields and/or larger areas for production (see Table 1.2). Because of some limited resources, low-input plants are an important prerequisite to solving future problems and to establishing sustainable agriculture. Such plants should be very efficient in their use of mineral plant nutrients (including N), fuel, water and arable land (high yields), but they should also be able to use the sun's energy more efficiently and unlimited plant nutrients from the air (such as N_2 and CO_2; see Table 1.4). Non-legumes should also be able to use N from the air for N-fixing symbiosis. Furthermore, the genetic pool available in plants, animals and microorganisms should contribute to optimizing plants and animals for a more efficient conversion of limited resources into feed and food. Maintaining the biodiversity of the available genetic pool is also a very important aspect of sustainable agriculture. Losses of biodiversity may have dramatic consequences in the future for plant breeding including plant biotechnology (HLPE, 2012; see also Table 1.4)

Subsequent animal feeding studies are necessary to demonstrate the digestibility/availability of the changed composition of the plants or the newly, or higher amounts of, expressed nutrients (see Fig. 1.3 and Chapter 5 for some examples).

Possible climate change may be an additional challenge for plant breeders and for sustainable development (Potthast and Meisch, 2012). Some authors (e.g. Easterling *et al.*, 2007; Reynolds, 2010) predict a 15–20% fall in global agricultural production by 2080 as a consequence of the expected climate change. The following climate change-related problems could be expected (Whitford *et al.*, 2010):

- Adaption to greater extremes in climate conditions and higher temperatures.
- The water supply may become limited or more variable; better adaptation of plants to drought resistance (Cominelli and Tonelli, 2010; Deikman *et al.*, 2012).
- Increasing soil salination.
- Higher disease infection and pest infestations (Wally and Punja, 2010).

A rapidly changing climate will require rapid development of new plant varieties. The negative effects of climate change could be greater than possible solutions by conventional plant breeding. Therefore, a large 'technology gap' between solutions by conventional breeding and the need for adaptation to climate change will result in adequate or lower plant yields (Whitford *et al.*, 2010). The UN (2010) expects an increase of greenhouse gas emissions from about 48 (2010) to about 66 Gt in 2030 and estimates a rise in global average temperatures of more than 5°C by the end of the century. Extreme weather situations such as thunderstorms, heavy rains, hailstorms, tornadoes, long dry periods or droughts, etc., as consequences of expected climate change, may have a dramatic influence on feed production and the feeding of food-producing animals. To achieve a 450 ppm CO_2 equivalent in the air, carbon dioxide emissions would need to be reduced from 48 Gt per year to 35 Gt in 2030. Plants undergo adaptive change to acclimatize to new environments. Drought-resistant, high water-use efficient, heat-tolerant and disease-resistant plants will be the important objectives of plant breeding under climate change. Therefore, techniques that are able to enhance the speed, flexibility and efficiency of plant breeding are required for the so-called 'second green revolution'.

In addition, insect-protected and herbicide-tolerant plants may also reduce the use of pesticides, with consequences for lower CO_2 emissions (lower carbon footprints) and a general reduction of global pesticide use (Phipps and Park, 2002). Life-cycle assessments to compare the environmental impact of isogenic and genetically modified herbicide-tolerant and/or insect-resistant plants are a great challenge for scientists working in the field (Persley, 2003; Bennett *et al.*, 2004).

Land and water are considered to be the greatest challenges on the supply side of food production. In 2030, Dobbs *et al.* (2011) estimate a 30% higher water need (an additional 1850 km^3) and between 140 and 175 million hectares (Mha) (about 10% of the present area) deforestation. Furthermore, the genetic pool available in plants, animals and microorganisms could also contribute to optimizing plants and animals for a more efficient conversion of limited resources into feed and food (see Table 1.4). Future strategies have to acknowledge the multifunctionality of agriculture and take into account the complexity of agricultural systems within different socio-economic situations. Farmers are not just producers; they are also managers of ecosystems. Therefore, different opinions and experiences on the impact of genetically modified (GM) plants on smallholder farmers in various regions should be expected (Kathage and Qaim, 2012; Kleemann, 2012).

Discussions on the potential of plant breeding by 'green biotechnology' are old (Persley, 1990; Hodges, 1999, 2000; Qaim, 2000; Borlaug, 2003; Avery, 2004) and they are not free from criticism (Altieri, 1998) and conflicts starting with the first steps of breeding and cultivating GM crops (Perlas, 1994; Altieri and Rosset, 1999). Nevertheless, there has been a dramatic increase in the cultivation of GM crops, starting with 1.7 Mha in 1996. In 2012, about 170 Mha of GM plants were cultivated worldwide (about 11% of total arable land; James, 2013). Most of these GM plants are tolerant of herbicides and/or resistant to insects (Fig. 1.2). Such plants do not contain higher amounts of desirable and undesirable substances and can be considered as substantially equivalent to their isogenic counterparts (OECD, 1993; see Chapters 4 and 6).

Currently, the interests of individuals or of some companies dominate, and these are not always in agreement with public interests, as discussed above (SCAR, 2008; Godfray *et al.*, 2010; Foley *et al.*, 2011). More fundamental and applied research should be conducted by independent, publicly sponsored research institutions (The Royal Society, 2009; Pardue, 2010) and the results should be made available to all those who are interested in such plants. Public–private partnerships should be formed with the mission to reach set goals in the coming decades (Arber, 2010).

1.3 Food-producing Animals as Part of the Food Chain

High portions of the yield of the most important GM plants (soybean, maize, cotton, rapeseed; see Fig. 1.2) are fed to

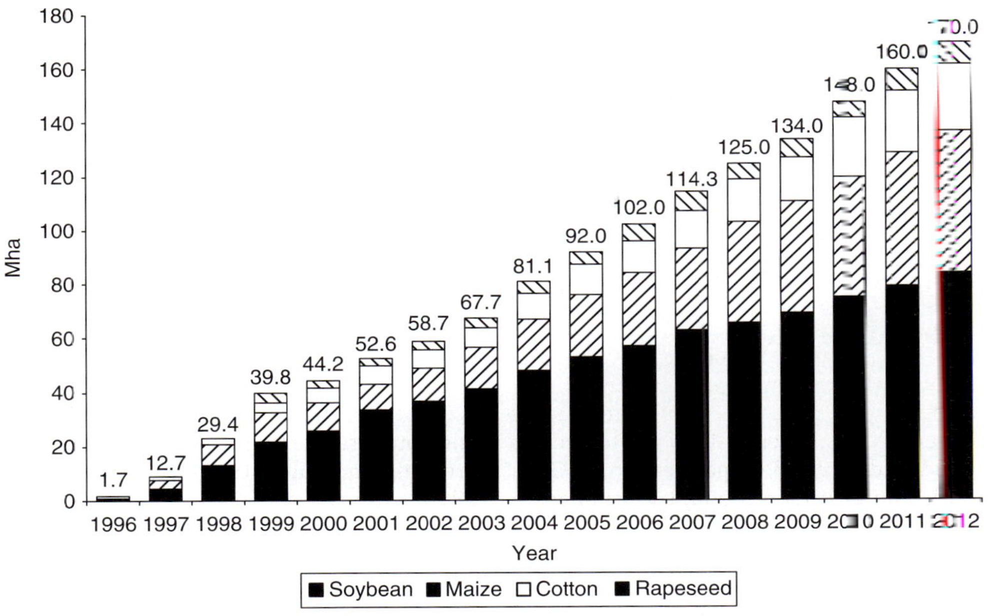

Fig. 1.2. Global area of transgenic crops (GM plants) in Mha (James, 2013).

food-producing animals (Table 1.5) and only small amounts are used for human nutrition.

Therefore, in the future, assessing the nutritive value of feeds and co-products from GM plants for food-producing animals will be a real challenge for animal nutritionists (Kleter and Kok, 2010; Flachowsky *et al.*, 2012; see Fig. 1.3 and Chapters 5, 6 and 7). In addition, GM animals will come on the market in the next few years (Golovan *et al.*, 2001a,b; Forsberg *et al.*, 2003; Niemann and Kues, 2007; Robi *et al.*, 2007; Fahrenkrug *et al.*, 2010; Niemann *et al.*, 2011; EFSA, 2012; FAO 2012b) and

nutritionists will have to deal with the energy and nutrient requirements of such animals, including animal clones (Fig. 1.3).

Various types of animal feeding studies are required in order to answer all the scientific and public questions and to improve the public acceptance of such food/ feed and animals (see Chapter 5). The current state and future challenges of the nutritional and safety assessment of feed from genetically modified plants will be analysed in the following chapters. The main objectives of those chapters are to consider the pros and cons of feeds from transgenic plants and

Table 1.5. Important food/feed from GM plants and the estimated proportions used as food or feed (author's estimation).

GM plant	Food	%	Feed	%
Soybean	Oil, proteins	25	Soybean (extracted oil) meal, full-fat soybean	75
Maize	Starch, maize meal, oil	15	Maize, oil, DDGS, gluten feed, silage, straw	85
Rapeseed	Oil	25	Rapeseed (extracted oil) meal, rapeseed expeller/cake, full-fat rapeseed	75
Cotton	Oil	15	Cotton seed (extracted oil) meal, expeller	85

Note: DDGS = dried distillers grain with added solubles.

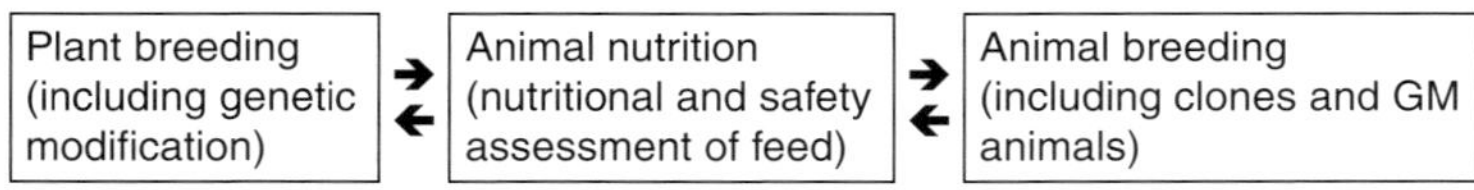

Fig. 1.3. Animal nutrition (nutritional assessment of feeds) between plant and animal breeding.

to demonstrate results in animal feeding. Different types of animal feeding studies for the nutritional assessment of GM feed will be assessed. Sometimes, it is impossible, and also not necessary, to strictly separate the nutritional and safety assessment of feed/food. Kleter and Kok (2010) and Davies and Kuiper (2011) consider the following aspects of risk assessments that also include nutritional aspects:

- characteristics of donor and recipient organism;
- genetic modification and its functional consequences;
- potential environmental impact;
- agronomic characteristics;
- compositional and nutritional characteristics;
- potential for toxicity and allergenicity of gene products, plant metabolites and whole GM plants;
- influence of processing on the properties of food and feed;
- potential for changes in dietary intake; and
- potential for long-term nutritional impact.

Some principles of the genetic modification of plants are demonstrated in Chapter 2.

1.4 Challenges and Developments

Resource productivity and/or resource efficiency measures are key challenges for the future, as shown in the two assessments below. Dobbs *et al.* (2011) integrated more than 130 potential resource measures in a resource productivity cost curve. Under the top 15 measures, accounting for roughly 75% of the total resource productivity, one may find many opportunities associated with agriculture. The following ranking shows the 15 opportunities (Dobbs *et al.*, 2011):

1. Building energy efficiency.
2. Increasing yield on large-scale farms.
3. Reduced food waste.
4. Reducing municipal water leakage.
5. Urban densification (leading to major transport efficiency gains).
6. Higher energy efficiency in the iron and steel industry.
7. Increasing yields on smallholder farms.
8. Increasing transport fuel efficiency.
9. Increasing the penetration of electric and hybrid vehicles.
10. Reducing land degradation.
11. Improving end-use efficiency.
12. Increasing oil and coal recovery.
13. Improving irrigation techniques.
14. Shifting road freight to rail and barge.
15. Improving power plant efficiency.

Another assessment has been carried out by KMPG International (2012). The authors analysed the global sustainability megaforces over the next two decades with an impact on every business and came to the following facts (no ranking):

- climate change;
- energy and fuel;
- material resource scarcity;
- water scarcity;
- population growth;
- wealth;
- urbanization;
- food security;
- ecosystem decline; and
- deforestation.

Both assessments contain similar elements concerning future developments and limitations. Of course, such assessments may be very helpful for the future, but they are man-made and not completely free of individual or group-influenced motions and expectations. For example, it is not possible to assess the consequences of new discoveries and developments.

In consequence, population growth with increasing age, arable land, fresh water and fuel limits, climate change and other developments require a radical rethinking of agriculture for the 21st century to meet this century's demands for feed, food, fibre and fuel, while reducing the environmental impact of their production (Fedoroff *et al.*, 2010; Tester and Langridge, 2010; Windisch *et al.*, 2013). Developments of plants under consideration of resources are mentioned in Table 1.4. The acceptance of such plants, as well as the farming systems that use them, are considered essential for the success of the new agriculture. Genetically modified organisms (GMOs) that are in the pipeline are described in Chapter 12. Furthermore, FAO (2012b) and Ruane (2013) provide information about the future of GM plants in developing countries.

More public investment will be needed, and new and imaginative public–private collaboration can also make the 'genetic revolution' beneficial for developing countries (Serageldin, 1999; Qaim, 2000; SCAR, 2008; The Royal Society, 2009; see also Chapters 12–15). Matten *et al.* (2008) recommend a global harmonization to decrease regulatory barriers. International organizations should play a key role in rationalizing regulatory systems. Furthermore, the public research sector will need to ensure that the risk assessment process is scientifically sound and transparent (Matten *et al.*, 2008; Miller, 2010; see also Chapter 3).

References

Aerts, S. (2012) Agriculture's 6 Fs and the need for more intensive agriculture. In: Potthast, T. and Meisch, S. (eds) *Climate Change and Sustainable Development – Ethical Perspectives on Land Use and Food Production*. Wageningen Academic Publishers, Wageningen, Netherlands, pp. 192–195.

Alexandratos, N. and Bruinsma, J. (2012) *World Agriculture Towards 2030/2050: The 2012 Revision*. ESA Working Paper 12-03. FAO, Rome (http://www.fao.org/docrep/016/ap106e/ap106e.pdf, accessed 26 May 2013).

Altieri, M.A. (1998) The environmental risk of transgenic crops: an agroecological assessment. *AgBiotech News and Information* 10, 405–410.

Altieri, M.A. and Rosset, P. (1999) Ten reasons why biotechnology will not ensure food security, protect the environment and reduce poverty in the developing world. *AgBioForum* 2, 155–162.

Arber, W. (2010) Genetic engineering compared to natural genetic variation. *New Biotechnology* 27(5), 517–521.

Avery, A.A. (2004) Farming into the future – people, pets and potables. In: *Future Shock Challenges and Opportunities*. The Australian Milling Conference, March 23–24, 2004, Melbourne, Victoria. FMCA Conference Secretariat, Hawthorn, Victoria, Australia, pp. 1–7.

Behl, R.K., Kubat, J. and Kleynhans T (2010) *Resource Management Towards Sustainable Agriculture and Development*. Agrobios (India), Jodhpur, India, 358 pp.

Bennett, R.M., Phipps, R.H., Strange A.M. and Grey, P.T. (2004) Environmental and human health impacts of growing genetically modified herbicide-tolerant sugar beet: a life-cycle assessment. *Plant Biotechnology Journal* 2, 273–278.

Borlaug, N. (1970) Nobel Lecture: The Green Revolution. Peace and Humanity. Oslo, 11 December 1970.

Borlaug, N. (2003) Not science, hunger is the foe. *Wall Street Journal*, Europe, 22 January 2003.

Casabona, C.M.R., Epifanio, L.E.S. and Cirion, A.E. (eds) (2010) *Global Food Security. Ethical and Legal Challenges*. Wageningen Academic Publishers, Wageningen, Netherlands, 532 pp.

Cominelli, E. and Tonelli, C. (2010) Transgenic crops coping with water scarcity. *New Biotechnology* 27, 473–477.

Davies, H.V. and Kuiper, H.A. (2011) GM-risk assessment in the EU: current perspective. *Aspects of Applied Biology* 110, 37–45.

DEFRA (2006) Determination of the environmental burdens and resource use in the production of agriculture and horticulture commodities. *Defra Project Report 2005*. Silsoe College, Cranfield, University, Bedford, UK (www.silsce.cranfield.ac.uk and www.defra.gov.uk, accessed 23 May 2013).

Deikman, J., Petracek, M. and Heard, J.E. (2012) Drought tolerance through biotechnology: improving translation from the laboratory to farmers' fields. *Current Opinion in Biotechnology* 23, 243–250.

D'Mello, J.P.F. (2011) *Amino Acids in Human Nutrition and Health*. CAB International, Wallingford, UK, 544 pp.

Dobbs, R., Oppenheim, J., Thompson, F., Brinkmann, M. and Zornes, M. (2011) Resource revolution: meeting the world's energy, materials, food and water needs. McKinsey sustainability and resource productivity practice. McKinsey Global Institute, November 2011, 20 pp.

Easterling, W.E., Aggarwal, P.K., Batima, P., Brander, K.M., Erda, L., Howden, S.M., *et al.* (2007) Food, fibre and forest products. In: Parry, M.L., Canziani, O.F., Palutikof, J.P., van der Linden, P.J. and Hanson, C.E. (eds) *Climate Change 2007: Impacts, Adaptation and Vulnerability.* Contribution of Working Group II to the Fourth Assessment Report of the Intergovernmental Panel on Climate Change, Cambridge University Press, Cambridge, UK, pp. 273–313.

EFSA (2012) Guidance of the risk assessment of food and feed from genetically modified animals and on animal health and welfare aspects. *EFSA Journal* 10, 2501 (43 pp.).

Fahrenkrug, S.C., Blake, A., Carlson, D.F., Doran, T., van Eenennaam, A., Faber, D., *et al.* (2010) Precision genetics for complex objectives in animal agriculture. *Journal of Animal Science* 88, 2530–2539.

FAO (2006) *World Agriculture: Toward 2030/2050. Interim Report.* Global Perspective Studies Unit, FAO, Rome.

FAO (2009a) *The State of Food and Agriculture – Livestock in the Balance.* FAO, Rome, 180 pp.

FAO (2009b) *How to Feed the World in 2050.* FAO, Rome, 120 pp.

FAO (2010) *Greenhouse Gas Emissions From the Dairy Sector. A Life Cycle Assessment.* FAO, Rome, 94 pp.

FAO (2012a) *The State of Food and Agriculture 2012 – Investing in Agriculture for a Better Future.* FAO, Rome, 180 pp.

FAO (2012b) GMOs in the pipeline: Looking to the next five years in the crop, forestry, livestock, aquaculture and agro-industry sectors in developing countries. Background Document to Conference 18 of the FAO Biotechnology Forum, 5 November to 2 December 2012 (http://www.fao.org/docrep/016/ap109e/ap109e00.pdf, accessed 26 May 2013).

Fedoroff, N.V., Battisti, D.S., Beachy, R.N., Cooper, P.J.M., Fischhoff, D.A., Hodges, P.C., *et al.* (2010) Radically rethinking agriculture for the 21st century. *Science* 327, 833–834.

Fink-Gremmels, J. (ed.) (2012) *Animal Feed Contamination – Effects on Livestock and Food Safety.* Woodhead Publishing, Cambridge, UK, 672 pp.

Flachowsky, G. (2002a) Evaluation of food security and food safety. *Landbauforschung Völkenrode* 52, 1–9. [In German]

Flachowsky, G. (2002b) Efficiency of energy and nutrient use in the production of edible protein of animal origin. *Journal of Applied Animal Research* 22, 1–24.

Flachowsky, G. (ed.) (2006) Possibilities of decontamination of 'Undesirable substances of Annex 5 of the feed law'. *Landbauforschung Völkenrode – FAL Agricultural Research*, Special Issue 294, 290 pp. [In German]

Flachowsky, G. (2008) What do animal nutritionists expect from plant breeding? *Outlook on Agriculture* 37, 95–103.

Flachowsky, G. (2010) Global food security: Is there any solution? *NovoArgumente* 105, 3–4, 64–68. [In German]

Flachowsky, G. (2012) Prospects of feed from genetically modified plants in livestock feeding. In: Mehra, U.R., Singh, P. and Verma, A.K. (eds) *Animal Nutrition – Advances and Developments.* Satish Serial Publishing House, Azadpur, Delhi, India, pp. 475–498.

Flachowsky, G. and Bergmann, H. (1995) Global feed potential and plants of moderate climate. In: Abel, H.J., Flachowsky, G., Jeroch, H. and Molnar, S. (eds) *Nutrition of Domestic Animals.* Gustav Fischer Verlag, Jena, Stuttgart, Germany, pp. 39–62. [In German]

Flachowsky, G. and Hachenberg, S. (2009) CO_2-footprints for food of animal origin – present stage and open questions. *Journal of Consumer Protection and Food Safety* 4, 190–198.

Flachowsky, G. and Kamphues, J. (2012) Carbon footprints for food of animal origin: What are the most preferable criteria to measure animal yields? *Animals* 2, 108–126.

Flachowsky, G., Dänicke, S., Lebzien, P. and Meyer, U. (2008) More milk and meat for the world. How is it to manage? *ForschungsReport* 2/2008, 14–17. [In German]

Flachowsky, G., Brade, W., Feil, A., Kamphues, J., Meyer, U. and Zehetmeyer, M. (2011) Carbon (CO_2)-footprints for producing food of animal origin – database and reduction potentials. *Übersichten zur Tierernährung* 39, 1–45. [In German]

Flachowsky, G., Meyer, U. and Schafft, H. (2012) Animal feeding studies for nutritional and safety assessments of feeds from genetically modified plants: a review. *Journal of Consumer Protection and Food Safety* 7, 179–194.

Flachowsky, G., Meyer, U. and Gruen, M. (2013) Plant and animal breeding as starting points for sustainable agriculture. In: Lichtfouse, E. (ed.) *Sustainable Agriculture Reviews 12.* Springer Science+Business Media, Dordrecht, Netherlands, pp. 201–224.

Foley, J.A., Ramankutty, N., Brauman, K.A., Cassidy, E.S., Gerber, J.S., Johnston, M., *et al.* (2011) Solutions for a cultivated planet. *Nature* 478, 337–342.

Forsberg, C.W., Phillips, J.P., Goölovan, S.P., Fan, M.Z., Meldinger, R.G., Ajakaiye, A., *et al.* (2003) The enviropig physiology, performance, and contribution to nutrient management advances in a regulated environment: the leading edge of change in the pork industry. *Journal of Animal Science* 81, Suppl. 2, E68–E77.

Giovannucci, D., Scherr, S., Nierenberg, D., Hebebrand, C., Shapiro, J., Milder, J., *et al.* (2012) Food and agriculture: the future of sustainability. A strategic input to the sustainable development in the 21st Century (SD21) project. United Nations Department of Economic and Social Affairs, Division for Sustainable Development, New York.

Godfray, H.C., Beddington, J.R., Crute, I.R., Haddad, L., Wawrence, D., Muir, J., *et al.* (2010) Food security: the challenge of feeding 9 billion people. *Science* 327, 812–818.

Golovan, S.P., Hayes, M.A., Phillips, J.P. and Forsberg, C.W. (2001a) Transgenic mice expressing bacterial phytase as a model for phosphorus pollution control. *Nature Biotechnology* 19, 429–433.

Golovan, S.P., Meidinger R.G., Ajakaiye, A., Cottrill, M., Wiederkehr, M.Z., Barney, D.J., *et al.* (2001b) Pigs expressing salivary phytase produce low-phosphorus manure. *Nature Biotechnology* 19, 741–745.

Grünberg, J., Nieberg, H. and Schmidt, T. (2010) Treibhausgasbilanzierung von Lebensmitteln (Carbon Footprints): Überblick und kritische Reflektion. *Landbauforschung – vTI Agriculture and Forestry Research* 60, 53–72. [In German]

Hall, D.C. and Hall, J.V. (1984) Concepts and measures of natural-resource scarcity with a summary of recent trends. *Journal of Environmental Economics and Management* 11, 363–379.

HLPE [High Level Panel of Experts on Food Security and Nutrition] (2012) Food security and climate change. A report by the High Level Panel of Experts on Food Security and Nutrition, June 2012. Committee on World Food Security, FAO, Rome, 98 pp.

HLPE (2013) Biofuels and fuel security (V0 Draft). A zero-draft consultation paper. 9 January 2013, 72 pp.

Hodges, J. (1999) Editorial: The genetically modified food muddle. *Livestock Production Science* 62, 51–60.

Hodges, J. (2000) Editorial: Polarization on genetically modified food. *Livestock Production Science* 63, 159–164.

Hoekstra, A.Y. and Champaign, A.K. (2007) Water footprints of nations: water use by people as a function of their consumption pattern. *Water Resource Management* 21, 35–48.

IPCC [Intergovernmental Panel on Climate Change] (2006) *IPCC Guidelines for National Greenhouse Gas Inventories*. IPCC, Bracknell UK.

IPCC (2012) Managing the Risks of Extreme Events and Disasters to Acvance Climate Change Adaption (SREX). IPCC, Geneva, Switzerland.

Jackson, A.A. (2007) Protein. In: Manr, J. and Truswell, S. (eds) *Essentials of Human Nutrition*, 3rd edn. Oxford University Press, New York, pp. 53–72.

James, C. (2013) *Global Status of Commercialised Transgenic Crops: 2012*. ISAAA, Ithaca, New York.

Kathage, J. and Qaim, M. (2012) Economic impacts and impact dynamics of Bt (*Bacillus thuringiensis*) cotton in India. *Proceedings of the National Academy of Sciences (PNAS)*. DOI:10.1073/pnas.1203647109.

Keyzer, M.A., Merbis, M.D., Pavel, L.F.P.W. and van Westenbeck, C.F.A. (2005) Diets shift towards meat and the effect on cereal use: Can we feed the animals in 2030? *Ecological Economics* 55, 187–202.

Kharas, H. (2012) The emerging middle class in developing countries. OECD Development Centre Working Paper No 285, January 2010. OECD, Paris, 61 pp.

Kleemann, L. (2012) Sustainable agriculture and food security in Africa: an overview. Kiel Working Paper, No 1812, 28 pp. (http://hdl.handle.net/10419/67346, accessed 25 May 2013).

Kleter, G.A. and Kok, E.J. (2010) Safety assessment of biotechnology used in animal production, including genetically modified (GM) feed and GM animals – a review. *Animal Science Papers and Reports* 28, 105–114.

KPMG International (2012) Expect the unexpected: building business value in a changing world. KPMG International, New York, 18 pp. (Excecutive summary; total document consists of 3 parts.)

Lee, R.-Y., Reiner, D., Dekan, G., Moore, A.E., Higgins, T.J.V. and Epstein, M.M. (2013) Genetically modified a-amylase inhibitor peas are not specifically allergenic in mice. *PLOS ONE* 8(1): e52972. DOI:10.1371/journal.pone.0052972 (accessed 25 May 2013).

Leip, A., Weiss, F., Monni, S., Perez, I., Fellmann, T., Loudjami, P., *et al.* (2010) Evaluation of the livestock sectors' contribution to the EU greenhouse gas emissions (GGELS) – final report. Joint Research Centre, European Commission, Brussels.

Matten, S.R., Head, G.P. and Quemada, H.D. (2008) How governmental regulation can help or hinder the integration of Bt crops within IPM programs. *Progress in Biological Control* 5, 27–39.

Miller, H. (2010) The regulation of agricultural biotechnology: science shows the better way. *New Biotechnology* 27(5), 628–634.

Morandini, P. (2010) Inactivation of allergens and toxins. *New Biotechnology* 27(5), 482–493.

Newman, J.A., Anand, H., Hal, M., Hunt, S. and Gedalof, Z. (2011) *Climate Change Biology*. CAB International, Wallingford, UK, and Cambridge, Massachusetts, 289 pp.

Niemann, H. and Kues, W. (2007) Transgenic farm animals: an update. *Reproduction, Fertility and Development* 19, 762–770.

Niemann, H., Kuhla, B. and Flachowsky, G. (2011) The perspectives for feed efficient animal production. *Journal of Animal Science* 89, 4344–4363.

OECD (1993) *Safety Evaluation of Foods Derived by Modern Biotechnology. Concepts and Principles.* OECD, Paris.

Pardue, S.L. (2010) Food, energy, and the environment. *Poultry Science* 89, 797–802.

Perlas, N. (1994) *Overcoming Illusions About Biotechnology.* ZED Books, London.

Persley, G.J. (1990) *Beyond Mendel's Garden; Biotechnology in the Service of World Agriculture.* CAB International, Wallingford, UK.

Persley, G.J. (2003) *New Genetics, Food and Agriculture: Scientific Discoveries – Societal Dilemmas.* International Council for Science, Paris (http://www.icsu.org).

Phipps, R.H. and Park, J.R. (2002) Environmental benefits of genetically modified crops: global and European perspectives on their ability to reduce pesticide use. *Journal of Animal Feed Science* 11, 1–18.

Pillai, R.R. and Kurpad, A.V. (2011) Amino acid requirements: quantitative estimates. In: D'Mello, J.P.F. (ed.) *Amino Acids in Human Nutrition and Health.* CAB International, Wallingford, UK, and Cambridge, Massachusetts, pp. 267–290.

Potthast, T. and Meisch, S. (eds) (2012) *Climate Change and Sustainable Development – Ethical Perspectives on Land Use and Food Production.* Wageningen Academic Publishers, Wageningen, Netherlands, 528 pp.

Powell, J.M., MacLeod, M., Vellinga, T.V., Opio, C., Falcucci, A., Tempio, G., Steinfeld, H., Gerber, P. (2013) Feed–milk–manure nitrogen relationships in global dairy production systems. *Livestock Science* 152(2), 261–272.

Qaim, M. (2000) Potential impacts of crop biotechnology in developing countries. In: Heidhues, F. and von Braun, J. (eds) *Development Economics and Policy*, Vol 17. Diss., Uni. Bonn, Peter Lang GmbH, Europäischer Verlag der Wissenschaften, Frankfurt am Main, Germany, 168 pp.

Rand, W.M., Pellett, P.L. and Young, V.R. (2003) Meta-analysis of nitrogen balance studies for estimating protein requirements in healthy adults. *American Journal of Clinical Nutrition* 77, 109–127.

Renault, D. and Wallender, W.W. (2000) Nutritional water productivity and diets. *Agricultural Water Management* 45, 275–296.

Reynolds, M.P. (2010) *Climate Change and Crop Production.* CAB International, Wallingford, UK, and Cambridge, Massachusetts, 292 pp.

Robi, J.M., Wang, P., Kasinathan, P. and Kuroiwa, P. (2007) Transgenic animal production and biotechnology. *Theriogenology* 67, 127–133.

Royal Society, The (2009) Reaping the benefits: science and the sustainable intensification of global agriculture. RS policy document 11/09, issued October 2009. RS 1608, ISBN: 978-0-85403-784-1.

Ruane, J. (2013) An FAO e-mail conference on GMOs in the pipeline in developing countries: the moderator's summary (http://www.fao.org/biotech/biotech-forum/, accessed 25 May 2013).

SCAR [EU Commission – Standing Committee on Agricultural Research] (2008) New challenges for agricultural research. Climate change, rural development, agricultural knowledge systems. The 2nd SCAR Foresight Exercise, Brussels, December 2008, 112 pp.

Schlink, A.A., Nguyen, M.-L. and Viljoen, G.J. (2010) Water requirements for livestock production: a global perspective. *Revue Scientifique et Technique-Office International des Epizooties* 29, 603–619.

Scholz, R.W. and Wellmer, F.-H. (2013) Approaching a dynamic view on the availability of mineral sources: what we may learn from the case of phosphorus. *Global Environmental Change* 23(11), 11–27.

Schwerin, M., Bongartz, B., Cramer, H., Eurich-Menden, B., Flachowsky, G., Gauly, M., *et al.* (2012) Climate change as a challenge for future livestock farming in Germany and Central Europe. *Zuechtungskunde* 84, 103–128. [In German]

Serageldin, I. (1999) Biotechnology and food

security in the 21st century. *Science* 285, 387–389.

Shade, R.E., Schroeder, H.E., Pueyo, J.J., Murdock, L.L., Higgins, T.J. and Chrispeels, M.J. (1994) Transgenic pea seeds expressing the α-amylase inhibitor of the common bean are resistant to bruchid beetles. *Nature Biotechnology* 12, 793–796.

Tester, M. and Langridge, P. (2010) Breeding technologies to increase crop production in a changing world. *Science* 327, 818–822.

UN (2010) The emissions gap report: Are the Copenhagen accord pledges sufficient to limit global warming to 2 degrees Celsius or 1.5 degrees Celsius: A primary assessment. November 2010. UN Environment Program (UNEP), Nairobi, 52 pp.

Verstraete, F. (2011) Risk management of undesirable substances in feed following updated risk assessments. *Toxicology and Applied Pharmacology*, DOI:10.1016/j.taap.2010.09.015 (accessed 25 May 2013).

Viljoen, A. and Wiskerke, S.C. (eds) (2012) *Sustainable Food Planing: Evolving Theory and Practice.* Wageningen Academic Publishers, Wageningen, Netherlands, 600 pp.

Wally, O. and Punja, Z.K. (2010) Genetic engineering for increasing fungal and bacterial disease resistance in crop plants. *Genetically Modified Crops* 1, 199–206.

Wals, A.E.J. and Corcoran, P.B. (eds) (2012) *Learning for Sustainability in Times of Accelerating Change.* Wageningen Academic Publishers, Wageningen, Netherlands, 550 pp.

Waterlow, J.C. (1999) The mysteries of nitrogen balance. *Nutrition Research Review* 12, 25–54.

Welzer, H. and Wiegandt, K. (eds) (2011) *Perspectives of Sustainable Development.* S. Fischer Verlag GmbH, Frankfurt/Main, Germany, 341 pp. [In German]

Wenk, C., Simon, O., Flachowsky, G. and Dupuis, M. (eds) (2009) Animal nutrition tomorrow: healthy animals – efficient and sustainable production – valuable products. Report on the meeting 2 April 2009. Vol 32. ETH Zürich, Zurich, Switzerland, 278 pp. [In German]

Wennemer, H., Flachowsky, G. and Hoffmann, V. (2006) *Protein, Population, Politics – How Protein Can Be Supplied Sustainably in the 21st Century.* Plexus Verlag, Mittenberg and Frankfurt/Main, Germany, 160 pp.

Whitford, R., Gilbert, M. and Langridge, P. (2010) Biotechnology in agriculture. In: Reynolds, M.P. (ed.) *Climate Change and Crop Production.* CAB International, Wallingford, UK, Cambridge, Massachusetts, pp. 219–244.

WHO (2007) Protein and amino acid requirements in human nutrition. Report of a Joint WHO/FAO/UNU Expert Consultation. *World Health Organization Technical Report Series 935,* 1–265.

Windisch, W., Fahn, C., Brugger, D., Deml, M. and Buffler, M. (2013) Strategies for sustainable animal nutrition. *Zuechtungskunde* 85, 40–53. [In German]

Young, V.R., Bier, D.M. and Pellett, P.L. (1989) A theoretical basis for increasing current estimates of the amino acid requirements in adult man, with experimental support. *American Journal of Clinical Nutrition* 50, 80–92.

Zoiopoulos, P.E. and Drosinos, E.H. (2010) *The Animal Feed Question in the Shadow of Contemporary Food Crisis: The European Challenge.* Nova Science Publishers, New York, 232 pp.

Zollitsch, W., Winckler, C., Waiblinger, S. and Haslberger, A. (eds) (2007) *Sustainable Food Production and Ethics.* Wageningen Academic Publishers, Wageningen, Netherlands, 550 pp.

2 Fundamentals of Plant Biotechnology

Marc De Loose[1]* and Ann Depicker[2]

[1]*Department of Plant Biotechnology and Bioinformatics, Ghent University, and Institute for Agricultural and Fisheries Research (ILVO), Merelbeke, Belgium; [2]Department of Plant Biotechnology and Bioinformatics, Ghent University, and Department of Plant Systems Biology, VIB, Ghent, Belgium*

2.1 The Importance of Biotechnology in Plant Breeding

Plant biotechnology is a general term describing a research domain, covering a broad spectrum of methodologies and techniques. Over the last decades, the output of the activities in this domain have resulted in an exponential increase of knowledge on the biological, biochemical and physiological processes of plant growth and development. This knowledge, in combination with the development of new breeding strategies and agricultural technologies, has led over the past 50 years to an enormous increase in crop yield and improvement in quality.

Plant biotechnology, sometimes also referred to as green biotechnology, is often perceived as a synonym for genetically modified (GM) plants, or transgenic plants. However, plant biotechnology is much broader than the applications of genetic modification, and in fact it is a continuum of different techniques and research domains, ranging from mutagenesis, organ and tissue culture and the use of molecular markers in breeding to transgenesis. In general, all these techniques in one way or another have resulted in broadening the deployable pool of useful genetic variation for the breeder, and transgenesis ultimately has the potential of an unlimited creation of genetic variation.

For more information, we refer the reader, at the end of this chapter, to some selected books covering this exciting area of plant research and technology in much greater detail (e.g. Chrispeels and Sadava, 2003; Slater *et al.*, 2003; Yunbi, 2010; Altman and Hasegawa, 2011).

2.1.1 Breeding and mutagenesis

Genetic variation is needed to enable new properties to be added to existing plant varieties in order to obtain new varieties that better suit the needs of society, farmers, industry and/or consumers. In the past, for the selection of plants and new traits, the breeder was restricted to the natural variation within the gene pool of crossable plants. Gene transfer and recombination between plant traits can be achieved by cross-hybridization after pollination (Fig. 2.1). However, with increasing phylogenetic distance, this way of gene transfer is rare or even impossible. Mutagenesis, making use of chemical agents, ultraviolet (UV) irradiation or isotope treatments, resulted in the random induction of mutations in the genome, leading to plants with modified phenotypes and added variability, although mostly a result of loss of function. Time-consuming screening of high numbers of

*E-mail: marc.deloose@ilvo.vlaanderen.be

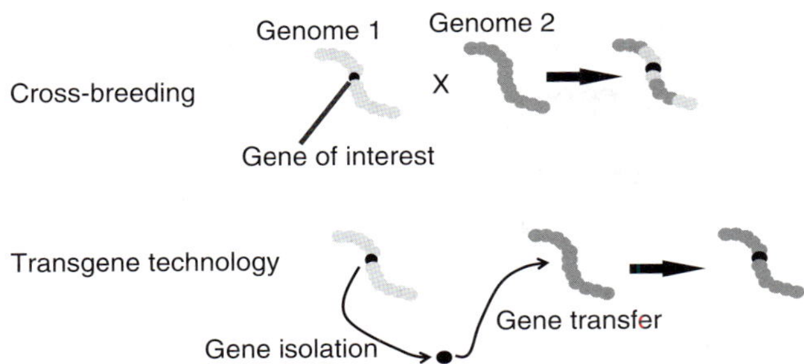

Fig. 2.1. Comparison of cross-breeding with transgene technology (or genetic modification technology) to introduce a gene of interest (encoding a trait of interest) present in genome 1 into a well-performing variety of genome 2. By breeding, crossing is needed between one parent of the well-performing variety and the other parent containing the trait of interest. Thereby, both genomes are mixed and many backcrosses are needed to obtain again the well-performing variety while keeping the trait of interest. By genetic modification, a trait from any organism can be isolated and introduced into the well-performing background without dragging along unwanted genetic information.

individuals is then needed to identify the mutant plants with the desired traits. Moreover, this is only possible for traits that are easy to measure at the phenotypic level.

2.1.2 Polyploidization

Chromosome doubling, by making use of colchicine, has been used intensively in the breeding of fodder crops, as often the derived polyploid varieties resulted in higher yield. Polyploidization is also a key technology for the creation of fertile interspecific hybrids, as it allows crosses between related plants with different ploidy levels. In this way, wild relatives could, after genome doubling, be crossed with crop species. This has been used, for example, for the introduction of pathogen-resistance genes from wild potato relatives into varieties of the cultivated potato.

2.1.3 Somatic hybridization

Somatic hybridization is a technique used to introduce novel genes into a crop genome from a donor species with which the crop will normally not interbreed. More sophisticated microinjection and cell fusion techniques allow whole cells or parts of cells to fuse to create composites or chimeras from unrelated species. The main problem has been the instability of the new genome combinations from two dissimilar species. Therefore, the use of somatic hybridization in breeding programmes has, in practice, been restricted to the introgression of genes from related plant species.

2.1.4 Transgenesis or genetic modification

Since the 1990s, the focus of breeding has shifted towards transgenesis and variants such as cisgenesis. Genetic modification (GM) is different from the previously described tools and traditional breeding in that any modification can be designed and tailored to achieve the desired effect (see Fig. 2.1). This GM methodology is more precise, has very limited problems with genome instability and the success rate is much higher than with mutagenesis, interspecific crosses and somatic hybridization. It is not only restricted to the introduction of new genes but can also be used to modify the expression of endogenous genes, or can shut down the expression of genes coding for undesired traits. Already in an early stage, primary transformants with a desired genetic configuration can be identified. Transgenic plants containing inadvertent fusions or disruptions of transgenes as well as of endogenous genes can be eliminated rapidly as the DNA sequence of the input DNA fragment is known. However, there are still limitations to breeding via transgene technology. For instance, only traits for which the corresponding genes are isolated and characterized can be transferred or modified (see Section 2.2.5). Moreover, agriculturally important traits, such as yield, are controlled by a whole set of genes and pathways, of which the expression is regulated as a complex network. Modifying these traits in crop plants by making use of the transgenic approach will depend largely on the knowledge gained on how the expression of these traits is regulated at the molecular level and on the availability of tools to regulate transgene expression in order to drive the desired metabolic pathways in the transgenic plant. On the basis of the current research in the domain of systems biology, one may also expect new breakthroughs for these traits soon.

The first barrier that needs to be circumvented in order to be able to develop a transgenic plant is the availability of a transformation procedure for the plant species to be transformed (see Section 2.2.5). Two ways of gene transfer can be distinguished: direct gene transfer, which makes use of physical forces to introduce the DNA to the nucleus of the accepting plant cell, or the *Agrobacterium tumefaciens* 'transport' system based on the soil bacterium gene transfer mechanism.

The second barrier is the access to genes coding for traits of interest. There are numerous transgenes developed today that code for both input and output traits. Input traits are those that potentially alter crop production. An example is Bt maize, which produces an insect toxic protein in the plant and which, as a result, does not need the application of a pesticide to control European corn borer infection. Another example is the introduction of resistance to the non-

selective herbicide, Roundup, trait in crops, allowing the spraying of Roundup for weed control without damaging crop plants. Output traits, on the other hand, are those that alter the harvested product. One example is the increase of the oleic acid content of soybeans, resulting in an improved product for food and industrial use. Another example is transgenic potatoes with an improved starch extractability. At the moment, most currently authorized and commercialized GM crops code for input traits, but a switch to more output traits is expected (see Chapter 12).

The third barrier is the huge costs required to commercialize a GM plant or derived product. During the development of GM crops, different categories of costs can be distinguished:

1. The development of the transgenic plant expressing the desired trait in a stable way, leading to the desired effect.
2. The development of commercial varieties containing the gene of interest by breeding.
3. The authorization dossier in order to allow the commercialization of the GMO event and its derived varieties.

The introduction of a transgenic crop in the field and of a transgenic plant derived product into the market thus requires huge investments, mainly for risk assessment and deregulation. Therefore, companies are focusing their investments mainly on crops that are cultivated on a large scale such as soybean, maize, canola and cotton, such that it becomes economically feasible. In research institutes and universities, on the other hand, tobacco, rice and *Arabidopsis thaliana* are most frequently used for genetic modification, due to well-developed transformation methods, easy propagation and well-studied genomes. There, they serve as model organisms for other plant species.

2.1.5 Molecular marker technology

Molecular marker technology is also considered as a plant biotechnological method. The development of molecular markers contributed to a higher efficiency in the screening and selection of particular hybrids after a breeding process. Restriction fragment length polymorphism (RFLP) was the first molecular marker, developed in early 1980. Another breakthrough was the emergence of polymerase chain reaction (PCR) in 1990. With this technology, a new generation of DNA markers was introduced into modern plant-breeding systems. Examples are randomly amplified polymorphic DNA (RAPDs), sequence characterized amplified regions (SCARs), sequence tagged sites (STS), single polymorphic amplification test (SPLAT), variable number of tandem repeats (VNTRs), amplified fragment length polymorphism (AFLP), DNA amplification fingerprinting (DAF), single-strand conformational polymorphism (SSCP), single-nucleotide polymorphism (SNP), microsatellites or short tandem repeats (STRs), cDNA, DNA microarrays and rDNA-internal transcribed spacers (ITS).

Molecular markers are DNA-based markers and they make it possible to select not for the trait but for the presence of the genetic information coding for the trait of interest. This is done by searching for DNA markers that have been proven to be genetically linked to the genes coding for the trait. It also allows the whole breeding process to be speeded up. After crossing a particular variety containing an elite genetic background with a plant containing the desired trait in an unfavourable genetic background, the use of molecular markers allows identification within the population of offspring plants, with maximal conservation of the elite genotype but additionally with the new trait of interest. Besides gaining time, this approach also makes it possible to carry out selection on small plants, even without the phenotypic expression of the genotype and the trait of interest. DNA markers are also very important for the fast introduction of a particular transgene-encoded trait in a variety of existing cultivars through conventional breeding, allowing valorization of GMO elite events.

2.2 GMO Technology: What, How and Its Importance to Plant Breeding

This section describes the different steps of developing a transgenic crop. The description will be general and rather theoretical. Depending on the crop and the gene of interest to be introduced, the procedure is slightly different. However, in general terms, it can be stated that the whole process of developing a commercial GMO variety consists of the following steps: isolation and functional analysis of genes encoding a particular trait; the assembly of a functional gene construct; transfer and integration of the gene construct into the genome; identification of elite events and breeding with the elite events; and preparation of the authorization dossier, needed for commercialization.

2.2.1 The first step: isolation and functional analysis of genes

Before discussing the process of gene isolation and characterization, it is relevant to describe what is meant by the term 'gene'. A gene is a molecular unit of heredity of a living organism. It is a name given to a stretch of DNA that codes for a polypeptide or for an RNA chain that has a function in the organism. Living beings depend on genes, as they specify all proteins and functional RNA chains. Genes hold the information to build and maintain an organism's cells and pass genetic traits to its offspring. All organisms have many genes corresponding to various biological traits, some of which are immediately visible and some of which are not. However, over the last decades the simple definition of a stretch of linked nucleic acids that code for a protein has been broadened and fine-tuned. New terms have been introduced, such as gene families, iso-proteins, as a result of alternative splicing, and ribozymes, which are RNA molecules that catalyse particular reactions.

In isolating a gene, the researcher makes use of the following characteristics of a gene to start his research: a gene has a certain primary structure (the sequence); a gene occupies a particular location in the genome; a gene encodes an RNA with a particular expression pattern; genes are coding for a protein or an mRNA with a defined function. This means that information on a gene can be collected at various points during development using different experimental approaches.

The discovery of the gene corresponding to the desired trait is the first step. In the context of developing transgenic crops, the choice of traits (genes) to be introduced will either be problem driven or opportunity driven. The first category is mainly input traits such as herbicide resistance, resistance to abiotic stresses and pest resistance. The second group depends on output traits, resulting in crops of which the harvested material has either new or improved characteristics. In the early days of plant genetic engineering, the capacity to identify and isolate the gene of interest was a significant restriction in applying transgene technology for crop improvement. The source of genes that were available for genetic engineering was rather limited. Often, these genes were discovered by accident. Many of the genes used were isolated from bacteria, such as antibiotic-resistance genes, herbicide-resistance genes or genes coding for insecticidal proteins.

Later, due to the huge and exponential efforts in plant molecular biology research, a broad range of genes, also of plant origin, became available. Important to the discovery and characterization of these genes was the introduction of *A. thaliana* as a model plant. The assembly of gene libraries and the characterization of the function of isolated genes using tools such as insertion mutagenesis resulted in a broad range of mutants of which it was known which trait was affected. Subsequently, mapping and sequencing of the mutant genomes allowed the identification of the gene that was encoding the trait of interest. Nowadays, the possibilities for discovering genes coding for particular traits is substantially increased by the introduction of bioinformatics and DNA sequencing, especially next-generation sequencing.

The main source of empirical information about gene function and structure has been, and still is, the sequencing of mRNA transcripts and corresponding cDNAs. Especially, the use of high-throughput methods applied on model plants, both monocots and dicots, has been very important. In this case, the DNA sequence is the starting point, and methods such as protein purification, complementation of mutant phenotypes and reversed genetics are examples of experimental approaches in identifying a gene's function. However, for 50% of the genes of most organisms, the functional or physiological properties of a gene product are still unknown. Recent improvements in sequencing methods have made it possible to generate huge data sets on the DNA sequences of whole genomes. In combination with the development of bioinformatic tools, new opportunities for identifying genes through exon discovery in genomic sequence data have been created. Two approaches can be used to identify candidate genes: making use of previously described gene sequences in other species or starting from the intrinsic characteristic of genes (e.g. nucleotide composition and sequence motifs).

Thus, as a result of fundamental research in the domain of gene function analysis over the last decades, enormous progress has been made in understanding how thousands of gene products interact with each other, resulting in a functional organism of which the developmental processes are co-ordinated. It has also provided insight on how an organism has the capacity to react towards environmental challenges.

The research output with the model species, Arabidopsis, allows the use of genes as such for genetic engineering, but it is also a bridge to isolate the homologous genes from the crop species. Molecular methods making use of cDNA libraries, gene libraries and the introduction of PCR technology have speeded up this process of isolating genes from crop plants on the basis of the knowledge obtained from the model species, Arabidopsis.

By the introduction of new analytical methods, such as metabolomic profiling and expression profiling, it is now possible to identify and characterize the function of genes and the resulting gene products in the metabolic pathways of living organisms. This knowledge opens up enormous possibilities for plant genetic engineering, allowing not only the expression of a new protein but also the ability to change complex metabolic pathways in the plant. This will allow the scope of input traits to be broadened, but especially it will open unlimited possibilities to improve the quality of plant crops with new or modified output traits.

2.2.2 The cloning step: from gene isolation towards the assembly of a functional transgene construct

After the discovery of the gene coding for the trait of interest and physically obtaining the DNA for this gene, the next step is to create a gene unit that will lead to the functional expression of the trait of interest in the acceptor plant. This means that the coding sequence for the trait needs to be linked to the necessary regulatory elements that are responsible for the transcription and translation of the coding sequence, finally resulting in a functional protein. In order to obtain a functional gene, different options are open, depending on the relationship between donor and acceptor organism and depending on the desired spatial and developmental expression pattern for the transgenic trait. Moreover, when specific demands on cell localization for the transgenic protein need to be taken into account, specific targeting signals need to be included in the construct.

The easiest and most straightforward way is when the intact gene is available and when the regulatory elements are functional in the acceptor plant and give rise to the desired expression profile. In this case, the gene as such can be used for the gene transfer: this means that the DNA sequence, as it is obtained from the donor organism, contains all the required information with the coding sequence as well as regulatory elements. However, in many cases, only the coding sequence of the gene of interest is

used, either because it is derived from a cDNA sequence and the regulatory sequences are not available or because the gene is derived from bacteria or another organism using different expression signals than in plants, or because the expression regulation of the original gene does not result in the desired spatial and developmental expression levels in the acceptor plant. In this case, extra cloning work needs to be done.

In short, a transgene requires the following necessary components:

1. The promoter, which is the on/off switch that controls the gene transcription during development in response to external biotic and abiotic circumstances. The promoter is responsible for tissue- or organ-specific expression or for constitutive expression of the coding sequence. An example of the latter is the widely used 35S cauliflower mosaic virus promoter. Genes under the control of the 35S constitutive promoter are highly transcribed throughout the whole life cycle and in most of the tissues and organs of the plant. The promoter is, in most cases, combined with the coding sequence by a transcriptional fusion in the 5′ untranslated region.

2. The coding sequence of the gene of interest. Often, this coding sequence is adapted in order to achieve a better expression in the transgenic plant. For example, in the case of certain bacterial genes, the adenine thymine (AT) content of the coding sequence is much higher than in plants and this results in aberrant premature transcription termination. This has been the case for the gene encoding the *Bacillus thuringiensis*-derived insecticidal protein (BT). By substituting A–T nucleotide pairs with G–C nucleotide pairs in the coding sequence without significantly changing the amino acid sequence, correct transcription could be obtained. Another reason to change the primary sequence is that the codon usage can be very different in the donor organism than in the acceptor plant. Then, the codon usage can be adapted by making use of the redundancy of the genetic code. In this way, enhanced translation efficiency and thus the production of transgenic protein in the plants, can be achieved.

3. The terminator sequence, which contains the signal for the end of transcription and for correct processing of the RNA.

4. Localization signals. In case there is a need for specific targeting of the transgenic protein towards particular organelles, the nucleus, the endoplasmatic reticulum, the vacuole or to the extracellular space, a signal peptide and cleavage signals need to be foreseen in the gene construct. In this case, a translational fusion between the signal or localization peptide and the coding sequence is made.

5. Translational fusion of tags. In particular cases, signals for post-translational modification or peptide tags are added, especially when the goal is to purify the recombinant protein produced in the transgenic plant. In such a case, it should be checked after cloning whether the fusion between the coding sequence and the tag are in frame.

In practice, transgene construction starts with the on-paper design of the complete gene sequence. Subsequently, the DNA fragment is made in a test tube by cutting and pasting different gene elements together; nowadays, the transgene encoding DNA fragment is made synthetically by polymerizing nucleotides in the desired sequence.

2.2.3 Selectable markers

Besides the transgene to be transferred to the plant cell, a selectable marker gene is linked to it most of the time, as expression of this selectable marker allows identification of the plant cells with the integrated transgene construct. Indeed, only in some systems is the transformation frequency high enough, meaning higher than 1%, to make simple PCR screening for transformed cells with the transgene construct possible. This is, for instance, the case after protoplast co-cultivation or after floral dipping of Arabidopsis flower stalks with the relevant *A. tumefaciens* strain. Thus, in case the efficiency of transformation/integration and regeneration of transformed cells is sufficiently high, one can avoid the use of selectable markers by screening the plants for the presence of transgenic DNA, and as high throughput PCR screening is now technically feasible, this is the preferred approach, if possible (see Section 2.2.7).

For tissue explant co-cultivation, however, the incorporation and expression of a transgene in a plant cell that subsequently regenerates into a transgenic shoot is a rare event. Therefore, a method is needed either to kill the non-transformed cells, such that only the transformed growing cells survive, or a visible marker is needed to distinguish the transformed from the non-transformed cells.

Selectable marker genes result in a selective advantage for the transformed cells: either (i) because of the expression of an enzyme that inactivates the selective agent (detoxification); or (ii) because of the expression of a resistant variant of the endogenous enzyme that is the target of the selection agent (tolerance). In the first case, the selectable marker genes encode enzymes that detoxify particular chemical products, such as antibiotics or herbicides, providing the transformed plant cell's resistance to these chemicals. In the second case, the selectable marker encodes an antibiotic- or herbicide-tolerant target enzyme.

An example of the first group is the antibiotic-resistance genes. Antibiotic-resistance markers detoxify the antibiotics by modifying them. For example, *neomycin phosphotransferase II* (NPTII) or *hygromycine phosphotransferase* (HPT) specifically phosphorylate neomycin/kanamycin/G418 and hygromicin, respectively. Kanamycin and hygromycin are taken up by the plant cells and they bind on to a subunit of the mitochondrial and chloroplast ribosomes, thereby inhibiting translation and thus blocking energy production and photosynthesis. The phosphorylated kanamycin and hygromicin compounds can no longer bind to the ribosomes. Thus, plant cells with the NPTII or HPT enzymes are resistant to kanamycin and hygromycin, respectively. In this way, these antibiotic-resistance selection markers are very often used to select transformed cells from a population of non-transformed plant cells, or to distinguish transgene-containing progeny plants from the segregants without the transgene (Fig. 2.2).

An example of the second category is 5-enolpyruvylshikimate-3-phosphate synthase (EPSPS). This plant enzyme plays a role in the biosynthesis of the aromatic amino acids, phenylalanine, tyrosine and tryptophan. The herbicide, glyphosate, is a competitive inhibitor of the endogenous enzyme. However, a variant EPSPS present in a special strain of *Agrobacteria* has a slightly altered shape. This alteration prevents glyphosate from binding, thus allowing the resistant EPSPS to catalyse the amino acid synthesis reaction. In this way, the expression of this gene in the transgenic plant gives a competitive advantage to the wild-type cells by bypassing the blocked biosynthetic pathway route and restoring the essential function.

Thus, only plants that have integrated the selectable marker gene will survive on tissue culture media complemented with the appropriate antibiotic or herbicide (see Section 2.2.7).

Similar to the transgene construct of interest, the selectable marker gene also needs the appropriate promoter and termination signals to allow functional expression of this trait. Selectable marker genes are driven mostly by promoters that result in constitutive expression such as the CaMV 35S and the nopaline synthase promoter for transformation of dicotyledonous plants and promoters of the ubiquitin gene of maize and the actin gene of rice for monocotyledonous plants.

Because antibiotics are used to combat human and animal pathogens, special care has been taken to study the spread of the resistance genes from the transgenic plants to the pathogens. Indeed, when the pathogens acquire these same resistance

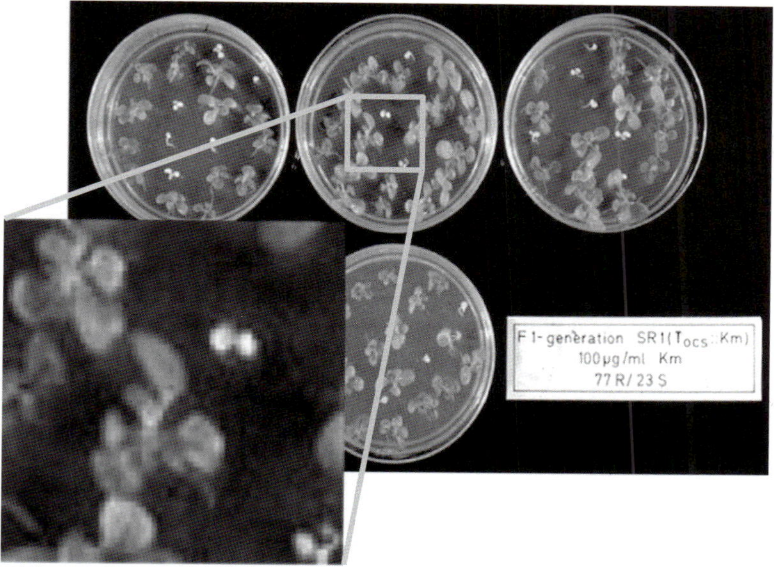

Fig. 2.2. Use of selectable markers, such as *nptII* conferring resistance to the antibiotic kanamycin. The progeny of a transformant segregate, the selectable marker and the transgene as a Mendelian marker in a 3 to 1 ratio: large seedlings that contain the antibiotic-resistance marker are green, and can make new leaves and roots; small seedlings that do not contain the selectable marker are sensitive to kanamycin, stop photosynthesis and are white.

genes, they become resistant to the antibiotics, and those antibiotics can then no longer be used to cure the disease. No evidence was found for the spread of the resistance genes and no generally accepted scientific arguments can be provided to underpin the concern in using antibiotic-resistance selection markers, but as a result of an increasing worldwide concern, several other selection markers have been developed for the selection of transgenic plants. One of them allows the transformed cells to metabolize a substrate that the wild-type cells cannot use as an energy source, and as a result, the transformed cells will grow out of the mass of non-transformed cells.

2.2.4 Cloning vectors for plant transgenes

A plethora of different ready-to-use cloning vectors for introducing the constructed transgene is available. These vectors are replicating in *Escherichia coli* and in *Agrobacterium* and already contain one of the available selectable markers or a marker for visual scoring of the presence of an expressed transgene (Fig. 2.3c). In a multicloning site or in a Gateway cassette, a fully assembled transgene can be introduced. Alternatively, these vectors contain the cloning site in between the regulatory elements, such that the coding sequence for the trait of interest can be inserted between regulation elements of choice. For specific goals such as for silencing a plant gene, a fragment of gene sequence is cloned in sense and antisense direction in between the regulatory elements, such that on transcription, a self-complementary transcript is formed and double-stranded RNA is generated.

Once the cloning vector with the DNA fragment to be transferred to the plant and carrying the (trans)gene and selectable marker is assembled, this vector is transferred to a disarmed *A. tumefaciens* strain containing the molecular machinery for transferring this DNA segment to the plant.

2.2.5 Transfer of the transgene construct into the genome of a plant cell

As discussed earlier, gene exchange between any organisms becomes possible by genetic transformation. This powerful tool enables plant breeders to broaden the genetic variation from which they can select in order to obtain new combinations of genes, leading to improved/adapted plant varieties. In other words, genetic transformation expands the possibilities for breeders beyond the limitations imposed by traditional cross-breeding and selection (see Fig. 2.1).

For most of the plant species, genetic transformation is carried out on tissue explants, of which a fraction of the cells is competent for regeneration to complete (fertile) plants after the transformation process. There are no universally applicable protocols for plant tissue culture, because experience has shown that protocols need to be optimized for each genus, species, cultivar, ecotype and different tissues used for transformation.

The following criteria need to be fulfilled in order to set up a successful transformation platform, and the different steps are discussed in the different paragraphs below:

1. Delivery of DNA to the plant genome without influencing cell viability negatively.
2. Selection of the transformants, the selectable marker gene and promoter.
3. Regeneration of intact plants.
4. Transmission of the transgenes into the next generations in fertile plants or stable maintenance and expression in vegetatively propagated crops.

In summary, transformation aims to create heritable changes in the plant as the result of the uptake and the stable integration of introduced DNA in one of the plant chromosomes.

Different approaches have been studied and developed to achieve DNA transfer into plants. A common feature is that in all cases the foreign DNA needs to enter the plant cell. Therefore, the DNA should first penetrate the cell wall and the plasma membrane before reaching the nucleus and integrating into the nuclear genome. In other words, the main goal of each of these methods is to transport the new gene(s) and deliver them into the nucleus of a cell without killing it.

Currently, there are two major techniques for transferring foreign DNA into an organism:

- The first method is based on the indirect physical transfer of foreign genes into target plant cells. The method used mostly is particle bombardment (biolistics). DNA is bound to tiny particles of gold or tungsten, which are subsequently shot into plant tissue or single plant cells under high pressure. The accelerated particles penetrate both the cell wall and the membranes. The DNA separates from the metal and is integrated into the plant genome inside the nucleus. Endogenously present DNA repair mechanisms play an important role in the stable integration of the foreign DNA into the genome. This method has been applied successfully for many cultivated crops, especially monocots like wheat or maize, for which transformation using A. tumefaciens has been less successful. The major disadvantage of this procedure is that serious damage can be done to the cellular tissue.
- The second method is making use of the 'machinery' of A. tumefaciens, which has the capacity to transfer DNA and proteins into the plant cell. A. tumefaciens is a soil bacterium and a natural plant parasite. To create a suitable environment for themselves, these Agrobacteria insert part of their genes into plant hosts, resulting in crown galls, which are a proliferation of plant cells near the soil level. The genetic information for tumour growth is encoded by the oncogenes that are transferred and located on the T-DNA, while the genetic information for the machinery to transfer this T-DNA, encoded in the vir genes, is found on a mobile, circular plasmid, the Ti plasmid (Fig. 2.3a). The T-DNA is delineated by a left border (LB) and a

right border (RB), consisting of ± 25 bp repeat sequence, and only the DNA segment in between the two border sequences is transferred to the plant cell. When A. tumefaciens attaches to a plant cell, it transfers this T-DNA from the bacterium through a cytoplasmatic bridge to the plant cell.

When A. tumefaciens is used for plant transformation, tumour induction is, of course, not wanted and therefore the bacterial oncogenes on the T-DNA are removed. The obtained A. tumefaciens strain is called disarmed, but still contains the virulence genes for DNA transfer. In such an A. tumefaciens strain, a replicating plasmid vector carrying whatever DNA fragment in between the LB and the RB can be introduced, and this strain will then transfer the new recombinant T-DNA to the plant cell (Fig. 2.3b).

The bacterium is thus providing a transport system, enabling transfer of foreign genes into plants. This method works especially well for dicotyledonous plants like potatoes, tomatoes and tobacco, but was originally less successful in crops like wheat and maize. However, recently, much progress on A. tumefaciens-mediated transformation has been made for monocot crops, and currently it has become the standard method. Several factors influence the A. tumefaciens-mediated transformation. The most important factors are the plant genotype and the explant type used as the target tissue for co-cultivation to be transformed. But also the A. tumefaciens strain, the binary T-DNA vector, the inoculation and co-culture conditions and the tissue culture/regeneration medium used influence the success rate.

2.2.6 Integration of the transgene construct into the genome of a plant cell

Irrespective of the transfer method, the introduced foreign DNA needs to be integrated in the genome of the acceptor plant cell. This is always a non-targeted process, meaning that the transferred DNA can integrate anywhere in any of the plant chromosomes (Fig. 2.3c). This is the result of the plant-mediated illegitimate recombination process. After the DNA fragment is transferred to the nucleus, the plant host enzymes, present to repair double-strand breaks, recognize the DNA ends of the bombarded fragment or the T-DNA and of broken chromosome ends. Those ends are then ligated in a non-homologous DNA end-joining process. When no double-strand breaks are available, the transferred DNA cannot integrate. Therefore, many DNA transfer processes result in transient expression because the introduced DNA is not recombined with the plant chromosome via illegitimate recombination. As a result, the presence of double-strand breaks and of the non-homologous end-joining process determine in part the competence of the plant cell to be transformed.

It should thus be stressed that the position of integration of the transgene will be different in every independently transformed cell. The integration process is random, but it is observed that there is a tendency that genes are inserted in genome regions that are actively transcribed. In the

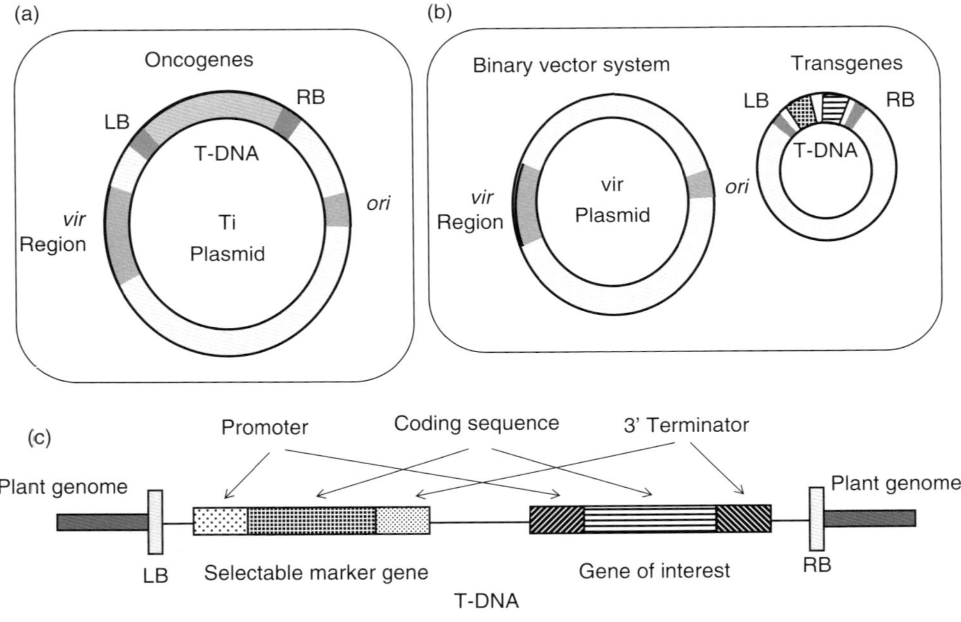

Fig. 2.3. *Agrobacterium tumefaciens*-mediated plant transformation. (a) The Ti plasmid contains the genetic information for transferring DNA to the plant cell in the *vir* region, while the genetic information that is transferred to the plant cells is present in the T-DNA. The T-DNA is delineated by the left border (LB) and the right border (RB) and contains the oncogenes, which induce crown galls on wounded surfaces of plants. (b) The *A. tumefaciens* vector system. The T-DNA is removed from the Ti plasmid, resulting in a vir plasmid, still containing the information to transfer DNA to plants. In this mutant strain, a binary vector plasmid is introduced containing a T-DNA in which the transgene construct (horizontally striped box) and the selectable marker (dotted box) are assembled. On contact with a plant cell, this *A. tumefaciens* strain will selectively transfer only the T-DNA segment with the gene of interest and selectable marker to the plant cell. (c) A simple transgene locus in a transformed plant, referred to as a transformation event, different in every independently obtained transformant. Basically, the inserted T-DNA segment contains a transgene encoding the trait of interest and a selectable marker gene. Both genes are assembled to contain a promoter where transcription starts, the coding sequence for the appropriate protein synthesis and a 3′ terminator to stop transcription.

case of co-transformation, i.e. when two gene constructs are mixed for particle gun bombardment or when two *A. tumefaciens* strains, each containing a different gene construct, are used, there is a high chance that both gene constructs are inserted in the same locus of the genome of the acceptor plant. This has consequences, as in such a case the two inserted traits will co-segregate during crossing experiments (see Section 2.2.10).

The advantage of using *A. tumefaciens* for transgene DNA delivery is that most of the integrated transgenes are present as intact constructs and, as predicted, inserted as a linear fragment from left to right T-DNA end. Another advantage is that most transformed cells contain only one or a limited number of integrated T-DNA copies. In the case of particle bombardment, more severe rearrangements occur both in the acceptor genome as well as in the transferred DNA segment. Moreover, frequently, multiple copies of scrambled DNA fragments of the transgene segment are observed in tandem arrays at the locus of insertion. As it has been shown that this can have an impact on the expression of the transgene, and on the stability of expression in subsequent generations (because the repeated tandem arrays become silenced), the *A. tumefaciens* transfer method is, in general, the preferred one.

2.2.7 Identification/selection of transformants

Most transformation methods make use of tissue culture to propagate plant cells from a tissue explant with regenerative properties. All the cells of the tissue explants are, in one step, subjected to the transformation process. The result is that a fraction of these cells will have taken up the foreign DNA, and only a fraction of these cells will also integrate the foreign DNA within the nuclear genome.

In order to identify the plant cells that have integrated the foreign gene, a selectable marker is used or a marker that allows transformed cells to be distinguished from non-transformed cells (see Section 2.2.3). When a selectable marker is linked to the gene of interest, cells that are transformed have a selective advantage over the non-transformed cells when grown on a medium containing the selection agent. The selection agent can be an antibiotic when the selectable marker is an antibiotic-resistance marker (such as the *nptII* gene or the *hpt* gene) or a herbicide when it is a herbicide-resistance gene such as biolaphos acetyltransferase (*bar*). The result of the selection is that cells that do not integrate or do not express the selectable marker will not survive on the medium to which the selectable agent is added. Without a selection marker, the small number of plant cells that have integrated the foreign DNA would be lost within the huge population of regenerating wild-type cells. Another claimed advantage is that the negative impact of position effects is immediately counterselected since only the transformants that express the selection marker well will survive the selection, and as the selection marker and the trait of interest are linked and integrated together in the genome, there is an indirect selection for good expression of the linked transgene. However, this is not always the case, and stable expression of the transgene should be controlled in several subsequent generations.

There are two major concerns raised by some stakeholders on the use of antibiotic-resistance markers: it might affect the therapeutic efficacy of the clinical use of these antibiotics and it might increase the potential horizontal gene transfer from the transgenic plant into intestinal and soil bacteria. The European Food Safety Authority (EFSA) evaluated these concerns and came to the conclusion that, according to information currently available, adverse effects on human health and the environment resulting from the transfer of the two antibiotic resistance marker genes, *nptII* and *hpt*, from GM plants to bacteria, associated with the use of GM plants, are unlikely. The EFSA also stated that mainly the presence of antibiotics in the environment and the frequent use of antibiotics were the key factors in driving the selection of dissemination of antibiotic-resistance genes. Moreover, they found that the use of antibiotic-resistance markers in transgenic plants did not add a substantial additional risk. Nevertheless, there is general agreement that antibiotic-resistance genes should be removed from transgenic plants on commercialization.

Therefore, to end up finally with antibiotic-resistant, marker-free GMOs, either an extra step is used to eliminate the selectable marker gene from the inserted DNA locus, or screening for DNA integration is performed by using PCR on many regenerated shoots in the absence of selection.

To remove the selectable marker, a technically simple approach is the use of co-transformation with two separate T-DNAs, of which one contains the transgene of interest and the second contains the selectable marker. In the case of unlinked integrations of both T-DNAs, the marker gene and the transgene of interest will segregate in the next generation, and thus marker-free GMO segregants can be selected. Another approach to remove the antibiotic selectable marker makes use of site-specific recombinases. When the selectable marker is flanked by the specific recognition sequences, transient expression of the site-specific recombinase will result in the removal of the DNA fragment containing the marker gene from the integrated T-DNA fragment. Best known is the Cre/Lox system.

However, it remains easier not to have to use a selection marker. In experimental conditions where the transformation frequency is sufficiently high, screening for uptake of DNA in the plant cells is feasible by using PCR. In this function, shoots need to be regenerated, allowing part of each individual plant to be sampled in order to extract genomic DNA for PCR analysis. Plants for which a positive signal is obtained in the PCR reaction have integrated the gene of interest. However, this does not guarantee expression of the trait. Therefore, the transgene mRNA or the transgene-encoded protein can be quantified.

For the herbicide-resistance marker, the major concerns focus on gene flow, development of weediness and the potential toxic or allergenic effects. Concerning gene flow and weediness, one should distinguish the impact on the environment on the one hand and agricultural management on the other hand. In case a functional herbicide-resistance gene is transferred to a wild relative, or if seeds from the herbicide-resistant crop are spread in nature, these plants will not have a competitive advantage over the natural population, as the selective agent is not used in nature and is thus not a problem. It is different in an agricultural context. Different management systems are available to cope with weeds, which can be wild species that have become resistant because of gene transfer or they can be volunteers being, for example, the result of a previous culture. The concern about toxicity and allergenicity has rather to do with the use of the herbicide than with the transformation event or the expression of the transgene.

2.2.8 Identification and characterization of an elite event

Once a set of transgenic plants has been selected, the identification of the best performing transgenic plant has to be started, referred to as the elite event. This is a step-by-step process, and depending on the goal, the trait, the experience and the available infrastructure, the order of these various steps can be different. In general, however, the criteria screened are the integration pattern, the expression profile and localization of the insertion in the genome.

To be useful for commercial application, first the expression levels should be sufficient in order to obtain the desired phenotype. Second, a simple integration pattern is desired, because complex integration patterns will make the breeding work with the selected event more complicated and might in some particular cases also lead to unstable expression in successive generations.

It is important to mention here that the primary transgenic plant will not be the variety that will be commercialized as such. As stated initially, this primary transformant allows a new trait to be introduced in the breeding programme. Therefore, it is necessary to check that the integrated fragment is stable over successive generations. Then, the transgene insert can be bred into different varieties.

In the regulatory process associated with the commercial release of a transgenic plant, the transgenic plant as such, and the transgene locus in particular, needs to be fully characterized (see Fig. 2.3). Commonly used methods for molecular characterization are PCR, Southern blot analysis and sequencing. These methods allow the amplification and sequencing of the inserted fragment and also the DNA region of the plant genome flanking the inserted fragment. These sequence data allow checking whether the DNA fragments that are to be inserted are indeed inserted in the expected configuration. In other words, it allows checking for deletions, insertions, repeats and mutations that could have occurred during the integration process. Based on the sequence data, a search is made for putative new open reading frames and whether these potentially newly formed open reading frames show homology to known toxic or allergenic proteins or peptides.

With Southern blotting, the structure of the inserted DNA can be unravelled. This analysis allows searching for multiple insertions, direct and indirect repeats of the transferred DNA fragment. Southern blotting is also used to scan the whole transgenic genome with a probe of the transferred DNA for potential secondary inserted small fragments. Also, with a probe of vector DNA, the absence of vector backbone sequences is controlled, as these vector backbone sequences may contain an antibiotic-resistance gene and prokaryotic origins of replication, which is unnecessary and not wanted.

In the past, Northern blotting and at present quantitative PCR are used to measure the transgene-derived mRNA accumulation levels in different parts of the transgenic plant. Western blotting and ELISA are used to check whether translation into proteins occurs. Western blotting also allows the size of the recombinant protein produced to be checked. For some transgene-encoded enzymes, functional assays are also available to check accumulation and the specific activity.

2.2.9 Risk evaluation of a transgenic event

Risk evaluation for food/feed use and environmental safety of GM plants is much broader than characterizing the transgene insert at the molecular level. Also, the impact on the nutritional value and the potential toxic and allergenic effects are analysed and evaluated (see Chapters 3–8).

In the case of plants of the so-called first generation (plants with input traits) the evaluation is based on studies of substantial equivalence with the original plant (see Chapters 5 and 6). In these studies, the choice of reference material with which the GMO is compared is very important. The primary transformant could be compared with the wild-type plant from which it is derived, but also a null segregant progeny plant of the transformant can be used as a reference/negative control. However, in real life these comparative studies are carried out on material that is harvested from plants that are the result of crossing and backcrossing the primary transgenic plant with a plant with a genetic background that has a potential economic value. In that case, the original plant is not the right comparator. The ideal comparator is the plant with a genetic background that is as identical as possible to the genetic background of the GMO plant that is being studied. Ideally, it is the isogenic material of the GMO plant that is the subject of the comparative analysis. When this material is not available, an alternative is to compare the different parameters in the function of the genetic variation that is at that moment representative of the genetic background at the basis of the commercialized GM varieties.

2.2.10 The elite event is the starting point for the development of GM plant varieties

Once an elite event is selected on the basis of its agronomic performance and/or product quality, the breeding work with this event can start. In fact, until this point a parallel can be seen with the selection of, and characterization work on, new candidate material collected by the breeder to use as parents. In both cases, the aim is to introduce 'new traits' in the breeding programme. Once these are identified, the next step is crossing and selecting plants with the new trait expressed in the earlier selected and combined genetic backgrounds (see Fig. 2.1).

Backcross breeding is used to introduce the transgene in an elite background. Transgenic plants are crossed with elite breeding lines used in traditional plant breeding in order to combine the desired traits of the elite parents and the transgene into a single plant. Therefore, offsprings are backcrossed repeatedly to the elite line in order to obtain a high-yielding transgenic variety. The result will be a GM plant with a yield potential and general characteristics close to current hybrids that express on top the extra trait encoded by the transgene. The backcrossing work can be speeded up substantially by making use of molecular markers. In practice, this means that as many molecular markers of the elite genetic background as possible spread all over the entire genome are combined with the markers specific for the transgene.

Sometimes, it can also be interesting to combine several transgenes in one new variety. These are often called stacked events. They can be obtained in two ways, either by crossing two single events, each containing one independent single insert, or by co-transformation. In the second case, an event can be selected where both transgenes are inserted in the same genetic locus. The advantage in this case is that both traits will co-segregate in the next generations, which makes breeding simpler. This is not the case when the two transgenes are combined via sexual crossing.

The entire genetic engineering process is basically the same for any plant. The length of time required to complete all steps from start to finish varies depending on the gene, the crop species and the resources available. It can take anywhere from 6 to 15+ years before a new transgenic variety is ready for release for growing in production fields.

2.3 Future Prospects

Genetic engineering and GM plants have the potential to address challenging topics in plant breeding. The technology allows a targeted broadening of the genetic variation from which the breeder can select at the start of the breeding programme, and it has the potential to tackle important biotic and abiotic constraints in agriculture. Both input and output traits can be dealt with, even if the genetic information is not available within the gene pool of the species to which the plant variety belongs. Another advantage is that it allows one specific new trait to be added to a plant while maintaining all the other desirable characteristics of the cultivar (see Fig. 2.1 and Chapters 7 and 12). However, we want to stress that this approach will not replace the conventional breeding programme, but it will broaden the possibilities and variety of the input material for breeding.

The development of a transgenic plant is labour-intensive and technologically very demanding, although not extremely difficult, but especially the identification of an elite event is a time-consuming process that needs technically skilled personnel with expertise in an environment with laboratory facilities and greenhouses suitable for growing transgenic plants in isolation. This last restriction is needed to fulfil the legal requirements for the cultivation of transgenic plants. Also, the breeding work is a substantial part of the process, including backcrosses and selfings in a controlled environment, where unwanted cross-pollination can be absolutely avoided.

Also, preparing the dossier for the authorization of cultivation/import/food and feed use is time-consuming and very expensive as compared to a new non-GM cultivar. Because of the complexity and because of different legal frameworks in different parts of the world, experienced collaborators are essential to prepare these dossiers.

Today, most of the GM plants that are commercialized code for input traits such as insect tolerance, herbicide tolerance, virus resistance and hybrid seed systems for heterosis. However, some output traits have also been developed, such as the production of oils, production of vitamin A precurser β-carotene and better digestibility and better extractability of starch (see Chapters 7, 10 and 12). Crops in which the GMO technology has been adopted successfully are maize, soybean, cotton, canola, potato, rice and papaya. But also for many other crops, one may expect to see products ready for commercialization in the coming years (see Chapter 12). This is a clear trend in North and South America, in India and China, and recently also in Africa. Successful introduction in Europe, however, will not depend on technological progress and investments but on the acceptance of these GM crops in society by different stakeholders; in the first place, by consumers (see Chapters 13–15). As it is known that the opinion of the consumer is influenced largely by the media, non-governmental organizations, retail, etc., it will be very interesting to see whether the controversy about GM crops will continue its stalemate of the past 10 years or whether acceptance of this technology will pave the way for European-targeted applications.

References

Altman, A. and Hasegawa, P. (2011) *Plant Biotechnology and Agriculture*, 1st edn. Academic Press, Amsterdam, Boston, Heidelberg, London, New York, Oxford, Paris, San Diego, San Fancisco, Singapore, Sydney and Tokyo.

Chrispeels, M.J. and Sadava, D.E. (2003) *Plants, Genes and Crop Biotechnology*, 2nd edn. Jones and Bartlett Publishers, Sudbury, Massachusetts.

Slater, A., Scott, N. and Fowler, M. (2003) *Plant Biotechnology: The Genetic Manipulation of Plants*. Oxford University Press, New York.

Yunbi, X. (2010) *Molecular Plant Breeding*. CABI International, Wallingford, UK.

3 Guidance Documents for Nutritional and Safety Assessment of Feeds from GM Plants

Yi Liu,* Anna Christodoulidou, Antonio Fernandez Dumont, Andrea Germini and Claudia Paoletti

European Food Safety Authority (EFSA), GMO Unit, Parma, Italy

3.1 Introduction

The use of genetically modified organisms (GMOs) – their deliberate release into the environment, their import and processing for food, feed and industrial uses – is regulated in the European Union (EU). Since the 1990s, different legislative instruments have been put in place to ensure their safety. Directive 2001/18/EC and Regulation (EC) No 1829/2003 are the main legal instruments regulating the cultivation and marketing of GMOs and derived products in the EU. A key requirement of these two legislations is a comprehensive and science-based risk assessment, which is a prerequisite for placing GMOs and their products on the EU market. This centralized risk assessment is performed by the European Food Safety Authority (EFSA), with the support of a panel of independent experts on GMOs (EFSA GMO Panel), a team of scientists from the EFSA GMO Unit and the input of EU member states. In the case of deliberate release into the environment by cultivation, the environmental risk assessment is carried out by a national competent authority of an EU member state. Such assessment is taken into account by the EFSA when preparing its scientific opinion.

The EFSA GMO Panel has published a set of guidance documents for the risk assessment of GM plants, their derived products for food and/or feed uses, non-food and/or non-feed uses and their cultivation. These guidance documents, explaining the strategy and defining the scientific criteria to conduct the risk assessment, are in line with risk assessment principles developed and agreed at the international level.

To date, applications for GMOs have included only plants and microorganisms. Since GM animals and derived food and feed products are still in the development phase, this chapter will focus on the safety and nutritional assessment of GM plants and their derived products in the EU. The nutritional and safety assessment of GM plants in other parts of the world is summarized briefly at the end of the chapter. The risk assessment of GM microorganisms is discussed in Chapter 11 of this book. There is a Russian textbook available on *Genetically Modified Food Sources – Safety Assessment and Control* (Tutelyan, 2013); however, the timing of this publication meant that this book could not be considered in the present volume.

3.2 EU Legislative Framework

The first legal instruments regulating GMOs in the EU (Council Directive 90/220/EEC and Council Directive 90/219/EEC) were developed in 1990 with the specific scope to protect human and animal health and the environment. Since then, the main legal instrument, considered as the horizontal

*E-mail: yi.liu@efsa.europa.eu

legal frame governing biotechnology in the EU, has become Directive 2001/18/EC (EC, 2001) of the European Parliament and of the Council of 12 March 2001 on the deliberate release into the environment of GMOs. Directive 2001/18/EC repeals Council Directive 90/220/EEC and strengthens previously existing rules on the release of GMOs into the environment, *inter alia* introducing principles for environmental risk assessment, mandatory post-market (environmental) monitoring, mandatory supply of information to the public, mandatory labelling and traceability at all stages of placing on the market and the establishment of a molecular register.

According to Directive 2001/18/EC, authorizations granted under Council Directive 90/220/EEC must be renewed in order to avoid disparities and to take full account of the conditions of consent of Directive 2001/18/EC. The authorization (renewable) is granted for a maximum period of 10 years, starting from its issue date. Following the placing on the market of a GMO, the notifier must ensure that post-market monitoring and reporting are carried out according to the conditions specified in the authorization. Directive 2001/18/EC, which is implemented in each member state by national regulations, deals with both small-scale field trials (voluntary releases carried out for experimental purposes, dealt with in part B of the Directive) and the marketing provisions of GMOs (dealt with in part C).

A 'sister' Directive, 'Council Directive 98/81/EC of 26 October 1998 Amending Council Directive 90/219/EEC on the Contained Use of Genetically Modified Microorganisms', governs the contained use of genetically modified microorganisms. Since this chapter focuses on GM plants, this Directive is not further detailed in this section.

In 2003, two regulations amending or repealing previous legal instruments were published: Regulation (EC) 1829/2003 (EC, 2003) of the European Parliament and of the Council of 22 September 2003 on genetically modified food and feed and Regulation (EC) 1830/2003 of the European Parliament and of the Council of 22 September 2003 concerning the traceability and labelling of genetically modified organisms and the traceability of food and feed products produced from genetically modified organisms, and amending Directive 2001/18/EC.

In Regulation (EC) 1829/2003, the rules for safety assessment have been strengthened and expanded. This regulation introduces, for the first time, specific rules on GM feed and enshrines labelling requirements for GM food and feed, previously covered only partially by Council Regulation (EC) 1139/98 and Commission Regulation (EC) 49/2000. As a main feature, this regulation implements the 'one key–one door' approach: one single authorization covers both food and feed uses, therefore filling the legal vacuum for feed product approval in the EU. The implementation of this regulation binds the risk assessment of GM food and GM feed together in one single process. Under Regulation (EC) 1829/2003 (in force since 18 April 2004), the applicant must submit an application containing the environmental and the food and feed safety risk assessment of the genetically modified event in question.

The scientific information provided in applications is evaluated by the EFSA, established by Regulation (EC) 178/2002 (EC, 2002) of the European Parliament and of the Council of 28 January 2002. In addition to evaluating these applications, the EFSA provides scientific advice to risk managers of the member states and of the European Commission, who are responsible for decision making with respect to product authorizations and inspection, and more in general for the management of food and feed safety.

The EU recognizes consumers' rights for information and labelling as a tool to make an informed choice. Since 1997, labelling to indicate the presence of GMOs as such or in a product is mandatory. Regulation (EC) 1830/2003 reinforced the labelling rules on GM food and feed: (i) mandatory labelling is extended to all food and feed irrespective of detectability; (ii) traceability is defined as the ability to trace GMOs and products produced from GMOs at all stages of their placing on the market and is implemented through their production and distribution chains.

The 1% threshold specified under Commission Regulation (EC) 49/2000 for the adventitious presence of approved GMOs was lowered in Regulation (EC) 1829/2003 to a new *de minimis* threshold of 0.9%. On 15 July 2011, Regulation (EC) 619/2011 on the low-level presence (LLP) of GMOs in food and feed imports came into force, regulating the adventitious presence of GM food and feed not approved in the EU. As a principle, this regulation follows the zero-tolerance policy for non-approved GM products. In order to implement this policy in realistic and operational terms, the LLP legislation defines the technical zero at the level of 0.1%. This is the lowest level of GM material that can be detected reliably by the EU Reference Laboratory during the validation of quantitative detection methods.

Recently, the European Commission has published 'Commission Implementing Regulation (EU) No 503/2013 of 3 April 2013 on applications for authorisation of genetically modified food and feed in accordance with Regulation (EC) No 1829/2003 of the European Parliament and of the Council and amending Commission Regulations (EC) No 641/2004 and (EC) No 1981/2006' (EC, 2013). This regulation came into force on 28 June 2013, with a transitional period until 8 December 2013, and enlists legally binding requirements that will need to be taken into account in the preparation and evaluation of GMO applications.

3.3 The European Food Safety Authority

Following a series of food crises in the late 1990s, Regulation (EC) No 178/2002 established the European Food Safety Authority (EFSA) in January 2002 as part of a comprehensive programme to improve EU food safety, to ensure a high level of consumer protection and to restore and maintain confidence in the EU food supply. The roles of the EFSA are to assess and communicate the risks associated with the food and feed chains, to advise risk managers and to address any scientific question or issue on food and feed safety within Europe. In the European food and feed safety systems, risk assessment is carried out independently from risk management. Being the responsible risk assessment body in the EU, the EFSA produces scientific opinions, guaranteeing a sound scientific foundation to European policies and legislation and supporting the European Commission, the European Parliament and EU member states in taking effective and timely risk management decisions. The EFSA's remit covers food and feed safety, nutrition, animal health and welfare, plant protection and plant health. In all these fields, the EFSA's most critical commitment is to provide objective and independent science-based advice and clear communication grounded in the most up-to-date scientific information and knowledge. The EFSA also consults EU national competent authorities on every GM plant application and addresses scientific concerns that are raised by these national authorities.

With respect to GMO risk assessment, the EFSA evaluates the data present in GMO applications and reviews all the scientific information relevant for the safety of any given GMO. This provides the scientific foundation necessary for the risk managers to authorize (or not) GM products into the EU market. The EFSA's work relies on the close collaboration between a panel of independent external experts on GMOs (the EFSA GMO Panel) and a team of scientists from the EFSA GMO Unit. The panel meets regularly in plenary sessions to discuss work in progress and to adopt scientific opinions. Each opinion results from a collective decision-making process, with every panel member having an equal say. As part of its remit, the GMO Panel also produces guidance documents to explain its approach to risk assessment, to detail scientific requirements and to ensure transparency in its work. The EFSA scientific evaluation of the risk assessment included in GMO applications is published in the form of 'EFSA scientific opinions', which are all available on the EFSA website (http://www.efsa.europa.eu/en/gmo/gmoscdocs.htm).

3.4 EFSA Guidance for Risk Assessment of Food and Feed from GM Plants

The European legal framework for GMO risk assessment requires the evaluation of any possible effect on human and animal health and on the environment that the release or the placing on the market of any given GMO may have. To do so, the EFSA has published two guidance documents: one focuses on food and feed safety – 'Guidance for risk assessment of food and feed from genetically modified plants' (EFSA, 2011a); the other focuses on environmental risk assessment (ERA) – 'Guidance on the environmental risk assessment of genetically modified plants' (EFSA, 2010a). Given the scope of this book, this chapter will focus primarily on the food and feed safety of GM products. We do, however, provide in Section 3.4.6 a short summary of the ERA approach adopted by the EFSA.

The current food and feed risk assessment strategy for GM plants and derived products (EFSA, 2011a) seeks to deploy methods and approaches to compare GM plants and derived products with their appropriate non-GM comparators. The underlying assumption of this comparative approach is that traditionally cultivated non-GM crops have gained a history of safe use for consumers and/or domesticated animals. These traditionally cultivated crops can thus serve as comparators when assessing the safety of GM plants and derived products.

The introduction of gene(s) in a plant by genetic modification has the objective to introduce, in the recipient plant, novel characteristics of interest. These characteristics are the so-called 'intended effects' or 'introduced traits' that fulfil the original objectives of the genetic modification. At the same time, the insertion of genes may result also in other, additional effects in the recipient plant, going beyond the intended effect(s) of the genetic modification: the so-called 'unintended effects'. Unintended effect(s) are potentially linked to genetic rearrangements or metabolic perturbations affecting different pathways (see Chapter 2).

Both intended and unintended differences are identified by comparing the agronomic, phenotypic and compositional characteristics of a GM plant with those of its non-GM comparator(s), grown under the same conditions. The relevance of the observed intended and unintended changes to human and animal health is then assessed by investigating the toxicological, allergenicity and nutritional properties of the GM crop. A detailed description on how this assessment is done in the EU is available in the EFSA guidance document (EFSA, 2011a).

The risk assessment includes four steps: hazard identification, hazard characterization, exposure assessment and risk characterization. According to Codex Alimentarius (CAC, 2007a), hazard identification is the identification of biological, chemical and physical agents capable of causing adverse health effects and which may be present in a particular food and feed or group of foods and feeds. This first step focuses on the identification of relevant differences between the GM plant and its non-GM comparator, taking into account natural variation. Codex Alimentarius (CAC, 2007a) defines hazard characterization as the qualitative and/or quantitative evaluation of the nature of the adverse effects associated with biological, chemical and physical agents, which may be present in food and feed. A dose–response assessment should be performed whenever data are obtainable. The exposure assessment aims to estimate quantitatively the likelihood of exposure of humans and animals to the food and feed derived from GM plants. In general, an exposure assessment characterizes the nature and size of the populations exposed to the food and feed derived from GM plants, together with the magnitude, the frequency and the duration of such exposure. It is necessary that every significant source of exposure is identified. Finally, the risk characterization is the qualitative and/or quantitative estimation of the probability of occurrence and severity of known or potential adverse health effects in a given population based on hazard identification, hazard characterization and exposure assessment (CAC, 2007a). A proper risk characterization should also identify and possibly quantify any uncertainty.

The hazard identification and characterization, as performed in the EU, includes the molecular characterization of the GM plant, its food and feed safety evaluation and its environmental impacts. These three components, together with the exposure assessment, allow characterizing the risk associated to any given GM plant and its derived products by addressing the following aspects:

1. Characteristics of the donor organisms and recipient plant.
2. Genetic modification and its functional consequences.
3. Agronomic and phenotypic characteristics of the GM plant.
4. Compositional characteristics of GM plants and derived products.
5. Potential toxicity and allergenicity of gene products (proteins, metabolites) of the whole GM plant and its derived products.
6. Anticipated intake and potential for nutritional impact.
7. Influence of processing and storage on the characteristics of the derived products.
8. Environmental impact of the GM plant.

3.4.1 Molecular characterization

The first step of the risk assessment of GM plants and derived products is the comprehensive molecular characterization of the GM plant in question. The objective of this characterization is to gather information on the structure and expression of the insert(s) and on the stability of the intended trait(s) to characterize the intended changes and to identify any potential unintended change. If present, these unintended changes, linked to the interruption of endogenous genes or to the possible production of new toxins or allergens, are further investigated in the relevant complementary part(s) of the risk assessment process.

The information required to perform the molecular characterization concerns both the genetic modification itself and the actual GM plant resulting from the transformation. This information focuses on the transformation method, the characteristics of the inserted nucleic acid sequence(s), the genomic location where the sequence(s) is/are inserted (insertion site(s)), the possible genomic alteration(s) due to the insertion of the sequence(s) and the expression and stability of the inserted sequence(s).

3.4.2 Food and feed safety evaluation

The comparative approach

As described above, the foundation of the food and feed safety evaluation is the comparative assessment which identifies differences in compositional, agronomic and phenotypic characteristics between the GM plant and derived products with respect to its non-GM comparator, taking into account the natural variation of the measured characteristics.

The EFSA 'Guidance for risk assessment of food and feed from genetically modified plants' (EFSA 2011a) provides detailed guidance on how to perform such comparative assessment. In particular, the EFSA's approach requires the simultaneous application of two complementary statistical tests: the test of difference and the test of equivalence (EFSA, 2010b). The test of difference is used to verify whether the GM plant, apart from the introduced trait(s), is different from its non-GM comparator. The test of equivalence is used to verify whether the agronomic, phenotypic and compositional characteristics of the GM plant fall within the range of natural variation, estimated from a set of non-GM reference varieties with a history of safe use included in the field trial. The combination of these two statistical tests allows the objective quantification of the degree of similarity between the GM plant and its non-GM comparator, taking into account natural variability.

The test materials (grain, forage, etc.) necessary to perform the comparative assessment (see Chapter 4) must be obtained from appropriately designed field trials in which the GM plant is grown together with its non-GM comparator and selected non-GM reference varieties. Until 2011, the

EFSA required the use of non-GM lines with comparable genetic background (near-isogeneic lines in the case of sexually propagated crops and isogenic lines in the case of vegetatively propagated crops) as comparators in the evaluation of GM plant applications. These non-GM comparators, defined as conventional counterparts, are derived from the breeding scheme used to produce the GM plant. In 2011, although the conventional counterpart remains the non-GM comparator of choice for the assessment of GM plants in the EU, the EFSA developed the 'Guidance on the selection of comparators for the risk assessment of genetically modified plants and derived food and feed' (EFSA, 2011b) to address the new challenges imposed by the increasing complexity of breeding schemes that did not always allow the production of a conventional counterpart but for which a fit-for-purpose alternative comparator could be identified.

Nevertheless, there will be cases where no suitable comparator exists. For instance, when the food and feed derived from a GM plant is not closely related to a food and feed with a history of safe use, or when a specific trait or a specific set of traits are introduced with the intention of significantly changing the composition or the physiology of the plant (see Chapters 5–7). In all these cases, a comparative risk assessment cannot be performed and a comprehensive safety and nutritional assessment of the GM plant and derived products needs to be carried out.

Compositional analyses

The comparative compositional analysis is usually performed on the raw agricultural commodity, as this represents the main point of entry of the material into the food and feed chain (see also Chapter 4). Additional analysis of processed products is conducted where appropriate, on a case-by-case basis. The compositional analysis is carried out on an appropriate range of compounds selected in accordance with the OECD consensus documents on compositional considerations for new plant varieties (OECD, 2012b). These compounds include, in general terms, the analysis of proximates, key macro- and micronutrients, anti-nutritional compounds, natural toxins and allergens, as well as other plant metabolites relevant for the specific plant species. The vitamins and minerals selected for analysis should be those present at levels which are nutritionally significant and/or those with a nutritionally significant contribution to the diet at the levels at which the plant or its derived products are consumed.

Depending on the intended effect(s) of the genetic modification, on the nutritional value and on the intended use of the GM plant, additional compositional analyses on specific compounds may be required. For example, in the case of oil-rich GM plants, a fatty acid profile for the main saturated monounsaturated and polyunsaturated fatty acids should be included. Whereas, in the case of GM plants intended to be used as an important protein source, an amino acid profile on individual amino acids and main non-protein amino acids should be included. The analysis of specific metabolites may also be relevant in case the genetic modification alters metabolic pathways affecting the physiology of the plant.

Depending on the outcome of the compositional analysis, and in particular whenever relevant changes between the GM plant and its non-GM comparator are identified, further toxicological and nutritional assessments are required.

Agronomic and phenotypic characteristics

The safety of the agronomic and phenotypic characteristics of the GM plant is also assessed following the comparative approach. A variety of endpoints are analysed, including: yield, flowering time, day degrees to maturity, duration of pollen viability, sensitivity to biotic and abiotic stress, etc. Where specific agronomic characteristics are the objective of the intended modification (e.g. a genetic modification conferring drought tolerance), additional studies under relevant selected conditions are required (e.g. in the case of drought tolerance, the performance of the GM plant should be tested under different water-limiting regimes).

Depending on the outcome of the comparative assessment of agronomic and phenotypic characteristics, and in particular whenever relevant changes between the GM plant and its non-GM comparator are identified, further toxicological and/or nutritional assessments are required.

3.4.3 Toxicological assessment

The scientific assessment of any risk that GM plants and derived products may pose to human and animal health includes a toxicological assessment, which is necessary to evaluate the safety of a GM food for human consumption and of a GM feed for animal intake (EFSA, 2011a). Such toxicological assessment begins with the evaluation of the newly expressed protein(s) produced by a GM plant as a consequence of its genetic modification. Further toxicological testing of the whole food and feed derived from the GM plant is hypothesis driven and is required on a case-by-case basis, depending on the evaluation of the novel protein(s) and the outcome of the comparative assessment.

As a guiding principle of the EFSA, the selection of a specific toxicological test follows internationally agreed test methods described by the OECD or by the European Commission (EC, 2004). The choice of test protocols depends on the type of GM plant and derived food and feed, on the genetic modification, on the intended and unintended alterations, on the intended use and anticipated intake and on the knowledge available. Since these test protocols have been developed for pure compounds, they may need adaptations before they can be applied to complex matrices like food and feed, including those derived from GM plants.

The toxicological assessment of the newly expressed protein(s) addresses not only the possible effects linked to its presence in the GM plant (the intended effect of the genetic modification) but also its possible effects on the potential presence of other new constituents, on the possible changes in the levels of endogenous constituents beyond normal variation and on the impact it may have on the composition and/or the phenotype of the GM plant (the potential unintended effects of the genetic modification). The information necessary to address these issues is generated from:

1. Molecular and biochemical characterization of the newly expressed protein(s), including amino acid sequence, molecular weight, post-translational modifications, description of the function, information on the enzymatic activities including the temperature and pH range for optimum activity, substrate specificity and possible by-products.
2. Up-to-date bioinformatics search for similarity to proteins known to be toxic to humans and animals.
3. Information on the stability of the newly expressed protein(s) under relevant processing and storage conditions for the food and feed derived from the GM plant and on the potential production of stable protein fragments generated through processing and storage.
4. Resistance of the newly expressed protein(s) to degradation by proteolytic enzymes (e.g. pepsin).
5. Twenty-eight-day repeated dose oral toxicity study in case the newly expressed protein(s) does not have a duly documented history of safe use.

According to the EFSA (EFSA, 2011a), the toxicological assessment of the whole food and feed derived from a GM plant is required only when the composition of the food and feed is substantially modified, or when there are indications for the potential occurrence of unintended changes based on the preceding molecular characterization and compositional or phenotypic analyses. In such cases, the testing programme includes a repeated dose 90-day oral toxicity study in rodents (EFSA, 2011c). Guidance on how to perform such a test is published by the EFSA (EFSA, 2011a), taking into account OECD guideline 408 (OECD, 1998). Depending on the outcome of the repeated dose 90-day oral toxicity study, further toxicological testing may be needed (e.g. studies on reproductive/developmental effects or on chronic toxicity; see Chapters 5 and 8).

The conclusion of the toxicological assessment provides information on: (i) the potential toxicity of the newly expressed protein(s) and, if present, of other natural constituents which might indicate adverse effects on human and animal health; and (ii) potential adverse effects of the whole food and feed derived from the GM plant.

3.4.4 Allergenicity assessment

Allergenicity is the potential of a substance to cause an allergy. An allergy is a pathological reaction of the immune response to that particular substance. Food allergy is an adverse immune response to food/feed and is different from toxic reactions and food intolerance. The majority of the substances causing an allergic response to food and feed are proteins.

Food allergy in humans, in food-producing animals and in companion animals has been described (EFSA, 2010c). Immune-mediated adverse reactions have been detected in pets (Verlinden *et al.*, 2006) as well as in food-producing animals, due to the replacement of animal proteins with vegetal proteins; for example, in young farm animals such as calves and pigs (Dreau and Lalles, 1999) and in intensively reared fish (Bakke-McKellep *et al.*, 2007). However, additional work is still needed for a more comprehensive understanding of the mechanisms involved in these immune-mediated adverse reactions.

Food allergy can be caused by various immune mechanisms and it is generally divided into two forms, IgE mediated (e.g. allergic reactions to peanuts or soybean) and non-IgE mediated (e.g. allergic eosinophilic gastroenteropathies). Since the form provoking the most severe allergic reactions, including anaphylaxis, is an IgE-mediated food allergy, this has been the focus of different international guidance documents

for allergenicity assessment of GMOs (CAC, 2009; EFSA, 2011a).

As indicated above, as Regulation (EC) 1829/2003 covers both GM food and feed, and therefore the potential for allergenicity in animals, both companion and livestock animals must be taken into account. In this context, the same principles apply to the allergenicity assessment of food as well as feed. The allergenicity assessment of GM plants and derived products, described in the EFSA's 'Scientific opinion on the assessment of allergenicity of GM plant and microorganisms and derived food and feed' (EFSA, 2010c) and implemented in the EFSA guidance (EFSA, 2011a), is divided into the following.

Allergenicity assessment of newly expressed proteins

There is not a single test that can predict the allergenicity properties of a protein. Therefore, both the EFSA (2010c, 2011a) and Codex Alimentarius (CAC, 2009) recommend a case-by-case, weight-of-evidence approach to assess the allergenicity of newly expressed protein(s). The cumulative body of evidence necessary for this assessment is based on the comparison of amino acid sequence similarity (bioinformatics search), specific serum screening, pepsin resistance test, other *in vitro* digestibility tests and, possibly, other studies (e.g. *in vitro* assays and *in vivo* models).

The EFSA (2010c, 2011a) also underlines the importance of including an adjuvanticity assessment of the newly expressed protein(s) in the allergenicity assessment. Adjuvanticity is the capacity of a substance to increase the immune response to an antigen when co-administered with that antigen. Strictly speaking, and unlike allergens, adjuvants do not have the capacity to trigger an allergic reaction per se, since they lack sensitizing potential. However, combined exposure to an adjuvant and an antigen may boost the immune response of an allergic individual animal to that particular antigen, causing more severe adverse reactions than when exposed to the antigen only.

Allergenicity assessment of the whole GM plant

When the plant receiving the new gene(s) is known to be allergenic, its allergenicity is compared with that of the appropriate non-GM comparator(s), taking into account natural variation. For this comparison, the approach to follow depends on the available information on the allergenicity of the recipient plant. Possible proposed approaches are: the use of analytical methodologies in conjunction with human sera; the inclusion of allergens among the endpoints tested in the compositional analysis; and immunological testing with sera collected from animals sensitized experimentally (EFSA, 2011a).

The conclusion of the allergenicity assessment provides information on: (i) the likelihood of the newly expressed protein(s) to be allergenic; and (ii) the likelihood of the GM plant to be more allergenic than the non-GM comparator.

3.4.5 Nutritional assessment

Food and feed derived from GM plants intended to be placed on the EU market should not be nutritionally disadvantageous to humans and animals. If the outcome of the comparative compositional assessment does not indicate a relevant difference between the GM plant and its non-GM comparator(s), except for the introduced trait(s), nutritional equivalence can be inferred and no further nutritional studies are needed (see also Chapter 4). If this cannot be demonstrated, animal feeding studies are necessary. For example, in the case of GM plants having an altered content of nutrients, animal studies with model or target species (e.g. poultry, pigs, ruminants, fish, etc.) are performed in order to determine the bioavailability of individual nutrients and their impact on animal performance and feed safety (see Chapters 5–8).

Properly designed animal feeding studies should span the growing or the finishing period to slaughter for chickens, pigs and cattle and should cover the major part of the lactation cycle for dairy cows and the laying cycle for hens or quails (see Chapter 5). Growth studies with aquatic species, such as carp, are preferable for feedstuff intended only for aquaculture.

The experimental design and statistical analysis of feeding studies depends on the choice of animal species, the type of plant trait(s) studied and the magnitude of the expected effect. Endpoint measurements vary according to the target species used in the study and should include data on animal health and welfare, animal losses, feed intake, body weight and animal performance (see Chapter 5).

The nutritional assessment addresses not only the nutritional relevance in the total diet for the consumers/animals of the newly expressed protein(s) (the intended effect of the genetic modification) but also the nutritional relevance of other possible new constituents, and the changes in the levels of endogenous constituents in the GM plant and derived food and feed (the potential unintended effects of the genetic modification).

The conclusion of the nutritional assessment of food and feed derived from GM plants provides information on: (i) the nutritional profile of the GM plant and derived products as compared to the non-GM comparator; and (ii) the altered nutrient levels in the GM-derived products that impact the anticipated intake and nutritional properties of the food and feed (EFSA, 2011a).

3.4.6 Environmental risk assessment

Environmental risk assessment (ERA) is an integral part of the safety assessment of GM plants and derived products. Although the ERA is always carried out for each GM plant application submitted to the EFSA, the amount of data requirement increases if the GM crop is expected to be cultivated in the EU (EFSA, 2010a). The ERA is based on the biological and ecological characteristics of the plant, the nature of the introduced trait(s), the receiving environment in which the plant will be introduced and the scale and frequency of the proposed introduction. Such an evaluation also considers the potential direct and indirect, as well as the immediate, delayed and cumulative long-term adverse effects.

Key elements for the ERA are the potential changes in the interactions between the GM plant and the biotic and abiotic factors, namely: changes in the persistence and invasiveness of the GM plant; potential for gene transfer; interactions between the GM plant and target organisms; interactions between the GM plant and non-target organisms; effects on biogeochemical processes and abiotic environment; and impacts of specific cultivation, management and harvesting techniques associated with the cultivation of the GM plant. As for food and feed safety assessment, the comparative approach is the guiding principle for the ERA of GM plants and it is applied to address each specific area of concern. Further details on the ERA strategy are provided by the EFSA (2010a) but are not addressed here since they are outside the scope of this chapter.

3.4.7 Exposure assessment

As already mentioned above, the aim of the food and feed exposure assessment is to estimate quantitatively the likelihood of exposure of humans and animals to products derived from GM plants. This is done first by determining the concentrations of the newly expressed protein(s) and of the endogenous constituents altered by genetic modification, and second, by identifying and quantifying any new constituents. In cases where the genetic modification results in an altered level of an endogenous constituent, or where a new constituent occurs naturally in other food and feed products, the anticipated change in total intake of this constituent must be assessed considering realistic scenarios. Ideally, intake levels should be estimated from representative consumption data and should take into account several factors such as the influences of processing and storage conditions and possible routes of exposure, as well as the characteristics

and dietary habits of the different consumer groups. However, since reliable consumption data may not always be available, data on import and production quantities may be helpful to assess worst-case scenarios. Probabilistic methods can also be valuable tools to determine plausible intake values.

In practice, intake estimates for any given food product can differ significantly depending on the data source. This is well illustrated by the following example: the intake of certain compounds has been estimated either from European or US population consumption data sets, or from the WHO Global Environmental Monitoring System - Food Contamination Monitoring and Assessment Programme (GEMS-Food Consumption Cluster Diets, accessable at http://www.who.int/foodsafety/chem/gems/en/index1.html) that include the European population and the FAO food balance sheet. The food balance sheets are based on the total amounts of food produced, imported, exported and utilized for various purposes, not on dietary surveys, and are generally considered to overestimate average consumption. These data sources are intrinsically different from each other and inevitably lead to different intake estimates, which cannot be evenly compared. As a first step to address this problem, the EFSA has consolidated 27 national dietary surveys into a so-called 'comprehensive European database' (EFSA, 2012a). The EFSA is currently undertaking further work to harmonize dietary data collections across EU member states (EFSA, 2012b).

3.4.8 Risk characterization and post market monitoring

The risk characterization of GM plants and derived products is based on data from hazard identification, hazard characterization and exposure assessment. A sound risk characterization considers all the available evidence collected during the assessment and highlights, whenever possible, any uncertainty identified at any stage of the risk assessment (EFSA, 2007). When addressing uncertainty, distinction should be made between uncertainty and variability. Uncertainty is linked to knowledge gaps or lack of data. Variability refers to variation that exists in reality; for example, individual variation in food consumption or varying concentrations of compounds in different food items. Uncertainty can be reduced through research, whereas variability cannot since it is an intrinsic property of a system.

As pre-market risk assessment studies cannot fully reproduce the diversity of consumers (human or animals), the possibility that unpredicted side effects may occur in some individuals – such as those with certain disease states, those with particular genetic/physiological character-istics or those who consume the products at high levels – remains. In such cases, a post-market monitoring (PMM) programme can be used to verify if the outcome of the pre-market risk assessment is confirmed in reality. Most GM plants that have been evaluated by the EFSA to date did not require PMM; however, this may change in the future with the increasing complexity of new generations of GM plants targeting significant compositional or physiological changes (see Chapter 7).

3.5 Applications of GM Plants for Food and Feed Uses

Currently, several GM plants and derived products are placed on the EU market, and an increasingly growing number of applications are entering the approval process (EC, 2012). At the time of writing of this chapter (February 2013), a total of 123 applications for GM plants have been submitted to the EFSA (EFSA, 2012c) under one of the two relevant legislations: Regulation (EC) No 1829/2003 and Directive 2001/18/EC. These applications cover a diversity of crops (mostly maize, followed by cotton and soybean) and traits (mostly herbicide tolerance, insect resistance, or a combination of the two). Other traits include: drought tolerance in maize, altered oleic acid content in soybean, or reduced amylose content in potato (Fig. 3.1). Most GM plant applications are for import and processing for food and feed or industrial uses (66%), implying that these GM plants are not going to be cultivated within the EU. Some GM plant applications are intended for deliberate release in the EU (15%), implying that these GM plants are going to be cultivated within the EU, if authorized.

If an applicant wishes to continue to market an authorized GM plant after the original 10-year approval decision, the GM plant must be reassessed by the EFSA prior to any renewal authorization decision by the European Commission and EU member states. At present, the EFSA has received a total of 23 renewal applications for GM plants (19%).

3.6 Nutritional and Safety Assessment of GM Food and Feed Outside of Europe

Modern biotechnology broadens the possibilities of genetic changes that can be introduced into organisms used for the production of human food and animal feed. All over the world, many countries have put in place different regulatory frameworks to assess the safety of GM plants and derived products. The statutory and non-statutory approaches regulating food and feed derived from GM plants may differ across countries, but the criteria used to assess the safety of these products is generally consistent from one country to another. This is probably attributable to the concerted efforts made by different forums created to develop internationally agreed approaches and standards to assess the safety and the impact of GM food and feed on human and animal health.

The Codex Alimentarius Commission (CAC), established by the Food and Agriculture Organization of the United Nations (FAO) and the World Health Organization (WHO), now with 180 member governments, published the 'Principles for the risk analysis of food derived from modern biotechnology' (CAC, 2011) and the 'Guideline for the conduct of food safety assessment of food produced using

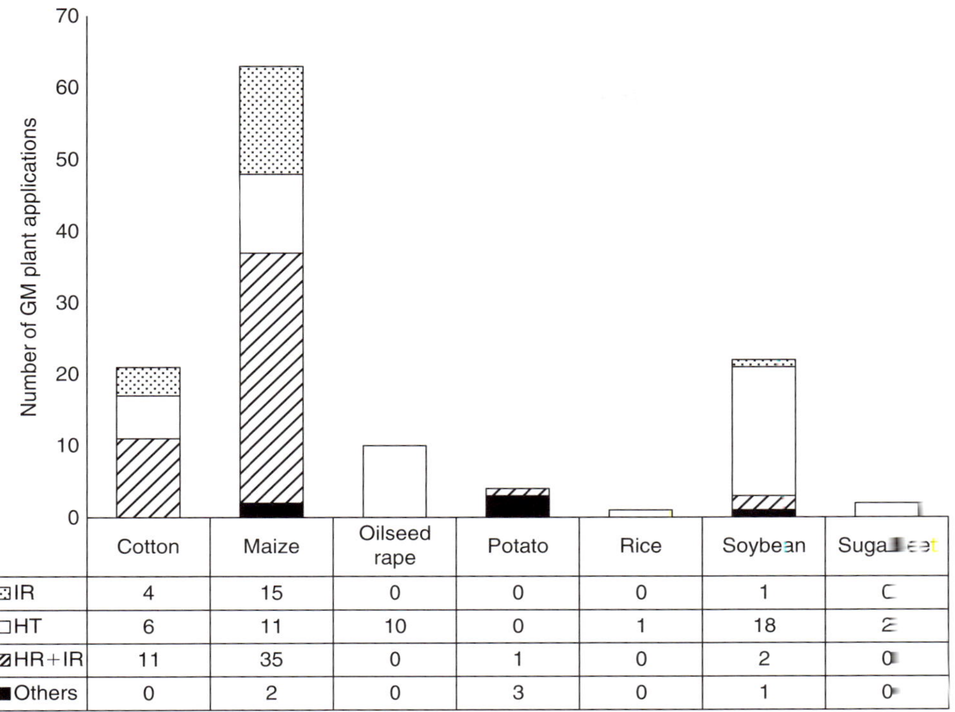

	Cotton	Maize	Oilseed rape	Potato	Rice	Soybean	Sugar beet
IR	4	15	0	0	0	1	0
HT	6	11	10	0	1	18	2
HR+IR	11	35	0	1	0	2	0
Others	0	2	0	3	0	1	0

Fig. 3.1. Number of GM plant applications submitted to the EFSA for risk assessment. The number of GM plant applications is divided by crop species in combination with their intended traits. The traits are denoted by pattern code: dotted = insect resistance (IR); white = herbicide tolerance (HT); dashed = herbicide tolerance and insect resistance (HR+IR); black = traits other than IR or HT (others).

recombinant-DNA plants' (CAC, 2008), as well as a large number of specific commodity standards. Codex standards, guidelines and more information can be found at http://www.codexalimentarius.net.

To support the CAC, ad hoc inter-governmental task forces on food derived from modern biotechnologies, joint FAO/WHO expert consultations, have published the 'Safety aspects of genetically modified foods of plant origin' (WHO, 2000) and the 'Evaluation of allergenicity of genetically modified foods' (WHO, 2001). Nevertheless, neither Codex nor WHO/FAO documents specifically address the use of GM plants for the production of animal feed.

The Organisation for Economic Co-operation and Development (OECD), with 34 member countries across the globe, plus the European Commission, have published four volumes on the 'Safety assessment of transgenic organisms' (OECD, 2010), a series of consensus documents for the work on the 'Harmonisation of regulatory oversight in biotechnology' (OECD, 2012a), a series of consensus documents for the work on the 'Safety of novel foods and feeds' (OECD, 2012b) and three consensus documents facilitating harmonization of the 'Work on the safety of novel foods and feeds' (OECD, 2012c), of which the 'Considerations for the safety assessment of animal feedstuffs derived from genetically modified plants' (OECD, 2003) specifically addresses the safety assessment of GM feed.

International treaties have also been put forward to enhance trade. Trade problems arise when countries have different legal requirements regarding the detection, labelling and approval procedures necessary to place GMOs and their products on the market. So far, no trade dispute over GMOs has been examined by the World Trade Organization's (WTO) dispute settlement panel. Several WTO agreements, such as the Agreement on the Application of Sanitary and Phytosanitary Measures (SPS), the Agreement on Technical Barriers to Trade (TBT), the Agreement on Trade-Related Aspects of Intellectual Property Rights (TRIPs) and the General Agreement on Tariffs and Trade (GATT), could be, in principle, applicable to the GMO issue. However, before doing so, further work needs to be undertaken to examine how the different WTO agreements can be tailored to address specific GMO issues (WTO, 2012).

An international agreement on living modified organisms (LMOs) was negotiated in 2000 under the convention on Biosafety of the United Nations. LMOs are basically GMOs that have not been processed and that could live if introduced into the environment (such as seeds). From 2003 onwards, 63 countries worldwide have ratified the international treaty, Cartagena Protocol on Biosafety to the Convention on Biological Diversity (commonly known as the Cartagena Protocol on Biosafety [CPB]), for handling the transboundary movement of LMOs resulting from modern bio-technologies from one country to another for direct use as food and feed or for processing (Cartagena Protocol, 2012). The CPB provides an international regulatory framework to reconcile the respective needs of trade and of environmental protection with respect to a rapidly growing bio-technology industry (Cartagena Protocol, 2010). As a result, domestic regulatory frameworks endorsing the CPB obtain the status of 'signature country' and are required to adhere to CPB requirements.

As an outcome of these multilateral efforts, despite some local differences in risk assessment approaches and in specific requirements, the trend is that more and more countries are joining the international agreement on trade to ensure sound pro-tection to consumers' health. Over time, this will lead to a clearer, more efficient and transparent authorization process, which will enhance international trade in products resulting from modern biotechnologies.

3.6.1 Australia and New Zealand

In Australia and New Zealand, the Food Standards Australia New Zealand (FSANZ) must approve any food produced from GM crops, or made using genetically engineered enzymes, before it can be marketed in Australia or New Zealand. In 2001, the Office of the Gene Technology Regulator (OGTR) and FSANZ took over the oversight of GMOs from the Genetic Manipulation Advisory Committee, which had been in force since 1987. The OGTR is a Common-wealth Government Authority in the Department of Health and Ageing, with legislative powers and which reports directly to Parliament through a Ministerial Council on Gene Technology. The OGTR decides on licence applications for the release of all GMOs, while regulation is provided by the Therapeutic Goods Administration for GM medicines or FSANZ for GM food. The individual state governments are then able to assess the impact of release on markets and trade and apply further legislation to control approved genetically modified products.

Stock feed legislation in Australia is under the responsibility of State and Territory jurisdictions, which have their own stock feed legislation. There are no labelling requirements for animal feed or feed ingredients derived from GM crops. However, when a GM crop is approved for commercial release, it is considered as safe as its conventional counterparts and can be used for the same purposes, including animal feed.

Further information on novel foods authorization in Australia and New Zealand can be found at http://www.foodstandards.gov.au/.

3.6.2 Asia: China, India and Japan

In response to research development and trade, several Asian countries have set up legislative frameworks and administrative measures to safeguard the commercial release of GM crops and their derived products.

In China, the State Council in 2001 passed the 'Regulations on administration of agricultural GMO safety', which was implemented via five administrative measures for the safety of (i) import, (ii) labelling, (iii) biosafety assessment, (iv) GM food hygiene and (v) inspection and quarantine of the import and export of agricultural GMOs (for details see Chapter 13). The Ministry of Agriculture (MoA) is responsible for the first three administrative measures. GMO developers legally based outside of the Chinese mainland must file an application to the MoA if they want to import GM material into China, either for commercialization or for experimental trials. The MoA twice a year entrusts institutions to perform environmental and edible safety tests and designates a national committee to review the application documents and the test reports. The MoA issues a safety licence within 270 days. The application must include a marketing approval certificate issued by the exporting country/region. So, only GM products approved for food and feed uses elsewhere can be imported into China. The 'Implementation regulations on safety assessment of agricultural genetically modified organisms' (MoA, 2002) define the two principles for the risk assessment of GMOs for agricultural uses (environmental release, food and feed uses) in China: the comparative assessment and the case-by-case approach. The MoA uses a class-based system (Classes I–IV categorizing unknown, low, medium and high risks, respectively) to determine the safety of the recipient organism, the type of genetic modification that influences the safety of the recipient organisms, the safety of the GMO, the production and/or processing activities that influence the safety of the GMO and the safety of the derived GM products. The Ministry of Health (MoH) is responsible for the administrative measures on GM food hygiene, dealing with GMOs for medical and pharmacological uses. The State Administration of Quality Supervision Inspection and Quarantine (SAQSQ) is responsible for the administrative measures for the inspection and quarantine of GM products. The latter two administrative measures, being outside the scope of this book, are not discussed further.

Further information on GMO authorization in China can be found at http://english.agri.gov.cn/hottopics/bt/201301/t20130115_9551.htm.

In India, the regulation of all activities related to GMOs and derived products is governed by 'Rules for the manufacture, use, import, export and storage of hazardous microorganisms, genetically engineered organisms or cells' (commonly referred to as Rules, 1989) under the provisions of the Environment (Protection) Act of 1935, through the Ministry of Environment and Forests (MoEF, 1989). The rules are implemented primarily by the MoEF, the Department of Biotechnology (DBT) and the Ministry of Science and Technology through six competent authorities: the Recombinant DNA Advisory Committee (RDAC); the Review Committee on Genetic Manipulation (RCGM); the Genetic Engineering Approval Committee (GEAC); the Institutional Biosafety Committees (IBSC); the State Biosafety Coordination Committees (SBCC); and the District Level Committees (DLC). The rules are very broad in scope and essentially capture all activities, products and processes related to or derived from biotechnology including GM food and feed, thereby making the GEAC the competent authority to approve or reject the release of GM food and feed in the marketplace.

Further information on GMO authorization in India can be found at http://dbtbiosafety.nic.in/.

In Japan, the safety assessment of foods and feedstuff produced by recombinant DNA techniques has been mandatory under the food sanitation law since April 2001. GM food and feed are dealt with by different ministries, all following similar scientific requirements. The Ministry of Health, Labour and Welfare (MHLW) receives applications for GMO marketing, and the

Food Safety Commission evaluates their safety in terms of human health (MHLW, 2012). The safety assessment of GM crops for feed uses is under the responsibility of the Ministry of Agriculture, Forestry and Fisheries (MAFF). The safety of GM crops used as livestock feed must be approved by the Feed Division of the Livestock Industry Department of the Agricultural Production Bureau in the MAFF (Tabei, 2003). Most of the items evaluated for feed safety are essentially the same as those for food safety described in the 'Standards for the safety assessment of genetically modified foods (seed plants)', published by the food safety commission of the Ministry of Health (MHLW, 2004).

Further information on GMO authorization in Japan can be found at http://www.fsc.go.jp/english/standardsforrisk assessment/geneticallymodifiedfoodfeed.html.

3.6.3 North America: USA and Canada

In the USA, the Food and Drug Administration (FDA) is the leading authority for the safety of food and animal feed. Within the FDA, the Center for Veterinary Medicine (CVM) is responsible for reviewing data on GM plants intended for use in animal feed. The Federal Food, Drug, and Cosmetic Act (FFDCA), which forms the legal basis for the regulation of foods by the FDA, defines a food as a product used for humans and animals. Therefore, requirements for food apply also to feed.

No specific law on GM plants and their derived products exists in the USA. In 1992, the FDA published a statement of policy in the Federal Register for foods derived from new plant varieties (FDA, 1992). This statement treats GM food and feed in the same way as conventional food and feed in the sense that a plant developed using recombinant DNA techniques (i.e. biotechnology) is not itself a regulatory trigger for food and feed safety oversight or pre-market approval. On the contrary, under US law, a regulatory oversight is triggered when a food is adulterated or misbranded

and therefore cannot be commercialized. An adulterated food is a food containing any poisonous or deleterious substance that may render it injurious to health, or containing an unsafe food additive or unsafe pesticide residue (Sec. 402, FFDCA). An unsafe food additive is one that has not been used according to an authorizing regulation (Sec. 409, FFDCA). An unsafe pesticide residue is one that has not been granted a tolerance or tolerance exemption (Sec. 408, FFDCA). The FDA has oversight of food additives and the Environmental Protection Agency (EPA) has oversight of pesticides. The implication for biotechnology-derived foods is that if they contain a food additive or pesticide, that food additive or pesticide must have gone through the relevant pre-market authorization procedure by the FDA or EPA before the GM food could be marketed. However, if the biotechnology-derived foods do not contain a food additive or a pesticide, they are not subject to any pre-market approval requirement. Therefore, the safety evaluation process is determined by the characteristics of the food or feed, not by its method of development. Apart from the presence of food additives or pesticides, foods (biotechnology-derived or otherwise) are still subject to post-market oversight.

Further information on GMO authorization in the USA can be found at http://www.fda.gov/Food/FoodScienceResearch/Biotechnology/.

In Canada, the safety assessment of GM food is separated from the safety assessment of GM feed. GM food is under the responsibility of the Food Directorate of Health Canada, whereas GM feed is under the responsibility of the Animal Feed Division of the Canadian Food Inspection Agency (CFIA). Nevertheless, both GM food and GM feed are considered categories of 'novel products' ('novel food' and 'novel feed', respectively) without a history of safe consumption by humans and animals. As far as novel foods are concerned, regulatory oversight is triggered by the new characteristics of the product rather than the process used to create the product. Potential food safety issues are those associated with toxins, contaminants and anti-nutritional factors that could be introduced into the food supply via the importation of new products, the introduction of a new species as a food source, the use of new processing techniques or changes in the genetic make-up of organisms.

In general terms, a novel feed is a feed that either has not been approved previously in Canada or that contains an intentional genetic modification resulting in a relevant phenotypic change as compared to a non-GM counterpart. The CFIA regulates the marketing of novel feeds in Canada through a pre-market notification and has published 'Regulatory guidance: feed registration procedures and labelling standards', which, in Section 2.6, include guidelines for the safety assessment of novel feeds (CFIA, 2012). The scientific requirements for such notification are specified under the Feeds Act, RSC 1985 (Department of Justice Canada, 1985) and the Feeds Regulations, 1983 (Department of Justice Canada, 1983). Based on the outcome of the safety assessment, the CFIA prepares the regulatory decision. Once a novel feed receives authorization, it is either enlisted in the Feeds Regulations or it is defined as substantially equivalent to another, already enlisted feed and is no longer considered as 'novel'.

Further information on novel foods authorization in Canada can be found at http://www.novelfoods.gc.ca.

3.6.4 South America: Argentina and Brazil

In Argentina, the risk assessment of GMOs is performed by the Agrifood Quality Directorate of the National Service for Agrifood Health and Quality (SENASA), a regulatory agency affiliated to the Ministry of Agriculture, Livestock and Fisheries. The risk assessment is performed by a scientific team with the advice of a Technical Advisory Committee composed of experts from different scientific disciplines, representing different sectors involved in the production, industrialization, consumption, research and development of GMOs. Guidelines for GMO risk assessment are approved under

Resolution 412/2002 setting the 'Principles and criteria for the assessment of food derived from genetically modified organisms' and the 'Requirements and rules of proceedings for the human and animal safety assessment of foods derived from genetically modified organisms' (SENASA, 2002).

Further information on novel foods authorization in Argentina can be found at http://www.senasa.gov.ar/contenido.php?to=n&in=731&io=7084.

In Brazil, the risk assessment of GMOs is under the responsibility of the National Biosafety Technical Commission. The National Biosafety Technical Commission is composed of 27 members, including scientists (with expertise in human health, animal health, plant health and environment), ministerial representatives and other experts. This commission also develops guidelines for the transport, importation and field experiments necessary for the approval of GM products in Brazil. Law 11.105, approved in 2005, constitutes the regulatory framework establishing safety rules and monitoring activities for GMOs (CTNBIO, 2005). The Council of Ministers is responsible for the risk management of GM products and evaluates the commercial and economic consequences of the release of a GM product into the market.

Further information on novel foods authorization, and opportunities and limitations for biotechnology innovation, in Brazil can be found at http://www.ctnbio.gov.br/ and in de Castro (2013).

3.6.5 Africa: South Africa

Only recently, in 2010, did the Common Market for Eastern and Southern Africa (COMESA) propose a draft policy on GM technology, which was sent to 19 national governments for consultation. Under the proposed policy, any new GM crops would first be scientifically assessed by the COMESA and, if deemed safe for the environment and human health, would receive authorization for cultivation in all 19 member countries, although final ratification would be left to each individual country.

Among the African countries, South Africa is the major GM crop grower. The regulation of GMOs in South Africa is defined in the Genetically Modified Organisms Act (GMO Act, Act No 15, 1997), its subsequent amendments and their applicable regulations (Act No 23 of 2006) (South African Government, 1997). The Act is intended to provide an 'adequate level of protection' during all activities involving GMOs that may have an adverse impact on the conservation and sustainable use of biological diversity and human and animal health. The GMO Act regulates the different authorization granted to GMOs in South Africa, namely contained use, field trials, import, commodity clearance (import for food and/or feed use) and general release (commercial plantings, food and/or feed use, importation and exportation). The GMO Act is implemented by the Directorate Biosafety of the Department of Agriculture, Forestry and Fisheries and is administered by the GMO Registrar.

Under the GMO Act, two regulatory bodies are involved in the risk assessment and risk management of GMOs in South Africa: the Advisory Committee, composed of independent scientists with different scientific backgrounds, and the Executive Council, composed of representatives from government departments. According to the regulations, the risk assessment of GMO applications is conducted in a scientifically sound manner and takes into account the following steps: (i) identification of any potential adverse effect resulting from the novel genotypic and/or phenotypic characteristics of the GMO; (ii) evaluation of the likelihood of these adverse effects, taking into account the level and kind of exposure to the GMO; (iii) evaluation of the consequences should these adverse effects be realized; and (iv) estimation of the overall risk posed by the GMO based on the evaluation of the likelihood and consequences of the identified adverse effects being realized.

Further information on the authorization of GMOs in South Africa can be found at http://www.biosafety.org.za/index.php.

3.7 Conclusions

Many countries and regions around the globe have, or are currently putting into place, regulatory frameworks, operational procedures and detailed guidelines for the safety assessment of GM plants and derived food and feed. Even in the frame of our limited comparison, it is evident that, despite general international agreement recognizing comparative assessment as the core principle for GMO risk assessment, differences in statutory and non-statutory approaches regulating food and feed derived from GM plants exist among countries. This calls for continuous efforts towards a global harmonization of regulatory frameworks and an international standardization of scientific requirements in order to enhance the comparability of risk assessments performed in different countries. Ultimately, such efforts would promote harmonization effectively, and consequently would boost trade in GM food and feed commodities.

Novel traits of GM plants in the developmental pipeline are those targeting metabolic or physiological pathways, either through a direct alteration of the existing pathways or through the insertion of new pathways. These 'second-generation' GM plants have been modified deliberately to enhance their nutritional properties or to improve their resistance to biotic and abiotic stresses. These novel traits pose new challenges to current risk assessment strategy based on comparative assessment, as there may be no comparator against which the food and feed products derived from these second-generation GM plants can be measured (CAC, 2007b). Effectively, this will make application of the substantial equivalence concept less straightforward.

In recent years, the development of new breeding technologies, allowing the insertion of foreign genes into crops' genomes at specific locations, has promoted the broadening of gene pools to levels unachievable with traditional breeding. These genomic changes are generally difficult to detect (Lusser *et al.*, 2011) and pose new challenges to the molecular characterization of GM plants, which will need to be addressed by the scientific community in the coming years.

We hope that this chapter achieves our initial objective; namely, to offer the reader an overview of the nutritional and safety assessment of GM plants and derived feed in the EU and to provide some useful insights on the strategies currently in place in some countries around the world. By no means do we expect to have addressed all aspects exhaustively, but if we have managed to stimulate your curiosity and to offer you guidance for further research in this area, our time has been well spent.

References

Bakke-McKellep, A.M., Penn, M.H., Salas, P.M., Refstie, S., Sperstad, S., Landsverk, T., *et al.* (2007) Effects of dietary soyabean meal, inulin and oxytetracycline on intestinal microbiota and epithelial cell stress, apoptosis and proliferation in the teleost Atlantic salmon (*Salmo salar* L.). *British Journal of Nutrition* 97(4), 699–713.

CAC (2007a) *Codex Alimentarius Commission Procedural Manual, Seventeenth edition.* Codex Alimentarius Commission, Joint FAO/WHO Food Standards Programme, Rome (http://www.fao.org/docrep/010/a1472e/a1472e00.htm, accessed 24 May 2013).

CAC (2007b) Report of the Working Group on the Proposed Draft Annex to the Codex Guideline for the conduct of food safety assessment of foods derived from recombinant-DNA plants: food safety assessment of foods derived from recombinant-DNA plants modified for nutritional or health benefits. CL 2007/18-FBT. CAC, Rome (ftp://ftp.fao.org/codex/circular_letters/cxcl2007/cl07_18e.pdf, accessed 21 June 2013).

CAC (2008) Guideline for the conduct of food safety assessment of foods derived from recombinant-DNA plants. CAC/GL 45-2003. Codex Alimentarius Commission (http://www.codexalimentarius.org/standards/list-of-standards/en/, accessed 24 May 2013).

CAC (2009) *Foods derived from modern biotechnology, Second Edition.* Codex Alimentarius Commission, Joint FAO/WHO Food Standards Programme, Rome (http://www.fao.org/docrep/011/a1554e/a1554e00.htm, accessed 24 May 2013).

CAC (2011) Principles for the risk analysis of food derived from modern biotechnology. CAC/GL 44-2003. Codex Alimentarius Commission (http://www.codexalimentarius.org/standards/list-of-standards/en/, accessed 24 May 2013).

Cartagena Protocol, The (2010) The Cartagena Protocol on Biosafety: full text (http://bch.cbd.int/protocol/text/, accessed 21 June 2013)

Cartagena Protocol, The (2012) Parties to the protocol and signature and ratification of the supplementary protocol (http://bch.cbd.int/protocol/parties/, accessed 24 May 2013).

CFIA (2012) Regulatory guidance: feed registration procedures and labelling standards (http://www.inspection.gc.ca/animals/feeds/regulatory-guidance/rg-1/eng/1329109265932/1329109385432, accessed 24 May 2013).

CTNBIO (2005) Biosafety Law 11.105/2005 (http://www.ctnbio.gov.br/index.php/content/view/1311.html, accessed 24 May 2013).

de Castro, L.A.B. (2013) *Opportunities and Limitations for Biotechnology Innovation in Brazil.* Bentham eBooks, Bentham Science Publishers, Oak Park, USA.

Department of Justice Canada (1983) Feeds Regulations (http://laws-lois.justice.gc.ca/PDF/SOR-83-593.pdf, accessed 24 May 2013).

Department of Justice Canada (1985) Feeds Act (http://laws-lois.justice.gc.ca/PDF/F-9.pdf, accessed 24 May 2013).

Dreau, D. and Lalles, J.P. (1999) Contribution to the study of gut hypersensitivity reactions to soybean proteins in preruminant calves and early-weaned piglets. *Livestock Production Science* 60(2–3), 209–218.

EC (2001) Directive 2001/18/EC of the European Parliament and of the Council of 12 March 2001 on the deliberate release of genetically modified organisms and repealing Council Directive 90/220/EEC. *Official Journal* L106, 17/04/2001, pp. 1–38.

EC (2002) Regulation (EC) No 178/2002 of the European Parliament and of the Council of 28 January 2002 laying down the general principles and requirements of food law, establishing the European Food Safety Authority and laying down procedures in matters of food safety. *Official Journal* L31, 1/02/2002, pp. 1–24.

EC (2003) Regulation (EC) No 1829/2003 of the European Parliament and of the Council of 22 September 2003 on genetically modified food and feed. *Official Journal* L268, 18/10/2003, pp. 1–23.

EC (2004) Directive 2004/10/EC of the European Parliament and of the Council of 11 February 2004 on the harmonisation of laws, regulations and administrative provisions relating to the application of the principles of good laboratory practice and the verification of their applications for tests on chemical substances. *Official Journal* L50, 20/2/2004, pp. 44–59.

EC (2012) EU register of genetically modified food and feed (http://ec.europa.eu/food/dyna/gm_register/index_en.cfm, accessed 24 May 2013).

EC (2013) Commission implementing regulation (EU) No 503/2013. *Official Journal of the European Union* L157, 8.6.2013 (http://eur-lex.europa.eu/LexUriServ/LexUriServ.do?uri=OJ:L:2013:157:0001:0048:EN:PDF, accessed 21 June 2013).

EFSA (2007) Opinion of the scientific committee related to uncertainties in dietary exposure assessment. *EFSA Journal* 438,1–54.

EFSA (2010a) Guidance of the environmental risk assessment of genetically modified plants. *EFSA Journal* 8(11), 1879.

EFSA (2010b) Statistical considerations for the safety evaluation of GMOs. *EFSA Journal* 8(1), 1250.

EFSA (2010c) Scientific opinion on the assessment of allergenicity of GM plants and microorganisms and derived food and feed. *EFSA Journal* 8(7), 1700.

EFSA (2011a) Guidance for risk assessment of food and feed from genetically modified plants. *EFSA Journal* 9(5), 2150.

EFSA (2011b) Guidance on the selection of comparators for the risk assessment of genetically modified plants and derived food and feed. *EFSA Journal* 9(5), 2149.

EFSA (2011c) Guidance on conducting repeated-dose 90-day oral toxicity study in rodents on whole food/feed. *EFSA Journal* 9(12), 2451.

EFSA (2012a) The EFSA Comprehensive European Food Consumption Database (http://www.efsa.europa.eu/en/datexfoodcdb/datexfooddb.htm, accessed 24 May 2013).

EFSA (2012b) EU Menu (http://www.efsa.europa.eu/en/datexfoodcdb/datexeumenu.htm, accessed 24 May 2013).

EFSA (2012c) EFSA's Register of Questions database (http://www.efsa.europa.eu/en/request/requests.htm, accessed 24 May 2013).

FDA (1992) Statement of Policy – Foods Derived from New Plant Varieties (http://www.fda.gov/Food/GuidanceRegulation/GuidanceDocuments/RegulatoryInformation/Biotechnology/ucm096095.htm, accessed 21 June 2013).

Lusser, M., Parisi, C., Plan, D. and Rodrigues-Cerezo, E. (2011) JRC scientific and technical report: new plant breeding techniques. State-of-the-art and prospects for commercial development. JRC63971. Publications Office of the European Union.

MLHW (2004) Standards for the safety assessment of genetically modified foods (seed plants) (http://www.fsc.go.jp/english/standardsforrisk assessment/geneticallymodifiedfoodfeed.html, accessed 24 May 2013).

MHLW (2012) Procedure for safety assessment of genetically modified foods/feeds (http://www.mhlw.go.jp/english/topics/foodsafety/dna/01.html, accessed 24 May 2013).

MoA (2002) Implementation regulations on safety assessment of agricultural genetically modified organisms. Ministry of Agriculture of the People's Republic of China (http://www.biosafety.gov.cn/image20010518/5421.pdf, accessed 21 June 2013).

MoEF (1989) Ministry of Environment and Forests – Rules for the manufacture, use, import, export and storage of hazardous microorganisms, genetically engineered organisms or cells. Official Gazette No 621 dt. 5-12-89 (http://bch.cbd.int/database/record.shtml?documentid=102938, accessed 21 June 2013).

OECD (1998) Test No 408: repeated dose 90-day oral toxicity study in rodents (http://www.oecd-ilibrary.org/environment/test-no-408-repeated-dose-90-day-oral-toxicity-study-in-rodents_9789264070707-en, accessed 24 May 2013).

OECD (2003) Considerations for the safety assessment of animal feedstuffs derived from genetically modified plants. Series on the Safety of Novel Foods and Feeds No 9 (http://www.oecd.org/dataoecd/16/40/46815216.pdf, accessed 24 May 2013).

OECD (2010) Safety assessment of transgenic organisms: OECD Consensus Documents (http://www.oecd.org/document/15/0,3746,en_2649_34385_37336335_1_1_1_1,00.html, accessed 24 May 2013).

OECD (2012a) Consensus documents for the work on harmonisation of regulatory oversight in biotechnology: biology of crops (http://www.oecd.org/document/15/0,3746,en_2649_34385_46726799_1_1_1_1,00.html, accessed 24 May 2013).

OECD (2012b) Consensus documents for the work on the safety of novel foods and feeds (http://www.oecd.org/document/39/0,3746,en_2649_34385_46805223_1_1_1_1,00.html, accessed 24 May 2013).

OECD (2012c) Consensus documents for the work on the safety of novel foods and feeds: facilitating harmonisation (http://www.oecd.org/document/55/0,3746,en_2649_34385_4681 4647_1_1_1_1,00.html, accessed 24 May 2013).

SENASA (2012) Principles and criteria for the assessment of food derived from genetically modified organisms and the requirements and rules of proceedings for the human and animal safety assessment of foods derived from genetically modified organisms (http://www.senasa.gov.ar//Archivos/File/File995-ingles412-.pdf, accessed 24 May 2013).

South African Government (1997) Genetically Modified Organisms Act (GMO Act, Act No 15, 1997) (http://www.info.gov.za/acts/1997/act15.htm, accessed 24 May 2013).

Tabei, Y. (2003) Risk assessment of GM crops, and the challenge of building public acceptance in Japan (http://www.agnet.org/library.php?func=view&id=20110718171149, accessed 24 May 2013).

Tutelyan, V.A. (2013) *Genetically Modified Food Sources: Safety Assessment and Control.* Elsevier, Academic Press, London.

Verlinden, A., Hesta, M., Millet, S. and Janssens, G.P.J. (2006) Food allergy in dogs and cats: a review. *Critical Reviews in Food Science and Nutrition* 46(3), 259–273.

WHO (2000) Safety aspects of genetically modified foods of plant origin. Report of a Joint FAO/WHO Expert Consultation. World Health Organization, Geneva, Switzerland.

WHO (2001) Evaluation of allergenicity of genetically modified foods. Report of a Joint FAO/WHO Expert Consultation. Food and Agriculture Organization of the United Nations, Rome.

WTO (2012) WTO trade topics (http://www.wto.org/english/tratop_e/tratop_e.htm, accessed 24 May 2013).

4 Compositional Analysis for Nutritional and Safety Assessment of Feeds from GM Plants

Gijs Kleter* and Esther Kok

RIKILT Wageningen UR, Wageningen, the Netherlands

4.1 Introduction

Since the first large-scale commercial introduction of genetically modified (GM) crops in the mid-1990s, they have increasingly become adopted around the globe, with the global area planted with GM crops reaching 170 million hectares (Mha) in 2012 (James, 2012). While GM crops are grown in an increasing number of countries outside the European Union (EU), the extent to which these crops are grown within the EU has remained limited. Most GM crop production occurs in Spain, where insect-resistant maize is grown (James, 2012). Notwithstanding this, the international trade in crop commodities will facilitate the import and export of GM variants of the commodity crops commonly used in animal feed, such as maize-, soybean-, cottonseed- and canola-derived products (e.g. gluten feed, distillers' grain, seed meals). The main modifications applied to these crops are of agronomic importance, namely insect resistance and herbicide tolerance (James, 2012). This involves the expression of the newly introduced genes, giving rise to proteins which impart these resistance traits, such as insecticidal proteins from the microorganism *Bacillus thuringiensis* (Bt) and herbicide-insensitive analogues of crop enzymes that are inhibited by herbicide-active ingredients, respectively. The genetic modifications of these crops have not targeted any modification of their nutritional value for the human and animal consumers of crop-derived products, yet a number of GM crops 'in the pipeline' bear characteristics such as altered contents of essential amino acids or fatty acids (see Chapters 7 and 12) that target improvement of the nutritional value of feed for livestock animals. In this chapter, we will discuss how the compositional analysis of GM crops carried out as part of the regulatory pre-market safety assessment can contribute to the nutritional assessment of these crops for livestock.

4.2 Principles of Safety Assessment

Compositional analysis is a key part of the nutritional and safety assessment of GM crops, which also comprises a range of other considerations, as reviewed by Kleter and Kok (2012). Before GM crops are allowed to be commercialized in many countries around the world, the companies that intend to commercialize them must obtain regulatory approval, i.e. these crops have to be admitted to the market as new products (see Chapter 3). This entails an approval procedure, part of which consists of an assessment of the safety of these products. This safety assessment is carried out by risk assessors, who will advise the policy makers on whether or not the products can be regarded as being as safe as comparable conventional varieties, so that the latter can take this into account when drafting a proposal for a policy

*E-mail: gijs.kleter@wur.nl

measure to approve the GM crop for marketing. Whereas regulations on GM crops may differ between nations, the approach followed for the safety assessment for the regulatory approval will be, by and large, the same from nation to nation (see Chapter 3). This is because the safety assessment of GM crops has been harmonized internationally through the guidelines published by Codex Alimentarius, a joint programme for food quality and safety of the United Nations' Food and Agriculture Organization (FAO) and the World Health Organization (WHO), involving the participation of representatives from most nations of the world. Documents developed by Codex Alimentarius, including standards, codes of conduct and guidance, serve as a point of reference in trade disputes over the safety of internationally traded foods should these be brought to the fore of international arbitration by the World Trade Organization (WTO) under the international WTO Agreement on the Application of Sanitary and Phytosanitary Measures (SPS agreement). The publication of the Codex Alimentarius guideline on the safety assessment of foods from plants developed through recombinant DNA technology was the culmination of years of effort by international organizations such as the FAO, the WHO, the Organisation for Economic Co-operation and Development (OECD) and others, which had already started before the first GM crops had even appeared on the market (reviewed by Kuiper *et al.*, 2001). While the Codex Alimentarius guideline applies to food, the same principles can, by and large, also be applied to animal feeds, as is done in many countries.

A key role in the safety assessment approach as developed through the international consensus-building efforts and described in the Codex Alimentarius guideline is the comparative approach, in which a GM plant is compared to its genetically nearest non-GM counterpart with a history of safe use. For example, for GM maize expressing a Bt protein, the counterpart for this comparison ideally would be near-isogenic non-GM maize. A comparative assessment commonly entails an analysis of the agronomic and phenotypic characteristics of the plant, such as its physiology, appearance, performance and reproductive traits, tested under field conditions representative of commercial cultivation. An extensive compositional analysis of the macronutrients, micronutrients, antinutrients, toxins and secondary compounds of the plant parts used for food or feed purposes is also performed, as well as a characterization of the molecular characteristics of the introduced genetic material and its expression (e.g. DNA construct used for the genetic modification procedure; structure of the inserted DNA; expression of introduced genes; levels of newly expressed proteins in different plant tissues and possibly also at different developmental stages of the plant). Within the EU, this will also comprise a molecular characterization of the place of insertion, i.e. the sequence of flanking regions of the inserted genetic construct, but these data may not be routinely required in other legislations.

The comparative approach is in line with the insight that feeds and foods are complex mixtures of compounds that may occur at variable levels and that have different kinds of health effects with different dose-response relationships. In addition traditional feeds and foods, while usually not having been tested for their safety per se, have a history of safe use based on experience and selection of, for example varieties of food crops containing admissible levels of intrinsic toxins and anti-nutrients (if present). In addition, processes (e.g heating) may have been established to remove these toxins and anti-nutrients from the crop-derived product prior to consumption. The differences observed in the comparison between the GM crop and its conventional counterpart will therefore have to be assessed for their potential effects on the basis of established knowledge on the safety and nutritional properties of the non-GM crop already being consumed by animals and/or humans. This comparative approach serves a twofold purpose. First, besides the intended effect, it aids the identification of any potential unintended effects of the genetic modification. Such an unintended effect might occur, for example, in a scenario in which an intrinsic gene of the host plant has been disrupted by the insertion of 'foreign DNA' by the genetic modification process into this gene, while a number of other mechanisms can also be envisaged (e.g. disruption of regulatory sequences). Second, comparison of the compositional data may also help to assess the possible nutritional impact should the observed changes in composition affect the nutritional value to an extent that reaches beyond the natural background variability within that specific crop.

4.3 Methodology

As noted above, the extensive comparative analysis that is usually carried out in the frame of the pre-market safety assessment of GM crops comprises a range of agronomic, phenotypic characteristics of the plant that, besides the intended effects of the genetic modification, help to identify potential unintended effects. In this section, we will focus on the compositional part of this analysis, as well as the design of the field trials used to produce samples for analysis and the interpretation of the outcomes.

4.3.1 Field trial design

For the purpose of the comparative compositional analysis, field trials are typically carried out in multiple locations during which the GM crop and its counterpart are grown together in the same localities and under the same conditions, as recommended by the Codex Alimentarius guideline (Codex Alimentarius, 2008). This will help to rule out all kinds of confounding environmental factors that might occur when the crops to be compared are grown in separate places and to ensure that the differences observed can be ascribed to the genetic modification. In the case of herbicide-tolerant GM crops, the Codex Alimentarius guideline indicates that an additional group of GM plants treated with the target herbicide to which the plant has been rendered tolerant (besides plants not treated with this herbicide but in the same way as the conventional counterpart) may be considered in some cases.

With regard to the field trial design, the Codex Alimentarius guideline states that the choice for the number and locations of the field trial sites, as well as the number of seasons, should represent the range of environmental conditions that the commercialized crop will be exposed to. Each location should be replicated, which means that it should contain multiple blocks, each with plots for the various test, counterpart and reference crops and their treatments (Codex Alimentarius, 2008).

4.3.2 Compositional analysis

The European Food Safety Authority's (EFSA) guidance for GM plants (EFSA, 2011) is further considered here besides the Code Alimentarius guideline because it expands on this guideline by providing more details and elaborating on methodologies. For example, the EFSA guidance notes that preferably the analysis is to be carried out on the raw agricultural commodity, as this will be the main source material entering the food and feed production chains. When appropriate, additional compositional analysis of the derived and processed food and feed products can be considered. This may be the case when processing or preserving steps that are typically used during production alter the properties of the GM end product as compared to the conventional counterpart (EFSA, 2011).

With regard to the parameters to be measured during the compositional analysis, the Codex Alimentarius guideline recommends that the 'key components' are analysed in the GM crop and in its counterpart (i.e. the control crop). It further explains that these key components include key nutrients and anti-nutrients that have a substantial impact on the overall diet and that are typical of the crop and the crop-derived feeds/foods (Codex Alimentarius, 2008). The EFSA guidance further explains that the choice of the vitamins and minerals to be analysed should indeed be those that are nutritionally significant based on the level at which they occur and the intake of the crop by consumers. Moreover, the choice for other parameters should also be guided by the characteristics of the product. For oilseeds, for example, analysis of the fatty acid profile of the oil in the plant should be carried out, while the amino acid composition is a key parameter to be analysed in products that are important sources of protein in human and animal nutrition (EFSA, 2011; Chapter 3).

With regard to the choice of compositional parameters to be analysed, a useful resource is the consensus documents on the key components of new crop varieties prepared by the OECD Task Force on the Safety of Novel Foods and Feeds, to which the EFSA guidance also specifically refers (EFSA, 2011). These documents have been prepared and adopted (by consensus procedure) by the members of this task force, which consists of representatives from OECD member countries, as well as observer nations and organizations. These consensus documents describe the role of the particular crop in nutrition, typical procedures for processing the crop to either food or feed, and which key compositional parameters are to be analysed in new crop varieties (OECD, 2013). These parameters can differ from crop to crop, so no generalized recommend-ations are possible and the parameters therefore have to be decided on a case-by-case basis. The range of recommended parameters usually include those that fall under the categories of gross composition parameters (e.g. protein, fat, ash), amino and fatty acid composition, micronutrients (particular minerals and vitamins), anti-nutrients (e.g. phytic acid in some crop seeds, including cereals), natural toxins (e.g. glycoalkaloids in potato tubers) and secondary metabolites (e.g. phenolic compounds in maize). Moreover, besides the parameters to be measured in the raw agricultural commodities, the OECD consensus documents also provide parameters for processed crop fractions typically used for food or feed purposes. A range of different crops are covered by the consensus documents, including, for example, soybean, canola (low-erucic acid rapeseed), maize, cotton, potato, sugar beet, rice, wheat, barley, forage legumes, sunflower, tomato, grain sorghum, sweet potato, sugarcane and papaya (OECD 2013). There is also an ongoing process of updating these documents, and updated versions of the canola and soybean documents, for example, were published in 2011 and 2012 respectively (OECD, 2011, 2012).

An example of the parameters recommended by the OECD consensus document for the analysis of soybean for food and feed purposes is provided in Table 4.1.

Table 4.1. Key compositional parameters recommended for analysis of new soybean varieties by the OECD Task Force on the Safety of Novel Foods and Feeds. (From OECD, 2012.)

Category	Parameter / Analyte	Crop part(s)	Food/feed
Proximates and fibre	Moisture, crude protein, crude fat, fibre (crude, ADF, NDF),[a] ash, carbohydrates	Seed, meal (feed), forage (feed)	Food, feed
Amino acids	Amino acids	Seed, meal (feed)	Food, feed
Fatty acids	Fatty acids	Seed, oil (food)	Food, feed
Minerals	Calcium, phosphorus	Seed, meal (feed)	Feed
Vitamins	Vitamin E (α-tocopherol), K_1	Seed, oil (food)	Food, feed
Anti-nutrients	Stachyose, raffinose	Seed, meal (feed)	Food, feed
	Phytic acid	Seed, meal (feed)	Food, feed
	Trypsin inhibitors	Seed, meal (feed)	Feed
	Lectins	Seed	Food, feed
Secondary compounds	Isoflavones	Seed	Food

Note: [a]ADF = acid detergent fibre; NDF = neutral detergent fibre.

4.3.3 Interpretation of outcomes

The results of the compositional analysis of the different parameters measured are to be analysed statistically, summarized and evaluated for possible changes in composition that may warrant a further assessment of the safety of the GM plant product. With regard to the interpretation of the outcomes, it is noteworthy that the Codex Alimentarius guideline discerns statistical significance from biological relevance in that statistically significant differences between the GM crop and its counterpart should be assessed in the light of the inherent background variability for the given parameter showing this difference (Codex Alimentarius, 2008). In other words, a difference in composition does not necessarily represent a hazard to food or feed safety. With regard to background variability, there are a number of information resources that are used, which may include literature data, database data and data from commercial reference varieties grown in the same field trials as the GM crop and the conventional counterpart. The latter is particularly required when a measure of the variability is already included in the statistical analysis according to EFSA guidance, which is well developed and elaborate on this point and treated in further detail below.

Literature and database data are summarized in table format by the above-mentioned OECD consensus documents, providing the ranges of values collected from these sources for the various macro- and micronutrients, anti-nutrients, toxins and secondary plant compounds. A more homogeneous range of input data is provided by the open-access ILSI Crop Composition Database, which contains compositional data for various crops grown during field trials under controlled conditions in different continents during recent years and for which additional data, such as the analytical methods used, are provided (ILSI, 2013). Data are entered by staff managing the website, following various quality checks before entering. Users can select data on a given parameter in a particular crop according to certain variables, such as geography, year and analytical methods used (ILSI, 2013). While the identity of the crop varieties is known and provided, they can neither be linked to specific values nor selected by the database users.

The inclusion of commercial reference lines into the field trial design so as to obtain information on the background variability of the compounds in these crops is required for the approach towards statistical equivalence testing recommended by the EFSA guidance, which provides detailed recommendations on how to apply statistics to the study outcomes. Two approaches are devised by the EFSA guidance for the analysis of the same parameters (end points), namely a test of difference and a test of equivalence, each having a different null hypothesis (EFSA, 2011).

In the case of difference testing, the GM plant is compared with its conventional counterpart and the differences identified may be considered as hazards based on the characteristics of the compositional parameter analysed (e.g. a toxin) linked to the pattern and extent of exposure of consumers to this parameter (e.g. whether the intake is sufficient to cause effects). The null hypothesis in difference testing is that there is no difference, which can be rejected or accepted based on the outcomes of the statistical difference test. Incorrect rejection of the null hypothesis is referred to as a 'type I error', i.e. concluding that there is a difference between GM and non-GM if there is actually none, in this case a false positive. The probability of a type I error is denoted as α, with a value typically set to 5%, while $[1 - \alpha]$ is called the 'specificity' of the test (hence 95% when α = 5%). Conversely, a type II error, denoted as β, is when the null hypothesis is incorrectly accepted (i.e. concluding that the parameter is not different between the GM and non-GM crops while it actually differs). The 'statistical power' (or 'sensitivity') of a given test design equals the value of $[1 - \beta]$, while a minimum power of 80% is commonly strived for. Power calculations are to be done in advance of a study so as to establish the magnitude of the statistical power to measure a given minimum difference (in this case, between the GM crop and its conventional counterpart) with a given variability in the parameter being analysed and in the light of the questions to be addressed by the study.

In the case of equivalence testing, the null hypothesis is non-equivalence and, in the approach devised by the EFSA guidance, a comparison is made between the GM crop and the various reference varieties. Equivalence limits are established on the basis of the variability of the values of the reference varieties. If the range of the values of the GM crop, expressed statistically as 90% confidence limits, lies within these boundaries, then the null hypothesis is rejected and the GM crop can be considered equivalent. In these cases, the type I error that is controlled (at the 5% level) is the false conclusion that the crops would be equivalent when they are actually not, while the type II error is the incorrect outcome that the null hypothesis is true and that the crops would therefore be non-equivalent.

The EFSA guidance foresees a number of scenarios for the comparison of the GM crop with its non-GM comparator based on eight possible combinations of outcomes; namely, two possible outcomes for the difference test (i.e. GM crop being statistically significantly different or not from the comparator) and four for the equivalence test (i.e. GM crop being equivalent; more likely to be equivalent than not; more likely to be non-equivalent than equivalent; or non-equivalent to its comparator) (EFSA, 2011). In the approach recommended by the EFSA, the reference crops grown during the same field trial have, in the first instance, replaced the other sources of information on background variability, such as the ILSI Crop Composition Database and literature data, providing an accurate indication of the variability of the particular crop grown under the same conditions as the GM crop and its conventional non-GM counterpart.

4.4 Composition in Relation to Nutrition

In the production of animal feeds for livestock, data on the composition of ingredients are commonly employed by producers to predict the nutritional value of the produced feed formulation. Hence, it can be argued that the data on the compositional analysis of the GM crop and its counterparts can provide insight into its possible nutritional impact.

A working group of the EFSA's Panel on Genetically Modified Organisms (GMO) reviewed the knowledge gathered through nutritional studies on GM crops and their counterparts, particularly with poultry, pigs and ruminants (beef and dairy cattle), as well as other species (EFSA, 2007). It was concluded that if the compositional data for the GM crop were comparable to those of non-GM crops, nutritional equivalence could also be established in these cases (EFSA, 2007), which was also in line with the conclusion by Flachowsky et al. (2005) that first-generation GM crops with agronomic traits did not influence the health and performance of livestock animals, nor the safety and quality of derived animal products (see also Chapter 6).

The EFSA's review also acknowledged that if the genetic modification affected the bioavailability of ingredients, i.e. affected their uptake by the animal during digestion of the GM-crop-containing feed, the impact might not be predicted solely with the aid of the compositional analysis (EFSA, 2007). In these cases, as well as for animal feeds with improved nutritional characteristics owing to either increased levels of nutrients or decreased levels of anti-nutrients, studies either in model laboratory animals or in target livestock animals under controlled conditions may be warranted (EFSA, 2007). Such tests could be focused either on assessing the digestibility of a certain nutrient or on the increased performance (related to improved nutrition) per se. The review by Flachowsky et al. (2012) gives an account of the requirements for such studies in laboratory and animal species, including the experimental design and physiological end points to be measured, which logically differ from one species to another and, conversely, also depend on the implicated feed nutrient/anti-nutrient. Bioavailability and other measures of nutritional impact (e.g. exposure, safe upper levels of nutrients) are similar to those assessed for the human nutritional impact of crops with nutritionally improved characteristics, as considered by a dedicated annex of the Codex Alimentarius guideline (Codex Alimentarius, 2008; see also Chapters 5–7).

4.5 Outlook

In order to prepare for the advent of future GM crops with more complicated modifications, which in turn may increase the likelihood of unintended effects, it has been suggested that advanced 'omics' techniques be developed and applied. The latter are holistic, non-targeted analytical techniques that can monitor the crop for changes in its constituents at different levels of cellular organization: including transcriptomics being used for measuring the levels of 'messenger' RNA (mRNA) indicative of gene expression activity; proteomics, measuring the diversity of expressed proteins; and metabolomics, analysing the diversity of chemical compounds (metabolites). Commonly used techniques include cDNA microarrays for transcriptomics (increasingly being replaced by next-generation sequencing (NGS) analysis); two-dimensional protein electrophoresis for proteomics (increasingly being replaced by mass spectrometry (MS) analysis of peptide structures); and nuclear magnetic resonance or gas or liquid chromatography coupled to mass spectrometry for metabolomics. A more detailed review of these different techniques and their potential application in GM crop safety assessment is provided by Kok *et al.* (2010). These authors also note that before omics can become mainstream in the risk assessment procedure, they should be standardized and validated, while databases also should be established to provide background data on the variability of the components measured under natural conditions in non-GM varieties of the specific crops. The outcomes should then help to identify differences, particularly those that are outside the boundaries of background variability, for

which it then needs to be determined if and which further tests need to be performed.

Using these omics techniques, various studies have been carried out in which GM crops were compared with their conventional non-GM counterparts. Several recent reviews highlight the findings that come out of such studies applying metabolomics as well as other omics techniques to GM crops. Davies *et al.* (2010), for example, reviewed a number of omics studies in maize, rice and potato, as well as some soft fruits, including studies that were carried out with various genotypes, in multiple locations and during more than one season. In many cases, the impact of environment and genotype was found to be substantial, while that of genetic modification appeared to be relatively minor.

In line with this finding by Davies *et al.* (2010) are the outcomes of a comprehensive review of the published literature on GMO safety, including studies on omics research on GM crops, by Ricroch (2013). With regard to the omics on crops with agronomic traits but without intended alteration of intrinsic crop metabolism, this author identified 36 studies on various cereals (barley, wheat, maize, rice) and non-cereals (cabbage, pea, potato, soybean). It was concluded that environmental factors had a greater impact than genetic modification, that there were also fewer impacts of it in comparison with conventional breeding techniques and that these outcomes did not raise concerns over the safety of commercialized GM crops (see Chapter 8).

4.6 Conclusions and Summary

A well-developed framework has been established for the regulatory pre-market safety assessment of GMOs based on international consensus as enshrined in the guideline of Codex Alimentarius (2008). While this guideline pertains to food, it also applies well to the safety assessment of animal feed. Central to the harmonized approach devised by the guideline is the comparative assessment, which entails a comparison of the compositional characteristics of a GM crop with its non-GM conventional counterpart and other comparators, of which the outcomes indicate whether or not there are differences between the GM crop and its comparators and therefore form the basis of decisions for further assessment of these differences. The extensive compositional analysis of nutrients, anti-nutrients, toxins and secondary metabolites therefore constitutes a key step in the safety assessment of GMOs. The consensus documents developed by the OECD Task Force on the Safety of Novel Foods and Feeds with recommendations on which crop-specific components are to be analysed for comparative safety are a valuable and widely used resource for this purpose.

In addition, data on the chemical composition of crops can help predict potential impacts on these crops' nutritional value. Recent reviews on state-of-the-art of studies on the composition of GM crops and their nutritional value in feeding studies in target livestock animals show that no additional findings are to be expected from animal feeding studies on the nutritional value of a GM crop if the crop's composition does not show any differences.

Before GM crops can be brought to the market, they need to undergo a pre-market safety assessment. An internationally harmonized approach has been established for this safety assessment based on the comparative assessment, which includes an extensive compositional analysis of the GM crop and its comparators, including a genetically close, non-GM conventional counterpart with a history of safe use, as well as other comparators. The crop materials to be analysed during this assessment should be obtained from field trials carried out in multiple locations according to an appropriate design, allowing for a robust statistical analysis and interpretation of the outcomes. Based on potential differences identified between the GM and the non-GM crops in this assessment, it can be decided whether a further assessment of these differences will be needed for their safety. The compositional data can also provide insight into the nutritional value of the GM crop to livestock animals and whether or not this has been changed compared with conventional non-GM crops. There have been promising advances in the fields of 'omics' analytical techniques that may be used as a supplementary tool to detect unintended changes in future GM crops with more complicated types of modifications, yet this will require standardization, validation and the establishment of databases on background variability.

References

Codex Alimentarius (2008) Guidelines for the Conduct of Food Safety Assessment of Foods Derived from Recombinant-DNA Plants. Codex Alimentarius, Joint FAO/WHO Food Standards Program, Food and Agriculture Organization, Rome (http://www.codex alimentarius.org/download/standards/1002/ CXG_045e.pdf, accessed 27 April 2013).

Davies, H.V., Shepherd, L.V.T., Stewart, D., Fraser, T., Röhlig, R.M. and Engel, K.H. (2010) Metabolome variability in crop plant species – when, where, how much and so what? Regulatory Toxicology and Pharmacology 53(3, Suppl. 1), S54–S61. DOI:10.1016/j.yrtph.2010. 07.004.

EFSA (2007) Safety and nutritional assessment of GM plants and derived food and feed: the role of animal feeding trials. Report of the EFSA GMO Panel Working Group on Animal Feeding Trials. Food and Chemical Toxicology 46, S2–S70. DOI:10.1016/j.fct.2008.02.008.

EFSA (2011) Guidance for risk assessment of food and feed from genetically modified plants. EFSA Journal 9(5), 2150 [37 pp.]. DOI: 0.2903/j.efsa.2011.2150 (www.efsa.europa.e / en/efsajournal/doc/2150.pdf, accessed 27 April 2013).

Flachowsky, G., Schafft, H. and Meyer, U. (2012) Animal feeding studies for nutritional and safety assessments of feeds from genetically modified plants: a review. Journal of Consumer Protection and Food Safety 7, 179–194. DOI:10.1007/ s00003-012-0777-9.

ILSI (2013) ILSI Crop Composition Database, Version 4.2. International Life Sciences Institute, Washington, DC (https://www.cropcomposition. org/query/index.html, accessed 27 April 2013).

James, C. (2012) Global Status of Commercialized Biotech/GM Crops: 2012 (ISAAA Brief 44-2012): Executive Summary. International Service for the Acquisition of Agri-Biotech Applications, Ithaca, New York (http://www.isaaa.org/resources/publications/briefs/44/executivesummary/default.asp, accessed 22 June 2013).

Kleter, G.A. and Kok, E.J. (2012) Safety of genetically modified (GM) crop ingredients in animal feed. In: Fink-Gremmels, J. (ed.) *Animal Feed Contamination; Effects on Livestock and Food Safety.* Woodhead, Oxford, UK, pp. 467–483.

Kok, E.J., Van Dijk, J.P., Prins, T.W. and Kleter, G.A. (2010) Safety assessment of GMO-derived foods. *Collection of Biosafety Reviews* 5, 82–111 (http://www.icgeb.org/~bsafesrv/pdffiles/Kok.pdf, accessed 27 April 2013).

Kuiper, H.A., Kleter, G.A., Noteborn, H.P.J.M. and Kok, E.J. (2001) Assessment of the food safety issues related to genetically modified foods. *Plant Journal* 27(6), 503–528. DOI:10.1046/j.1365-313X.2001.01119.x.

OECD (2011) Revised Consensus Document on Compositional Considerations for New Varieties of Low Erucic Acid Rapeseed (Canola): Key Food and Feed Nutrients, Anti-Nutrients and Toxicants (Series on the Safety of Novel Foods and Feeds No 24; ENV/JM/MONO(2011)55).

Organisation for Economic Co-operation and Development, Paris (http://www.oecd.org/env/ehs/biotrack/49343153.pdf, accessed 22 June 2013).

OECD (2012) Revised Consensus Document on Compositional Considerations for New Varieties of Soybean [Glycine Max (L.) Merr.]: Key Food and Feed Nutrients, Antinutrients, Toxicants and Allergens (Series on the Safety of Novel Foods and Feeds No 25; ENV/JM/MONO(2012)24). Organisation for Economic Co-operation and Development, Paris (http://search.oecd.org/officialdocuments/displaydocumentpdf/?cote=ENV/JM/MONO(2012)24&doclanguage=en, accessed 27 April 2013).

OECD (2013) Consensus Documents for the Work on the Safety of Novel Foods and Feeds. Organisation for Economic Co-operation and Development, Paris (http://www.oecd.org/science/biosafety-biotrack/consensusdocumentsfortheworkonthesafetyofnovelfoodsandfeeds.htm, accessed 27 April 2013).

Ricroch, A.E. (2013) Assessment of GE food safety using '-omics' techniques and long-term animal feeding studies. *New Biotechnology* 30(4), 349–354. DOI:10.1016/j.nbt.2012.12.001.

5 Types of Feeding Studies for Nutritional and Safety Assessment of Feeds from GM Plants

Gerhard Flachowsky*

Institute of Animal Nutrition, Friedrich-Loeffler-Institute (FLI), Federal Research Institute for Animal Health, Braunschweig, Germany

5.1 Introduction

Feeding studies in laboratory animals and targeted livestock species may be useful for assessing the nutritional and safety impact of intended genetic changes. This information is more important for plants with substantial changes in composition (plants with output traits, or second-generation genetically modified (GM) plants) than for substantially equivalent plants (first-generation plants). In approaching the evaluation of nutritional value and safety of feed from GM plants, two questions must be answered.

The first question is how the feed will be used. Is it intended to be consumed as a whole feed replacing a traditional feed, or is it intended that the product of genetic modification will be separated from the plant and consumed as an ingredient or as a co(by)-product? The approach to nutritional and safety assessment will be different in both cases. Feeding studies should be done with those components used in animal nutrition.

The second key question deals with the extent of consumption of the genetically modified feed or ingredients. The intake level must be known or predictable in advance of performing animal feeding studies. The highest possible amount of feed

from transgenic plants should be used in animal feeding studies.

Various types of animal feeding studies are required to answer all the scientific and public questions, and to improve the public acceptance of such food/feed and animals (see ILSI, 2007; EFSA, 2008; Flachowsky *et al.*, 2012). The current state of the nutritional and safety assessment of feed from modified plants and the future challenges will be analysed in this chapter. The necessity of feeding studies with food-producing animals depends also on the outcome of the compositional analysis (see Chapter 4).

The main objective of the present chapter is to consider the pros and cons of various animal feeding studies for the nutritional assessment of GM feed. Sometimes, it is impossible, and also not necessary, to separate strictly the nutritional and safety assessment of feed/food.

The principles of substantial equivalence assessment have developed (OECD, 1993) into a systematic approach that focuses on comparing a particular GM plant to the nearest isogenic relative using agronomic metrics and compositional analysis (see Chapter 4) to determine if genetic modification has produced unintended pleiotropic effects. On the other hand, it must be asked, in the case of first-generation plants, if such rigorous analyses as mentioned in Chapter

*E-mail: gerhard.flachowsky@t-online.de

3 and 4 as well as feeding studies are necessary, because unregulated crops produced by other breeding methods also undergo genetic changes and contain unintended effects (Chassy, 2010). However, it is possible that the process of genetic modification could result in unintended, potentially adverse pleiotropic changes (Cellini *et al.*, 2004) that might not be detected analytically (Delaney, 2007). Therefore, feeding studies in laboratory and target animals are considered necessary for nutritional and safety assessment.

Kleter and Kok (2010) and Davies and Kuiper (2011) consider the following aspects, which also include nutritional aspects for risk assessments:

- Characteristics of donor and recipient organisms (see Chapter 2).
- Genetic modification and its functional consequences (see Chapter 2).
- Potential environmental impacts (see Chapter 3).
- Agronomic characteristics.
- Compositional and nutrit.ional characteristics (see Chapter 4).
- Potential for toxicity and allergenicity of genetic products, plant metabolites and whole GM plants (see Chapters 3 and 4).
- Influence of processing on the properties of feed (see Chapter 4).
- Potential for changes in dietary intake (see Chapters 6 and 7).
- Potential for long-term and multi-generational nutritional impact (see Chapter 8).

5.2 Types of Feeding Studies

Before commencing feeding studies, compositional analysis (including nutrients and undesirable substances) of all feeds used in the study (see Chapter 4) and of mixed feed given to experimental animals, as well varicus *in silico*, *in vitro* or *in sacco* studies (see ILSI, 2003, 2007; FSANS, 2007; DBT, 2008; EFSA, 2008, 2011a) are the key elements of animal feeding studies and contribute substantially to the nutritional and safety assessment of GM plant-derived

feed and food. Unfortunately, there are some feeding studies with high public interest but with large weaknesses in the field of clear characterization of feeds and mixed feeds and experimental design (e.g. Velimirow *et al.*, 2008; Seralini *et al.*, 2012).

Feeding studies with laboratory animals, but much more with target animals, are key elements for the nutritional and safety assessment of feed/food from GM plants. Depending on the scientific questions, the following types of feeding studies (see Table 5.1) are well established and may be carried out:

- Laboratory animal models for the toxicity testing of single substances (single-dose toxicity testing; repeated-dose toxicity testing; reproductive and developmental toxicity testing; immunotoxicity testing, etc.; OECD, 1998a; DBT, 2008; EFSA, 2008; Ladics *et al.*, 2010).
- Laboratory animal models for the safety and nutritional assessment of whole GM feed and food (in general 90-day feeding studies for safety assessment) to detect unintended effects, subchronic animal tests, for margins of safety, etc. (OECD, 1998b; DBT, 2008; EFSA, 2011b).
- Studies to measure the digestibility/ bioavailability of nutrients from GM plants and to analyse the influence of GM products on the metabolism of target animals/categories (ILSI, 2004, 2007; Flachowsky and Böhme, 2005; DBT, 2008; EFSA, 2008).
- Tolerance studies to analyse the influence of maximal amounts of GM feeds on animal health and welfare (ILSI, 2007; DBT, 2008; EFSA, 2008).
- Efficacy studies to measure the influence of GM feed on animal yield/performance, feed conversion rate (FCR) and slaughtering performance, as well as the safety and composition/quality of food of animal origin (ILSI, 2007; DBT, 2008; EFSA, 2008; see Chapters 6 and 7).
- Long-term studies to find out the long-term effects of GM feed (e.g. whole growing period in the case of growing animals, whole laying period in the case of laying animals, or one or more

lactations in the case of lactating animals; whole lifespan of animals; Flachowsky *et al.*, 2012; Snell *et al.*, 2012; see Chapter 8).

- Multi-generational studies to analyse the influence of GM feed on the fertility/reproductive performance of animals (Flachowsky *et al.*, 2005b; BEETLE, 2009; Snell *et al.*, 2012; see Chapter 8).
- Studies with GM animals to determine the energy and nutrient requirements of modified animals and to analyse the quality and safety of food of animal origin, which are beyond the scope of this book (EFSA, 2012).

In general, the expense of the studies mentioned above increases from the top to the bottom of Table 5.1. Therefore, long-term studies and multi-generational experiments with target animals are very rare (see Chapter 8). Limited feed amounts in earlier breeding stages may also restrict animal numbers and the duration of studies with target animals, especially with large animals such as ruminants and pigs. In summary, the following factors (see also Table 5.2) may influence the types of animal feeding studies:

- Scientific question(s).
- Availability of GM feeds (especially in the early stages of breeding) and adequate comparators.

- The highest possible amounts of GM feed included in the diets.
- Financial budget.
- Availability of equipment, animals and qualified personnel.

Feeding studies with target animals will be considered in more detail in the following paragraphs. Product quality (e.g. milk, meat, fish, eggs, etc.) and the carry-over of ingredients of feed into food of animal origin (e.g. fatty acids, minerals, vitamin , undesirable substances) should also be measured in feeding studies or after slaughtering the animals (see Chapters 7 and 10). All animal feeding studies should be conducted with consideration of the principles of animal health and welfare (Russell and Burch, 1959; EFSA, 2012).

5.3 Experimental Design and Animal Feeding

The objective of the animal feeding study is to compare the GM variety, or co-products from the GM variety, with a near-isogenic conventional variety as a control treatment.

During the first years of animal experimentation with GM feed, only two groups (GM feed and non-GM counterpart) were fed in most studies (summarized by Clark and Ipharraguerre, 2001; Flachowsky and Aulrich, 2001; Aumaitre *et al.*, 2002 .

Table 5.1. Important types of feeding studies with animals, for nutritional and safety assessment of feed from GM plants and animals.

Type of studies	Laboratory animals	Target animals
Testing of single substances (28-day study)	X	
90-day rodent feeding study	X	
Long-term feeding study	X	X
Multi-generation feeding study	X	X
Determination of digestibility/availability (including rumen fermentation and metabolism)	X	X
Efficiency (performance) study		X
Product study (composition and quality of food of animal origin)		X
Tolerance study		X
Studies with GM animals[a]		X

Note: [a]Not covered in this book.

Table 5.2. Advantages of feeding studies with GM plant-derived feed/food with laboratory or target animals.

Laboratory animals	Target animals
Internationally agreed study protocols	Representative for target animal species/categories (extrapolation of data possible)
Small amounts of feed, higher number of repetitions	Higher amounts of GM products are fed to animals (recommendations for practical feeding, direct transfer of results)
Lower costs for feed and equipment	All 'control' animals fed with comparators (isogenic, commercial) are available for the market (no waste animals)
	Studies on the transfer of valuable and undesired substances in food of animal origin (composition and quality of the food)

Because of the normal biological ranges, the differences between two groups should not be overestimated. If two treatments are significantly different, that does not mean that the difference is large enough to be of any biological importance or any practical significance. Therefore, one or more commercial reference lines were later included in such studies to help compare the data with commercial lines. Some statistically significant differences between the GM plant and the near-isogenic line may occur by chance and may not be biologically relevant. Reference lines may help to delineate the range of values typical for the crop type (ILSI, 2003; EFSA, 2011a). For experimental design, the following parameters/criteria should be stated clearly:

- Animal species/category.
- Number of animals; number of groups.
- Termination of experiment.
- Initial body weight.
- Highest portion of GM feed should be included or better: dose–response studies.
- Clear characterization/analysis of all feed used in the study.
- Balanced diets, adjusted diets according to the scientific recommendations of the National Research Council (NRC) or the Society of Nutrition Physiology (GfE) for animal species/categories.
- Health and welfare; keeping of animals.
- Feed intake, weight gain or animal yield.
- Removal of animals; animal losses.

All studies should be conducted according to internationally accepted protocols (OECD, 1998a,b; ILSI, 2003, 2007; EFSA, 2008, 2011a). Statistical analysis and interpretation of results should be done according to adequate publications and under consideration of the EFSA's comments (2011c).

5.4 Laboratory Animals

Usually, the OECD guideline tests (OECD TG 407 and 408; OECD, 1998a,b) for chemicals are used for the safety testing of single substances including new products resulting from genetic modification (e.g. newly expressed proteins; EFSA, 2006, 2008). Generally, rodents (rats or mice) are used over a period of 28 days/1 month for single-dose or repeated-dose toxicity testing. The detailed testing strategy should be selected on a case-by-case basis, based on prior knowledge regarding the biology of the products, so that relevant end points are measured in the test (for more details see OECD, 1998a,b; EFSA, 2006, 2008, 2011b; FDA, 2007).

A 90-day rodent feeding study should be carried out, as indicated by molecular, compositional, phenotypic, agronomic or other analysis (e.g. changes in metabolic pathways). Such toxicity studies should only be performed on a case-by-case basis to provide additional information for the risk assessment (EFSA, 2008). It seems to be nearly impossible from the nutritional point of view to adapt toxicity studies for testing whole feed. The OECD (1998b) guideline has been developed to assess the safety of additives and not to test whole feed/food. The EFSA (2011b) states that the purpose for a repeated-dose, 90-day oral toxicity study on whole food/feed is to reassure the public that the GM food/feed is just as safe and nutritious as its traditional comparator. In such cases, high portions of the whole feed/food should be supplemented to a basal diet, knowing that the energy and nutrient requirements of laboratory animals (NRC, 1995) are not met and imbalances in some nutrients, especially amino acids, could be expected (EFSA, 2011b). This statement seems to be important from the view of many GM feeds rich in protein (e.g. soybean, cotton, rapeseed). It is nearly impossible to make scientific conclusions under imbalanced conditions (NRC, 1995). Adjusted diets – if possible – should be fed under those conditions.

Another point of criticism of the 90-day feeding study with rodents is the duration of the experiments for safety assessment. Some authors (e.g. Seralini *et al.*, 2011, 2012) consider 90 days to be too short for various parameters such as fertility and reproduction, histopathology of some organs (e.g. liver, kidney), hormone status, etc. The authors propose 2-year studies with mature rats or other animals (at least three species; de Vendomois *et al.*, 2009), which should include sexual hormone assessment and reproductive, developmental and trans-generational studies. Details about the expenses of such studies and the potential scientific yield, as well the higher levels of safety, are not given (see Chapter 8 for long-time and multi-generation studies). Recently, such a 2-year study (Seralini *et al.*, 2012) with rats was carried out, with confusing results because of many weaknesses (e.g. no characterization of the composition of feedstuffs and mixed feed, low number of animals) in experimental design. More details about the necessity of studies with laboratory animals and useful end points have been described by the EFSA (2003, 2008, 2011b) and OECD (1998b).

As already mentioned, the scientific output to contribute to the nutritional and safety assessment of feed for the nutrition of target species with such studies in laboratory animals would be very small or negligible. In some cases, appropriately designed animal toxicology studies can provide an additional measure of safety assurance. In general, however, such studies in laboratory animals are unlikely to reveal unintended minor compositional changes that have gone undetected by targeted analysis because they lack adequate sensitivity. Specific studies with target animals may contribute more substantially to the nutritional assessment of feed and could be useful for safety assessment. This conclusion also seems to be very important from the view of GM feed used in animal nutrition (see Table 5.2).

Model animals (mice, rats, rabbits or small target animals (chicks, quails, piglets) are also used to measure the digestibility/ bioavailability of nutrients in GM crops (see Section 5.5.1) because of the high costs and the limited feed amounts available in some cases, especially in early plant-breeding stages (see Chapters 6 and 7 for more details).

5.5 Target Animals

Studies with target animals (food-producing animals) are focused mainly on nutritional concerns. Up to now, such studies have paid less attention to safety aspects (EFSA, 2003, 2011a). Future feeding studies with target animals should also be used for the safety assessment of GM plants because of the high proportion of these plants used for animal nutrition (see Chapter 1) and the cultivation of second-generation GM plants (see Chapters 7 and 12). The type of studies depends on the type of genetic modification in the plants or animals and the availability of GM feed or GM animals.

Animal feeding studies should be conducted in target animal species to demonstrate the nutritional properties that may be expected from the use of the modified crop, specific crop components or co-products from GM crops (ILSI, 2007). The feeding period should cover the whole lifespan (see Table 5.3) or a representative period (e.g. in the case of laying hens or dairy cows).

5.5.1 Measuring digestibility/ bioavailability (nutrient availability)

Animal feeding studies play an important role in testing the nutritional value of the introduced trait in a nutritionally improved crop. Analyses of nutrient composition provide a solid foundation for assessing the nutritional value of feeds (see Chapter 4); however, they do not provide information on nutrient availability.

In the case of substantial changes in plant composition (GM plants with output traits, or second-generation GM plants; see Section 5.5.3 and Chapter 7), studies measuring the digestibility/availability of some nutrients or nutrient precursors are necessary (ILSI, 2004, 2007; Flachowsky and Böhme, 2005; EFSA, 2008).

Such studies should be done on laboratory animals or, preferably, on target animals. Studies with laboratory animals need less feed and lower costs (see Table 5.2): sometimes, the application of results to target animals is questionable, but such studies are already possible during early plant-breeding stages.

5.5.2 Efficiency studies including transfer of nutrients with GM plants with input traits (first generation)

The objective of efficiency trials is to measure the effect of feed from GM plants on the performance of food-producing animals and to compare the results with an isogenic counterpart and some commercial products.

Many feeding studies have been carried out during the past years to show the substantial similarities (OECD, 1993) of feed derived from first-generation GM plants (without substantial changes in their composition or GM plants with input traits; see Chapter 6). Most of the studies were done as efficiency trials and GM feed was compared in adjusted diets with their isogenic counterparts and some conventional commercial varieties (one to ten varieties in some cases). The experimental designs were done according to the recommendations by ILSI (2003; Table 5.4) and the EFSA (2006, 2008, 2011a). Questions concerning the tolerance of some feeds in animals (tolerance studies to determine the maximum inclusion level of the feed in diets) may also be included in efficiency trials. Statistical significance, but not biological relevance, is a conflicting subject in some publications and also with the public. Recently, the EFSA (2011b,c) contributed to solving this conflict and explored the concept of biological relevance. It is recommended that the nature and size of biological changes or differences should

Table 5.3. Examples of lifespans for growing/fattening animals (in days).

Animal species/categories[a]	Conventional/ more intensive	Organic/more extensive
Chickens for fattening (broilers)	30–42	56–84
Turkeys for fattening	56–168	70–112
Growing/fattening pigs	150–300	200–400
Veal calves	80–200	–
Growing/fattening bulls	300–500	400–600

Note: [a]Laying hens and dairy cattle are usually used for longer periods. Laying hens: about 126–140 days for growing (pullets) and about 300–360 days (1 year) for the laying period. Dairy cattle: about 22–36 months for growing (heifers) and 1–10 years for lactation (average in Europe, two to five lactations).

be defined before studies are initiated. A pre-defined relevant biological effect should be used to design studies with sufficient statistical power to be able to detect such effects if they truly occur. Such conclusions may be very helpful, but their realization depends on the experimental equipment of research institutions, especially in cases of large animals (ruminants, pigs). Nevertheless, there is a certain subjective component in this recommendation, which will require some discussion in the future.

During the past few years, many efficiency studies and some reviews on the nutrition and safety assessment of feed from GM plants (mostly first-generation plants) have

Table 5.4. Some recommendations from the 'Best practices for the conduct of animal studies (efficiency studies) to evaluate crops genetically modified for input traits' (first-generation GM plants) (adapted from ILSI, 2003).

Animals (species/ categories)	Number of animals (coefficient of variation 4–5%)	Duration of experiments	Composition of diets[a]	Measurements/end points
Poultry for meat production	10–12 pens per treatment, with 9–12 birds per pen	5 weeks or more	Balanced diets	Feed intake, gain, feed conversion, metabolic parameters, body composition
Poultry for egg production	12–15 replications per treatment, with 3–5 layers per pen	18–40 weeks of age, at least three 28-day phases	Balanced diets	Feed intake, egg production, feed conversion, egg quality, metabolic parameters
Swine	6–9 replications per treatment, with 4 or more pigs per replication	Piglets (7–12 kg) 4–6 weeks Growers (15–25 kg) 6–8 weeks	Balanced diets	Feed intake, gain, feed conversion, metabolic parameters, carcass quality
Growing and finishing ruminants	6–10 replications per treatment, with 6 or more cattle per replication	90–120 days	Balanced diets	Feed intake, gain, feed conversion, carcass data, metabolic parameters
Lactating dairy cows	12–16 cows per treatment	Latin square 28-day periods Randomized block design	Balanced diets	Feed intake, milk performance and composition, body weight, metabolic parameters, body condition score (BCS), cell counts in milk, animal health

Note: [a]Efficiency studies to evaluate feed from GM plants with output traits (second-generation GM plants) should be done under consideration of recommendations by the EFSA (2008, 2011a) and ILSI (2007); feed from GM plants should be included in high portions in the diets and compared with near-isogenic counterparts (commercial varieties) to show the biological range.

been published (Clark and Ipharraguerre *et al.*, 2001; Flachowsky and Aulrich, 2001; Aumaitre *et al.*, 2002; Flachowsky *et al.*, 2005a, 2007; CAST, 2006; Spiekers *et al.*, 2009) and are described in more detail in Chapter 6. Furthermore, the ILSI (2003) and EFSA (2006, 2008, 2011a) documents also summarize the present state of knowledge in the feeding of GM plant-derived feed of the first generation to target animals. In the ILSI document (ILSI, 2003), protocols for evaluating feedstuffs from GM plants with input traits in poultry for meat and egg production, pigs, lactating dairy cows and growing and finishing ruminants are given.

The necessity of animal feeding studies with feed from first-generation GM plants, or of substantial equivalent plants, is often questioned concerning their sensitivity and scientific output. According to various guidance documents (EFSA, 2006, 2008, 2011a), such studies are not needed urgently for nutritional assessment and are not required for safety assessment. No animal feeding studies are required if the differences in compositional analyses between isogenic and transgenic plants are small or negligible (first-generation GM plants; see Chapter 4). The scientific yield of such studies is considered as negligible by some authors (EFSA, 2008, 2011a).

On the other hand, feeding experiments with target animals with first-generation GM plants may contribute towards showing the public the nutritional equivalence and safety of the feed, and therefore they could improve the public's acceptance of GM feed. Furthermore, recommendations for optimal amounts of feed from GM plants in target animal feeding may be deduced.

Another point is the so-called wastage of animals (the '3Rs'; Russell and Burch, 1959). Under the present regulations, only animals fed with non-permitted GM feed cannot be used in the food chain. This means that if a GM feed with its isogenic counterpart and four commercial varieties are tested in a feeding study, more than 80% of target animals can be used for human nutrition. Therefore, efficiency feeding studies with first-generation GM

plants could be useful in some cases (see Tables 5.3 and 5.4).

More details on conducting animal feeding studies to evaluate GM crops modified for input traits (first-generation GM crops) are described in detail by ILSI (2003), the EFSA (2011a) and in Chapter 6.

5.5.3 Efficiency studies including transfer of nutrients with GM plants with output traits (second generation)

More and other feeding studies are necessary for the nutritional assessment of food/feed from second-generation GM plants (so-called plants with substantial changes in composition, plants with output traits, or plants with improved nutritional quality or so-called 'biofortified' plants; see Chapters 7 and 12). The principle of substantial equivalence as used for the safety and nutritional assessment of feed from first-generation GM plants does not have adequate relevance for feed with substantial changes in composition (ILSI, 2004, 2007; EFSA, 2008; Llorente *et al.*, 2011). ILSI (2004) mentioned various examples of modifications by biotechnology and distinguished in the following groups (see also Chapters 7 and 12):

- Proteins and amino acids.
- Carbohydrates.
- Fibre and lignins.
- Oils/lipids/fatty acids.
- Vitamins and minerals.
- Nutraceuticals.
- Anti-nutrients.
- Allergens and substances causing food/feed intolerance.
- Toxins.

Specific studies are necessary to demonstrate the effects of genetic modification on the nutritional value of feeds. Experimental designs for such studies with second-generation GM crops are described in detail by the EFSA (2008, 2011a), Flachowsky and Böhme (2005) and ILSI (2007). King (2002) described a three-step process for evaluating plant biofortification in human nutrition (see also Chapter 7). Flachowsky and Böhme

(2005) submitted models for studies to assess biofortified plants as well as plants with decreased concentration of undesirable substances in GM crops. Different experimental designs are necessary to demonstrate the efficiency of changes or of expressed nutrients/constituents in GM crops, such as:

- Bioavailability or conversion of nutrient precursors into nutrients (e.g. β-carotene; see Table 5.5).
- Digestibility/bioavailability of nutrients (e.g. amino acids, fatty acids, minerals, vitamins; see Table 5.6).
- Efficiency of substances which may improve nutrient digestibility/availability (e.g. enzymes; see Table 5.7).
- Efficiency of substances with surplus effect(s) (e.g. prebiotics).
- Improvement of sensory properties/palatability of feed (e.g. essential oils, aromas).

Balance studies with laboratory/target animal species/categories are necessary to assess the conversion of nutrient precursors (e.g. β-carotene) into nutrients (e.g. vitamin A). At least two groups of animals are necessary to assess the bioconversion of the precursor into the nutrient (Table 5.5 and Chapter 7). All animals should be fed with balanced diets except the substance under investigation.

Other studies are necessary to assess the effect/s of the increased content of one or more essential nutrients in the GM plant (Table 5.6). Answers should be provided on its digestibility/bioavailability, but also on the effects on feed intake, animal yield and composition/quality of food of animal origin.

Dose–response studies seem to be helpful in some cases. Studies with restricted (adequate to control; pair feeding to control animals) and *ad libitum* feed intake are recommended if any influence of the component on the feed intake level is expected.

The evaluation of the bioavailability of essential nutrients should be measured in appropriate species. For example the precaecal digestibility of essential amino acids of pigs should be measured in specific prepared pigs (e.g. GfE, 2005); those for avian species in cecectomized cockerels. Similar models should be used to assess the effects on non-essential substances such as enzymes, pre- and probiotics, essential oil or substances with influence on the sensory properties or palatability of feed intake and animal yield (see Flachowsky and Böhme, 2005; EFSA, 2008).

Table 5.5. Proposal to assess the conversion of nutrient precursors of the second generation into nutrients (GM plant with output traits; e.g. conversion of the precursor β-carotene into vitamin A). (From EFSA, 2008.)

Groups[a]	Composition of diets	Measurements; end points[b]
1[c]	Balanced diet with typical amounts of the isogenic counterparts (unsupplemented control)	Depend on genetic modification of plants, for example: • feed intake, animal's growth • concentration of specific/converted substance(s) in most suitable indicator organs (e.g. vitamin A in the liver)[d] • additional metabolic parameters such as depots in further organs or tissues, activities of enzymes and hormones
2	Balanced diet with adequate amounts of the transgenic counterpart (e.g. rich in β-carotene)	
3	Diet of Group 1 with β-carotene supplementation adequate to Group 2	
4	Diet of Group 1 with vitamin A supplementation adequate to expected β-carotene conversion into vitamin A	

Notes: [a]Some animal groups are fed with commercial/isogenic control feed to find out the biological range of the parameter(s). [b]Depletion of specific nutrient in experimental animals could be necessary. [c]Adequate feed amounts (pair feeding) for all animals; depletion phase for all animals before experimentation. [d]Up to the steady state in the specific target organ.

Table 5.6. Proposal for nutritional assessment of a GM feed ingredient in which the concentration of an essential nutrient has been increased. (From EFSA, 2008.)

Groups	Diet composition	Measurements
1	Typical levels of near-isogenic parenteral line fed *ad libitum*	Depends on claim of genetic modification: • precaecal digestibility of amino acids • indicator values for minerals and vitamins • feed intake • metabolic parameters • animal performance, feed efficiency • incorporation of substances in animal tissues Quality of food of animal origin
2	Balanced diet including typical levels of transgenic line (e.g. increased levels of amino acids, minerals, vitamins, etc.) fed *ad libitum*	
3	Balanced diet including typical levels of the isogenic line with supplementation of adequate amounts (to Group 2) of changed nutrient (e.g. amino acids, minerals, vitamins, etc.) pair feeding to Group 2	
4	Inclusion of further groups with other commercial varieties	

Table 5.7. Proposal to assess the effects of enhanced nutrient utilization (e.g. by enzymes).

Groups	Diet composition	Measurements
1	Balanced diet including typical levels of near-isogenic parenteral line fed *ad libitum*	Depends on claim of genetic modification: • feed intake • animal performance(s) • digestibility of specific nutrient(s) • metabolic parameters Quality of food of animal origin
2	Balanced diet including typical levels of transgenic line (e.g. content of specific enzyme); feeding level of Group 1	
3	Diet of Group 1 plus enhancer adequate to transgenic line (or dose–response studies); feeding level of Group 1	
4	Diet of Groups 2 and 3, fed *ad libitum*	

Other types of experimental designs are necessary to assess the effects of decreased concentration of undesirable substances in GM crops, such as:

- Lower concentration of inhibiting substances (e.g. phytate).
- Lower concentration of toxic substances (e.g. mycotoxins, native plant toxins such as gossypol, glucosinolates, glyco-alkaloids, glycoproteins).
- Lower concentration of cell wall constituents (e.g. lignin, silicate).

Such undesirable substances may affect feed intake and animal health and performance negatively, but also nutrient availability and metabolic parameters. The reduction of undesirable substances in crops is one of the most important objectives of genetic modification.

Phytate is an anti-nutritive factor that reduces the phosphorus availability from plants. Table 5.8 shows a model that may contribute towards demonstrating the effects of lower phytate content in animals (see also Chapters 7 and 12, as well as Spencer *et al.*, 2000, for an example).

After a description of the production, handling, storing and processing of GM and isogenic counterparts, as well as the sampling and analysis of harvested and processed material, ILSI (2007) distinguished feeding studies for specific-trait groups and protocols for evaluating whole feedstuffs with genetically modified output traits. Proteins and amino acids, carbohydrates, lipids and fatty acids, vitamins and antioxidants, minerals, enzymes and anti-nutrients are mentioned as specific-trait groups. Protocols for animal feeding studies

Table 5.8. Proposal to assess the effects of inhibitors of nutrient availability (e.g. phytate).

Groups	Diet composition	Measurements
1	Balanced diet including typical levels of isogenic parenteral line fed *ad libitum*	Depends on claim of genetic modification:
2	Balanced diet including typical levels of transgenic line adequate to Group 1 (e.g. low phytate crop) fed *ad libitum*	• feed intake • animal performance, feed and nutrient (e.g. P) efficiency
3	Diet of Group 1 and supplemented with inhibited nutrient (e.g. phosphorus) fed in adequate amounts to Group 2	• digestibility of inhibited nutrien • concentration of inhibited nutrien in indicator organs/tissues
4	Inclusion of further groups with other commercial varieties (unsupplemented and supplemented with inhibited nutrient)	

to assess GM crops with output traits are given for poultry meat production, egg production, pigs, lactating ruminants, growing and finishing ruminants and aquaculture.

Furthermore, efficiency studies with second-generation GM plants may also be used or combined with studies to measure the digestibility/bioavailability of the newly expressed substances. Changed composition of GM plants and derived feed may also influence the composition of food of animal origin, as has been demonstrated exemplarily for soybeans with a modified fatty acid pattern (for more details, see Chapter 10).

5.5.4 Long-term and multi-generation feeding studies

Long-term feeding studies cover the whole lifespan, or a very long period of this, of the animals, for example, in the case of laying hens or dairy cows (see Table 5.3 and Chapter 8). Apart from the animals' performance, the answers expected from such studies include their fertility and health when fed with high amounts of GM feed. Can animal feeding trials contribute to the assessment of long-term effects? This was the main question of the BEETLE study (BEETLE, 2009). The assessment of the data and results from the online survey of BEETLE (2009) on animal health did not show any new aspects. Some participants of the online survey expected only potential long-term effects in relation to allergenicity in humans, but all other possible adverse

long-term effects were assessed as b⸗i⸗g negligible. In general, a methodical im⸗r⸗⸗e- ment of the risk assessment procedⵣre ⸗a⸗ been recommended, including a ⸗i⸗⸗⸗r number of replications and add⸗ti⸗ra⸗ control groups to demonstrate the bi⸗⸗o⸗i⸗a⸗ range of measured parameters.

In addition to long-term feeding s u⸗i⸗s multi-generation studies (mostl⸗ ⸗⸗⸗e generations; see Flachowsky *et al.*, ⸗0⸗⸗b Snell *et al.*, 2012) should be carried ⸗⸗t ⸗c test the influence of GM feed on r⸗⸗o- duction, long-term health and me⸗a⸗di⸗ effects in laboratory and target a⸗in⸗⸗s More details about long-term and ⸗⸗⸗ⵣi generation feeding studies are descr⸗be⸗⸗r Chapter 8.

5.6 Conclusions

The objective of animal feeding stu⸗i⸗ ⸗i⸗ *vivo* studies) is to characterize the ef⸗e⸗⸗ o⸗ specific feeds in animals. Feeding ⸗t⸗⸗⸗e start with the compositional analysi⸗ a⸗c i⸗ *silico* and/or *in vitro* measurements.

Feeding studies can be don⸗ ⸗⸗t⸗ laboratory and target animals. Lab⸗r⸗t⸗r⸗ animals are used for testing single sub⸗t⸗⸗⸗e⸗ and toxicological parameters in anim⸗l⸗.

The measuring of digestibility/ava⸗la⸗i⸗t⸗ of nutrients, long-term feeding stu⸗⸗e⸗ ⸗n⸗ multi-generation studies are don⸗ ⸗⸗t⸗ laboratory and target animals. T⸗⸗e⸗a⸗⸗ studies to measure the optimal d⸗⸗a⸗e ⸗ specific feed in animal diets, e⸗i⸗⸗e⸗⸗ experiments and the influence of f⸗e⸗⸗ o⸗

body composition or composition and quality of milk and eggs should be measured in target animals.

More details of results with feeds from GM plants in animal nutrition are described in the Chapters 6–8 and 10.

References

Aumaitre, A., Aulrich, K., Chesson, A., Flachowsky, G. and Piva, G. (2002) New feeds from genetically modified plants: substantial equivalence, nutritional equivalence, digestibility, and safety for animals and the food chain. *Livestock Production Sciences* 74, 223–238.

BEETLE (2009) Long-term effects of genetically modified (GM) crops on health and the environment (including biodiversity), Executive Summary and Main Report. Federal Office of Consumer Protection and Food Safety (BVL), Berlin, 132 pp. + 6 Annexes.

CAST [Council for Agricultural Science and Technology] (2006) Safety of meat, milk, and eggs from animals fed crops derived from modern biotechnology. Issue Paper 34, 8 pp. CAST, Ames, Iowa.

Cellini, F., Chesson, A., Coquhonn, I., Constable, A., Davies, H.V., Engel, K.-H., *et al.* (2004) Unintended effects and their detection in genetically modified crops. *Food Chemistry and Toxicology* 42, 1089–1123.

Chassy, B.M. (2010) Food safety and consumer health. *New Biotechnology* 27, 534–544.

Clark, J.H. and Ipharraguerre, I.R. (2001) Livestock performance: feeding biotech crops. *Journal of Dairy Science* 84, E. Suppl., E9–E18, 237–249.

Davies, H.V. and Kuiper, H.A. (2011) GM-risk assessment in the EU: current perspective. *Aspects of Applied Biology* 110, 37–45.

DBT [Department of Biotechnology] (2008) Protocols for food and feed safety assessment of GE crops. Department of Biotechnology, Ministry of Science and Technology, New Delhi, 34 pp.

Delaney, B. (2007) Strategies to evaluate the safety of bioengineered foods. *International Journal of Toxicology* 26, 389–399.

EFSA (2006) Guidance document of the Scientific Panel on Genetically Modified Organisms for the risk assessment of genetically modified plants and derived food and feed. *EFSA Journal* 99, 1–100.

EFSA (2008) Safety and nutritional assessment of GM plant derived food and feed. The role of animal feeding trials. *Food and Chemical Toxicology* 46, 2–70.

EFSA (2011a) Guidance for risk assessment of food and feed from genetically modified plants. *EFSA Journal* 9(5), 2150, 37 pp.

EFSA (2011b) Guidance on conducting repeated-dose 90-day oral toxicity study in rodents on whole food/feed. *EFSA Journal* 9(12), 2438, 21 pp.

EFSA (2011c) Statistical significance and biological relevance. *EFSA Journal* 9(10), 2372, 17 pp.

EFSA (2012) Guidance on the risk assessment of food and feed from genetically modified animals and on animal health and welfare aspects. *EFSA Journal* 10(1), 2501, 43 pp.

FDA [Food and Drug Administration] (2007) Guidance for industry and other stakeholders, toxicological principles for the safety assessment of food ingredients. Redbook 2000; July 2000, revised July 2007. US Department of Health and Human Services, Food and Drug Administration, Center for Food Safety and Applied Nutrition (http://www.cfsan.fda.gov/guidance.html, accessed 3 June 2013).

Flachowsky, G. and Aulrich, K. (2001) Nutritional assessment of GMO in animal nutrition. *Journal of Animal Feed Science* 10, Suppl. 1, 181–194.

Flachowsky, G. and Böhme, H. (2005) Proposals for nutritional assessments of feed from genetically modified plants. *Journal of Animal Feed Science* 14, 49–70.

Flachowsky, G., Chesson, A. and Aulrich, K. (2005a) Animal nutrition with feeds from genetically modified plants. *Archives of Animal Nutrition* 59, 1–40.

Flachowsky, G., Halle, I. and Aulrich, K. (2005b) Long-term feeding of Bt-corn – a ten-generation study with quails. *Archives of Animal Nutrition* 59, 449–451.

Flachowsky, G., Aulrich, K., Böhme, H. and Halle, I. (2007) Studies on feeds from genetically modified plants (GMP). Contributions to nutritional and safety assessment. *Animal Feed Science and Technology* 133, 2–30.

Flachowsky, G., Meyer, U. and Schafft, H. (2012) Animal feeding studies for nutritional and safety assessments of feed from genetically modified plants: a review. *Journal of Consumer Protection and Food Safety* 7, 179–194.

FSANS (2007) The role of animal feeding studies in the safety assessment of genetically modified foods. Report of Workshop of Food Standards Australia New Zealand, 15 June 2007, Canberra, 19 pp.

GfE [Society of Nutrition Physiology] (2005) Standardised praecaecal digestibility of amino acids in feedstuffs for pigs – methods and concepts. *Proceedings of Society of Nutrition Physiology* 14, 185–205.

ILSI (2003) Best practices for the conduct of animal studies to evaluate crops genetically modified for input traits. International Life Sciences Institute, Washington, DC, 62 pp.

ILSI (2004) Nutritional and safety assessments of foods and feeds nutritionally improved through biotechnology. *Comprehensive Reviews in Food Science and Food Safety* 3, 36–104.

ILSI (2007) Best practices for the conduct of animal studies to evaluate crops genetically modified for output traits. International Life Sciences Institute, Washington, DC, 194 pp.

King, J.C. (2002) Evaluating the impact of plant fortification on human nutrition. *Journal of Nutrition* 132, 511S–513S.

Kleter, G.A. and Kok, E.J. (2010) Safety assessment of biotechnology used in animal production, including genetically modified (GM) feed and GM animals – a review. *Animal Science Papers and Reports* 28, 105–114.

Ladics, G.S., Knippels, L.M. and Penninks, A.H. (2010) Review of animal models to predict the potential allergenicity of novel proteins in genetically modified crops. *Regulation Toxicology and Pharmacology* 56, 212–224.

Llorente, B., Alonso, G.D., Bravo-Almonacid, F., Rodriguez, V., Lopez, M.G., Carrari, F., *et al.* (2011) Safety assessment of non-browning potatoes: opening the discussion about the relevance of substantial equivalence on next generation biotech crops. *Plant Biotechnology Journal* 9, 136–150.

NRC [National Research Council] (1995) *Nutrient Requirements for Laboratory Animals.* 4th revised edn. National Academy Press, Washington, DC, 173 pp.

OECD (1993) *Safety Evaluation of Foods Derived by Modern Biotechnology. Concepts and Principles.* OECD, Paris.

OECD (1998a) OECD guidelines for the testing of chemicals – repeated dose 28-day oral toxicity study in rodents, 407 (http://browse.oecdbookshop.org/oecd/pdfs/free/9740701e.pdf, accessed 3 June 2013).

OECD (1998b) Guideline for the testing of chemicals – repeated dose 90-day oral toxicity study in rodents, 408 (http://browse.oecdbookshop.org/oecd/pdfs/free/9740801e.pdf, accessed 3 June 2013).

Russell, W.M.S. and Burch, R.L. (1959) *The Principles of Humane Experimental Technique.* Special Edn. Universities Federation for Animal Welfare, Potters Bar, UK.

Seralini, G.-E., Mesnage, R., Clair, E., Gress, S., Vendomois, J.S. de and Cellier, D. (2011) Genetically modified crops safety assessments: present limits and possible improvements. *Environmental Science Europe* 23, 10 (http://www.enveurope.com/content/23/1/10, accessed 3 June 2013).

Seralini, G.-E., Clair, E., Mesnage, R., Gress, S., Defarge, N., Malatesta, M., *et al.* (2012) Long term toxicity of a Roundup herbicide and a Roundup-tolerant genetically modified maize. *Food and Chemical Toxicology* (http://dx.doi.org/10.1016/j.fct.2012.08.005, accessed 1 May 2013).

Snell, C., Bernheim, A., Berge, J.-P., Kuntz, M., Pascal, G., Paris, A., *et al.* (2012) Assessment of the health impact of GM plant diets in long-term and multigenerational animal feeding trials: a literature review. *Food and Chemical Toxicology* 50, 1134–1148.

Spencer, J.D., Allee, G.L. and Sauber, T.E. (2000) Phosphorus bioavailability and digestibility of normal and genetically modified low-phytate corn for pigs. *Journal of Animal Science* 78, 675–681.

Spiekers, H., Meyer, H.H.D. and Schwarz, F.J. (2009) Effects of long term utilization of genetically modified maize (MON 810) in dairy cattle feeding on performance and metabolism parameters (in German). *Schriftenreihe Bayerische Landesanstalt für Landwirtschaft* (LfL), Vol 18, 105 pp.

Velimirow, A., Binter, C. and Zentek, J. (2008) Biological effects of transgenic maize NK603 × MON 810 fed in long term reproduction studies in mice. Research Report of Section IV, Vol §/2008 of the Bundesministerium für Gesundheit, Familie und Jugend, Vienna.

Vendomois, J.S. de, Roullier, F., Cellier, D. and Seralini, G.-E. (2009) A comparison of the effects of three GM corn varieties on mammalian health. *International Journal of Biological Sciences* 5, 706–726.

6 Feeding Studies with First-generation GM Plants (Input Traits) with Food-producing Animals

Gerhard Flachowsky*

Institute of Animal Nutrition, Friedrich-Loeffler-Institute (FLI), Federal Research Institute for Animal Health, Braunschweig, Germany

6.1 Introduction

Nutritionists distinguish genetically modified (GM) plants mostly into first-generation plants and second-generation plants. This designation is purely pragmatic or historical; it does not reflect any particular scientific principle or technological development.

The first generation of GM plants is generally considered to be crops carrying simple input traits such as increased resistance to pests or tolerance against herbicides. Other inputs, such as more efficient use of water and/or nutrients or an increased resistance against heat and drought are not expected to cause any substantial change in composition and nutritive value. Such plants could also be considered from the nutritional point of view as plants of the first generation. The newly expressed proteins that confer these effects occur in modified plants at very low concentrations (see Chapters 4, 5 and 9) and do not change their composition or feeding value significantly when compared with isogenic lines.

The following are currently considered to be feed/food safety issues for new (GM) crops:

- the newly introduced DNA (see Chapters 2, 4 and 9);
- the safety of the newly introduced genetic products (see Chapter 9);
- the potential toxicity of the newly expressed protein(s) (see FDA, 2007, and Chapter 9);
- potential changes in allergenicity, unintended effects giving rise to allergenicity or toxicity (see Chapter 3 and EFSA, 2010); and
- changes in nutrient composition, undesirable substances and feeding value (see Chapters 4, 5 and 7).

Most of these issues are not considered in detail here. The paradigm of the so-called 'substantial equivalence' (OECD, 1993) is the frame of the assessment and the first step of the assessment based on composition, but the substantial equivalence is not relevant for biotech crops of the next generation (crops with output traits; Llorente *et al.*, 2011).

Some authors consider this traditional assessment (comparison with 'known' traditional/historical counterparts) as not really science based and propose a registration and assessment of newly expressed criteria.

*E-mail: gerhard.flachowsky@t-online.de

In the case of plants with input traits, the gene products are functional proteins that affect a plant pest adversely or confer herbicide tolerance. These genes are not normally expected to affect biochemical pathways or cascades (Herman *et al.*, 2009). Therefore, some authors (e.g. Matten *et al.*, 2008; Herman *et al.*, 2009; Giddings *et al.*, 2012) criticize the present regulations and suggest that compositional assessment and feeding studies with feed from first-generation GM plants are no more necessary for evaluating the safety of transgenic crops than they are for plants bred traditionally.

6.2 Composition

Currently, the compositional equivalence of GM plants with input traits is considered a cornerstone of the case-by-case safety and nutritional assessment of such plants in the EU (see Chapter 4). Following this procedure, the composition of a transgenic crop is compared with that of non-transgenic comparators with a history of safe use. When compositional equivalence is established between the endogenous components of transgenic and non-transgenic plants, the safety assessment can focus on the properties of the products newly expressed by the transgenes (Kuiper *et al.*, 2001; Chassy, 2002) and no additional feeding studies seem to be needed. Such routine feeding studies generally add nothing or little to a nutritional assessment of feed (ILSI, 2003; EFSA, 2006, 2008; Davis and Kuiper, 2011). But, nevertheless, many feeding studies have been done to compare feed from a GM plant with its isogenic counterpart and some commercial varieties.

However, there are some discrepancies between transgenic plants and their isogenic counterparts. For example, transgenic Bt maize contains a gene from the soil bacterium, *Bacillus thuringiensis* (Bt), which encodes for the formation of a specific protein (Cry-protein) that is toxic to common lepidopteran maize pests, such as the European corn borer or corn rootworm (Alston *et al.*, 2002; Al-Deeb and Wilde, 2003; Magg *et al.*, 2003; Siegfried *et al.*, 2005). Maize plants that are less severely weakened by the corn borer might be expected to show better resistance to field infections, particularly infection by *Fusarium* subspecies. As a consequence of the lower level of fungal infection in the field, reduced mycotoxin contamination is to be expected, as summarized by Wu (2006a) and demonstrated with respect to various mycotoxins, but not in all cases (see Table 6.1). In studies made over several years, Dowd (2000) and Wu (2006a) investigated the influence of various levels of infestation with corn borers on isogenic and Bt hybrids with respect to mycotoxin contamination and came to the conclusion that, overall, a lower level of mycotoxin contamination was detected in the transgenic hybrids despite the considerable geographical and temporal variation observed.

Barros *et al.* (2009) compared the fungal and mycotoxin contamination in Bt maize and non-Bt maize grown in Argentina. The authors found significant lower total fumonisin levels in Bt maize in all seven locations in two harvest seasons compared to non-Bt maize (see Table 6.1). There was no significant difference in deoxynivalenol levels between Bt and non-Bt maize.

Application of the fungicide Tebuconazole did not alter either the infection or the toxin levels in Bt and non-Bt maize hybrids. Wu (2006b) and Brookes and Barfoot (2008) analysed the impact of mycotoxin reduction in Bt maize on economy, health, environmental and socio-economic effects and reported some important advantages for both the consumer and the producer (see Chapters 13 and 14).

6.3 Digestion Trials

After determination of composition (see Chapter 4) and various *in vitro* studies digestion experiments are the first step to check substantial equivalence under *in vivo* conditions.

Some studies have been carried out at the Institute of Animal Nutrition, Braunschweig, Germany (Table 6.2), and also by other authors as follows:

Table 6.1. Concentration of selected *Fusarium* toxins in isogenic and transgenic (Bt) maize grains (concentration in the transgenic hybrids expressed as a per cent of the isogenic foundation hybrid), by various authors.

		Mycotoxin					
		Deoxynivalenol		Zearalenone		Fumonisin B$_1$	
Author	Growing season/ region	Isogenic (ng/g)	Bt (%)	Isogenic (ng/g)	Bt (%)	Isogenic (µg/g)	Bt (%)
Munkvold *et al.*, 1999	1995	n.r.[a]	n.r.	n.r.	n.r.	8.8	54
	1996	n.r.	n.r.	n.r.	n.r.	7.0[b]	24
	1997	n.r.	n.r.	n.r.	n.r.	16.5[b]	13
Cahagnier and Melcior, 2000	France	350	79	n.r.	n.r.	1.0	20
	Spain	176	11	n.r.	n.r.	6.0	10
Pietri and Piva, 2000	1997 (*n* = 5)					19.8	10
	1998 (*n* = 11)	n.s.d.[c]		n.s.d.		31.6	17
	1999 (*n* = 30)					3.9	36
Valenta *et al.*, 2001	Corn borer:						
	infested (*n* = 15) not	873	18	256	13	n.r.	n.r.
	infested (*n* = 15)	77	70	19	15	n.r.	n.r.
Bakan *et al.*, 2002	France	472	154	3	<d.l.	n.r.	n.r.
	France	751	44	33	12	n.r.	n.r.
	France	179	101	3	133	n.r.	n.r.
	Spain	82	20	7	43	n.r.	n.r.
	Spain	271	7.4	4	75	n.r.	n.r.
Reuter *et al.*, 2002	1999:						
	Germany	343	<d.l.[d]	3	<d.l.	n.r.	n.r.
Papst *et al.*, 2005	Infested plots	1990	67	n.r.	n.r.	4.849	<0.1
	Protected plots	1294	63	n.r.	n.r.	16	n.d.[e]
Barros *et al.*, 2009	2003	1800	89	n.r.	n.r.	0.173	25
	2004	2400	79			0.633	32
	Argentina					Total	Fumonisin

Notes: [a] Not reported; [b] total fumonisin; [c] no significant difference (very low concentration); [d] below the detection limit; [e] not detected.

1. Maize and maize products: Barriere *et al.* (2001), with Bt maize silage in sheep; Donkin *et al.* (2003), with Roundup Ready (RR) maize in dairy cattle; Custodio *et al.* (2006), with Bt maize in pigs; Scheideler *et al.* (2008a), with maize in laying hens.
2. Rapeseed and co-products: Stanford *et al.* (2003), with canola meal in sheep.
3. Beets and co-products: Hartnell *et al.* (2005), with sugarbeets, fodder beets and beet pulp in three studies in sheep.

As expected, none of the authors found significant differences in the digestibility of dry matter/organic matter and various nutrients between feed from isogenic plants or their transgenic counterparts.

The results of the proximal analyses and the digestibility trials (Table 6.2) with feeds or co-products of feeds of first-generation GM plants (GM plants with input traits) show the similarity of such plants to their isogenic, non-transgenic counterparts. Some authors investigated the degradation of tDNA and newly expressed proteins during silage making and feed processing (e.g. Berger *et al.*, 2003; Aulrich *et al.*, 2004; Lutz *et al.*, 2006), as well as in the digestive tract and the effects of tDNA and newly expressed proteins on intestinal microbiota in ruminants (e.g. Chowdhury *et al.*, 2003a; Alexander *et al.*, 2004; Lutz *et al.*, 2005; Wiedemann *et al.*, 2006; Guertler *et al.*, 2008, 2009) and non-ruminants (e.g. Ash *et al.*, 2003; Chowdhury *et al.*, 2003b; Reuter and Aulrich, 2003; Buzoianu, 2011; Walsh *et al.*, 2011). Most authors describe a nearly complete degradation of tDNA and newly

Table 6.2. Summary of studies to measure the apparent digestibility of feeds from first-generation GM plants in comparison with isogenic counterparts, conducted at the Institute of Animal Nutrition, Braunschweig, Germany.

GM plant	Analytical measurements	Animal species/ categories	Animal number (isogenic/ transgenic)	Duration (days)	Composition	Digestibility	References
Bt maize Grain	Crude nutrients, amino acids, fatty acids, NSP, minerals, mycotoxins	Growing and fattening pigs	3 periods (39/60/80 kg body weight) 6/6 pigs	14	No significant differences	No significant differences	Reuter *et al.*, 2002
	Crude nutrients, starch, NSP, amino acids, fatty acids, minerals	Laying hens	6/6	10	No significant differences	No significant differences	Aulrich *et al.*, 2001
	Crude nutrients	Broilers	6/6	5	No significant differences	No significant differences	Aulrich *et al.*, 2001
	Crude nutrients	Sheep	4/4	24	No significant differences	No significant differences	Aulrich *et al.*, 2001
Pat maize Grain	Crude nutrients, starch, sugar, NSP, amino acids, fatty acids	Pigs	5/5	14	No significant differences	No significant differences	Böhme *et al.*, 2001
Pat sugarbeets Roots	Crude nutrients, sugar	Sheep	4/4	24	No significant differences	No significant differences	Böhme *et al.*, 2001
	Crude nutrients, sugar	Pigs	5/5	14	No significant differences	No significant differences	Böhme *et al.*, 2001
Top silage	Crude nutrients	Sheep	4/4	24	No significant differences	No significant differences	Böhme *et al.*, 2001

Note: NSP = non-starch polysaccharides.

expressed proteins by silage making and the processing of feeds and in the digestive tract, and found no significant influence on the apparent digestibility of nutrients or total feed of animals. Some authors (e.g. Mazza *et al.*, 2005; Sharma *et al.*, 2006; Tudisco *et al.*, 2010) detected traces of tDNA in animal bodies (see Chapter 9 for more details).

Buzoianu *et al.* (2012a,b) investigated the effects of Bt maize on the intestinal microbiota of pigs and found a higher cecal abundance of Enterococcaceae, Erysipelotrichaceae and Bifidobacterium and a lower abundance of Blautia in pigs fed with Bt maize than in those fed with an isogenic maize diet. A lower enzyme-resistant starch content in Bt maize, which is most likely a result of normal variation and not due to genetic modification, may account for some of the differences observed within the cecal microbiota.

6.4 Animal Feeding Studies

Feeding studies with target animals (food-producing animals) are recommended if studies with laboratory animals (see Chapter 5) are not able to answer all the specific questions, if unintended effects could be expected and/or if a nutritional evaluation seems to be necessary.

In most cases, studies about composition (see Chapter 4) and feeding studies with laboratory animals do not require animal feeding trials with feed of first-generation GM plants.

Nevertheless, many feeding studies with food-producing animals were done to compare feeds from first-generation GM plants with their isogenic counterparts (see Table 6.3). Later (since about 2000), one or more commercial varieties were included in the studies to help explain any unexpected differences or to confirm any expected differences observed between feed from GM plants and the control (e.g. McNaughton *et al.*, 2007). Commercial conventional varieties should typically be produced in the region where GM plants and their isogenic counterparts come from.

The following experimental design should be used to test feeds from first-generation GM plants in target animals (see Chapter 5):

- feed from GM plants;
- feed from the isogenic counterpart; and
- feed from typical commercial/conventional (non-GM) varieties (1–5 groups; ILSI, 2007; EFSA, 2008, 2011a).

The investigated feed should be applied in balanced diets in the highest possible amounts.

Most of the studies were done with broiler chickens because of the low costs of such experiments (see Chapter 5), the short experimental periods and the small amounts of feed required for such studies. Surprisingly, many studies were also done with dairy cattle (see Table 6.3), despite the high costs and the large amount of feed and equipment needed for such studies. Table 6.3 shows some examples of feeding studies with various GM plants and different traits from first-generation GM plants in food-producing animals.

One example to demonstrate the importance of additional groups fed with 'local' feed is shown in Table 6.4. No significant differences were determined between transgenic maize (DAS-59122-7) and its isogenic counterpart, apart from a significantly higher relative liver weight (p <0.05) of female broilers fed with GM maize in the diet. All values of the broilers were in a normal physiological range of all measurements (see Table 6.4). The inclusion of commercial non-GM varieties in the field and in animal feeding studies may contribute to avoiding an overestimation of experimental data and to distinguishing between statistical significance and biological relevance (EFSA, 2011b).

During the last few years, about 150 feeding studies have been done with food-producing animals, reported in scientific peer-reviewed papers and summarized in some reviews (e.g. Clark and Ipharraguerre, 2001; Flachowsky and Aulrich, 2001; Aumaitre *et al.*, 2002; Flachowsky *et al.*, 2005, 2007, 2012; CAST, 2006; Alexander *et al.*, 2007). An update of all published feeding studies can be found in FASS (2013).

Table 6.3. Some results of feeding studies with feed from various GM plants with different traits of the first generation in food-producing animals (in comparison with isogenic counterparts and mostly with commercial varieties).

GM plant/trait	Animal species/category (animals per treatment)	Duration (days) or living span (kg)	Portion in diet (% or kg)	Parameters	Main results	References
Maize						
Insect protected (Bt maize); silage	Dairy cattle (12)	91 days	70%	Feed intake, milk yield, composition	No significant influence on milk yield and composition	Barriere *et al.*, 2001
Insect protected (Bt 11 maize); silage/grain	Dairy cattle (16)	4 × 21 days	40/28% of DM; silage/grain	Feed intake, milk yield, composition, rumen fermentation	No significant effects on intake, yield and rumen fermentation	Folmer *et al.*, 2002
Glyphosate-tolerant or insect-protected (European corn borer) maize; silage and grain (3 trials)	Dairy cattle (6/8/8)	63/84/84 days	42–60% silage; 20–34% grain	Feed intake, milk yield, composition, ruminal degradability	No significant differences in all measured parameters	Donkin *et al.*, 2003
Glyphosate-tolerant (RR) and rootworm-resistant maize (Bt); grain, silage	Dairy cattle; 2 studies; 2 × 4 × 4 Latin square	28/21 days	40/23% silage/grain (1); 26.7% grain (2)	Feed intake, milk yield, composition	No effect on milk yield and composition	Grant *et al.*, 2003
Glyphosate tolerant; silage and grain (RR-NK603 maize)	Dairy cattle; 4 × 4 Latin square (8)	56 days	30% silage; 27.3% grain	Feed intake, milk yield, composition	No significant effect on milk yield and milk composition	Ipharraguerre *et al.*, 2003
Insect protected (Bt), herbicide tolerant; silage	Dairy cattle (8)	28 days	45%	Milk yield, composition, tDNA, Cry1Ab	No influence on composition of silage and milk, milk yield, no tDNA and Cry1Ab in milk	Calsamiglia *et al.*, 2007
Insect protected (Bt); silage, grain	Dairy cattle (15)	765 days	63% roughage, 41% concentrate	Feed intake, milk yield, composition, tDNA, Cry1AB, metabolic parameters	No significant influence on measured data	Guertler *et al.*, 2009; Spiekers *et al.*, 2009; Steinke *et al.*, 2010
Insect-protected (rootworm) and herbicide (glufosinate)-tolerant maize	Dairy cattle (15; switchback design)	28 days	23% grain; 21% silage	Feed intake, milk yield, composition	No significant influence of GM feed on all parameters	Brouk *et al.*, 2011
Insect protected (Bt maize); silage	Growing/finishing bulls (20)	246 days	75%	Feed intake, weight gain, slaughtering data, fate of tDNA	No significant influence on growing and slaughtering data, no tDNA in body tissues	Aulrich *et al.*, 2001
Insect-protected Bt maize: 2 studies; grazing residues (1) or silage (2)	Growing/fattening steers ((1) 33; (2) 64)	(1) 70 days; (2) 101 days	Ad libitum/90%	Feed intake, growth of animals	Some small differences between treatments, but not consistent	Folmer *et al.*, 2002
Glyphosate tolerant (IIII); grain	Steers; 3 studies (49/49/50)	[illegible] days	70%	Feed intake, growing and slaughtering data	No effect on animal performances and carcass characteristics	Erickson *et al.*, 2003

Continued

Table 6.3. Continued

GM plant/trait	Animal species/ category (animals per treatment)	Duration (days) or living span (kg)	Portion in diet (% or kg)	Parameters	Main results	References
Insect protected (Bt, rootworm), 3 studies; grazing residues (1) or grain (2/3)	Growing/fattening steers (32/50/49)	60/112/102 days	Residues *ad libitum* + protein suppl.	Feed intake, growth, slaughtering data, meat quality	No significant effects on animal yield and meat quality	Vander Pol *et al.*, 2005
Insect-protected (Bt 11) maize; grain	Calves (6)	Age: 2 months Duration: 3 months	43%	Feed intake, weight gain	No significant differences	Shimada *et al.*, 2006
Insect-protected (Bt) and glufosinate (pat)-tolerant maize; grain	Finishing steers (20)	109 days	82%	Feed intake, weight gain, slaughtering data	No significant differences to growing and slaughtering data	Huls *et al.*, 2008
Insect-protected (Bt 176) maize, grain	Growing/finishing pigs (36)	91 days	70%	Feed intake, growing/ slaughtering data	No significant differences in growing and slaughtering data; no tDNA in body tissues	Reuter *et al.*, 2002
Glyphosate-tolerant (RR); grain	Growing/finishing pigs, 2 studies (36/40)	22–116; 30–120 kg in studies (1) and (2)	68/74; 78/82% in grower/finisher feed	Feed intake, growing and slaughtering data	No significant effects on animal performances and carcass characteristics	Hyun *et al.*, 2004
Insect protected (Bt, rootworm); grain	Growing/finishing pigs; 2 studies (18/40)	(1) 23–117 kg; (2) 30–115 kg	(1) 69/75 in grower 1/2; 79/82 in finishers 1/2; (2) 65/72/76% in grower/finisher 1/2	Feed intake, growing/ slaughtering data, meat composition	No significant differences in growing/slaughtering data and body composition	Hyun *et al.*, 2005
Insect-protected (Bt 11) maize, grain	Weaner/growing/ finishing pigs (60)	17–120 kg (110 days)	70/76% (grower/ finisher)	Feed intake, growing, slaughtering results, meat quality	No significant differences in weight gain, higher feed intake, influence on muscle colour (more intense yellow)	Custodio *et al.*, 2006
Insect-protected (Bt 11) maize; grain	Finishing pigs (32)	60–110 kg (50 days)	77–83%	Feed intake, growing/ slaughtering results, meat quality	No differences in weight gain and carcass characteristics, influence on muscle colour (less intense yellow)	Custodio *et al.*, 2006
Insect-protected (Bt, rootworm) and glufosinate (pat)-tolerant maize; grain	Growing/finishing pigs (36)	37–127 kg	69/75/82%; (37–60/60–90/90–127 kg)	Feed intake, growing/ slaughtering data, carcass quality	No significant differences in growing/slaughtering data and body composition	Stein *et al.*, 2009a
Insect-protected (Bt) and glufosinate (pat)-tolerant maize; grain	Growing/finishing pigs (24)	23.5–120 kg	65/74/81%; (23.5–60; 60–90; 90–120 kg)	Feed intake, growing/ slaughtering data	No significant differences in growing and slaughtering results	Stein *et al.*, 2009b

Insect-protected (Bt MON810) maize; grain	Male weanling pigs (16)	7.5 kg; 31 days	38.9%	Feed intake, weight gain, organ morphology	No significant effects on growth and slaughtering data, no histopathological changes or alterations in blood biochemistry	Walsh *et al.*, 2012
Insect-protected (Bt) maize (grain, 2 trials)	Laying hens (18)	26 weeks	60%	Feed intake, laying performance, fate of tDNA	No significant effects on laying performances, no tDNA in organs, meat or eggs	Aeschbacher *et al.*, 2005
Insect-protected (Bt; rootworm) and glufosinate (pat)-tolerant maize (grain)	Laying hens (72)	84 days (3 × 28 days)	64.75%	Feed intake, body weight gain, egg production and quality	No significant differences in laying performances and egg quality	Jacobs *et al.*, 2008
Insect-protected (Bt, rootworm resistant) maize; grain (2 trials)	Laying hens (48/12)	56/28 days	51.1% in both studies	Feed intake, performances, fate of transgenic protein	No significant effects on feed intake and animal yields; transgenic protein was digested extensively, similar to other feed proteins; no transgenic protein fragments in eggs and tissues	Scheideler *et al.*, 2008a
Insect-protected (Bt) maize; grain	Laying hens (84)	112 days	60%	Feed intake, egg production and quality	No significant differences in laying performance and egg quality	Scheideler *et al.*, 2008b
Insect-protected (Bt 176) maize; grain	Broilers (640)	38 days	58/64% (starter/ finisher)	Feed intake, weight gain, carcass characteristics	Better feed conversion, increased weight of pectoralis minor muscle; no further differences	Brake and Vlachos, 1998
Glyphosate-tolerant (RR) maize; grain	Broilers (80)	40 days	50–60%	Feed intake, growth, carcass characteristics	No significant differences between RR maize and conventional maize	Sidhu *et al.*, 2000
Insect-protected (Bt 11) maize; grain	Broilers (800)	47 days	48–63%	Feed intake, growth, carcass characteristics	No significant differences	Brake *et al.*, 2003
Insect protected (Bt) maize; grain	Broilers (27)	35 days	73.6%	Feed intake, growth, fate of DNA	No significant differences, no tDNA in body tissues	Tony *et al.*, 2003
Insect-protected (Bt, rootworm) maize (2 trials); grain	Broilers (100)	42/42 days	55/60% (starter/ grower-finisher)	Feed intake, growth, slaughtering data	No significant differences between all treatments	Taylor *et al.*, 2003
Insect-protected (Bt) maize (3 trials); grain	Broilers (23)	39 days	60%	Feed intake, growing and slaughtering data, fate of tDNA	No significant differences between all treatments	Aeschbacher *et al.*, 2003
Insect-protected (MON810) maize; grain	Broilers (216)	42 days	48–62%	Feed intake, growth	No significant differences	Rossi *et al.*, 2005

Continued

Table 6.3. Continued

GM plant/trait	Animal species/category (animals per treatment)	Duration (days) or living span (kg)	Portion in diet (% or kg)	Parameters	Main results	References
Insect resistant (Bt, rootworm, European corn borer) and herbicide tolerant (glyphosate, RR) (2 trials); grain	Broilers (120)	42 days	55/60% (starter/grower-finisher)	Feed intake, growth, carcass composition and quality	No significant differences between all treatments	Taylor *et al.*, 2005a,b
Insect resistant (Bt, European corn worm) and herbicide tolerant (glyphosate, RR); grain (2 trials)	Broilers (120/100)	42 days	55/59% for starter/grower	Feed intake, growth, carcass composition and quality	No significant differences between all treatments	Taylor *et al.*, 2007a
Insect resistant (European corn borer) and glyphosate tolerant (RR)	Broilers (100)	42 days	57/59% for starter/grower	Feed intake, growth, carcass composition, meat quality	No significant differences between all treatments	Taylor *et al.*, 2007b
Insect-resistant (Bt) and glufosinate-tolerant (pat) maize; grain	Broilers (120)	42 days	53/58/70% (starter/grower/finisher)	Feed intake, growth, carcass yield	All the values within the tolerance intervals; GM maize nutritionally equivalent to non-GM control maize	McNaughton *et al.*, 2007
Insect-resistant (Bt) and herbicide-tolerant (glyphosate and acetolactate synth.) maize; grain	Broilers (120)	42 days	58.5/64/71.5% (starter/grower/finisher)	Feed intake, growth, carcass yield	No significant differences between all treatments	McNaughton *et al.*, 2008
Insect-resistant (Bt) and herbicide-tolerant (pat) maize; grain	Broilers 120)	42 days	63/67.5/74% (starter/grower/finisher)	Feed intake, growth, carcass yield	No significant differences between all treatments	McNaughton *et al.*, 2011c
Soybeans						
Glyphosate-tolerant (RR) soybean meal	Dairy cattle (12)	28 days	10.2%	Milk yield, composition	Higher FCM yield of GM group; weaknesses in experimental design (see Flachowsky and Aulrich, 1999)	Hammond *et al.*, 1996
Glyphosate-tolerant (RR) soybean meal	Pigs (50)	100 (24–111 kg)	24/19/14% (grower/early and late finisher)	Feed intake, weight gain, carcass parameters	No significant influence on animal performances, slaughtering results and meat quality	Cromwell *et al.*, 2002

Glyphosate-tolerant (RR) soybean meal	Pigs (12)	100 days	24/19/14% (grower, early and late finisher)	Feed intake, weight gain, fate of tDNA	No significant influence on animal performances, no tDNA or fragments of transgenic proteins in animal tissues	Jennings et al., 2003
Glyphosate-tolerant (RR) soybean meal	Laying hens (84)	84 days	23.5%	Feed intake, laying performance	No significant effects on performance and egg quality	Mejia et al., 2010
Glyphosate-tolerant (RR) soybean meal	Laying hens (24)	84 (3 × 28 days)	13.6–19.6%	Feed intake, laying performance	No significant effects on performance and egg quality	McNaughton et al., 2011a
Glyphosate-tolerant (RR) soybean meal	Broilers (120)	42 days	33/27% (starter/grower)	Feed intake, weight gain, carcass parameters	No significant influence on animal performances and slaughtering results	Hammond et al., 1996
Insect-protected (Bt soya) soybean meal	Broilers (90)	41 days	34%	Feed intake, weight gain, carcass parameters	No significant influence on animal performances and slaughtering results	Kan and Hartnell, 2004b
Glyphosate-tolerant (RR) and sulfonylurea/imidazolinona-tolerant soybean meal	Broilers (120)	42 days	30/26/21.5% (starter/grower/finisher)	Feed intake, weight gain	No significant influence on animal performance and slaughtering results	McNaughton et al., 2007
Glyphosate-tolerant (RR) soybean meal	Broilers (100)	42 days	33/30% (starter/grower)	Feed intake, weight gain, carcass parameters	No significant effects on animal performance and carcass parameters	Taylor et al., 2007c
Glyphosate-tolerant (RR) soybean meal	Broilers (120)	42 days				McNaughton et al., 2011b
Glyphosate-tolerant (RR) soybean meal	Catfish (100)	70 days	45–47%	Feed intake, weight gain	No significant influence on animal yields	Hammond et al., 1996
Glyphosate-tolerant (RR) soybean meal	Atlantic salmon	210 days	25%	Growth, body composition, organ development, haematological and clinical parameters, fate of tDNA	Some differences in intestinal development, but no other diet-related zootechnical and morphological differences, no tDNA in body tissues	Sissener et al., 2009a,b; Sanden et al., 2011
Canola Glyphosate-tolerant (RR) canola meal	Sheep (lambs) (8 wethers, digestibility; 15 lambs, feeding)	21 (digestibility), about 70	6.5%	Digestibility, feed intake, growth, carcass composition	No significant effect on digestibility, feed efficiency, growth, carcass characteristics and meat quality	Stanford et al., 2003
Glyphosate-tolerant (HH) canola meal	Broilers (100)	42 days	25/20% (starter/finisher)	Feed intake, weight gain, carcass composition	No significant differences in feed intake, performance and carcass composition	Taylor et al., 2004

Continued

Table 6.3. Continued

GM plant/trait	Animal species/ category (animals per treatment)	Duration (days) or living span (kg)	Portion in diet (% or kg)	Parameters	Main results	References
Glyphosate-tolerant (RR) canola meal	Rainbow trout (2 studies) (45 per level and treatment)	(1) 10–529 g; (2) 16–145 g	5/10/15/20%; 2 studies	Feed intake, weight gain, body composition	Some differences between groups depending on inclusion level, but in summary, GM canola meal is equivalent to parenteral line; no effect on body composition	Brown *et al.*, 2003
Cottonseed						
Insect protected (Bt) and glyphosate tolerant (RR); whole seeds (2 studies)	Dairy cows (4 × 4 Latin square)	4 × 28 days	2.5 kg per day; about 10% of DM	Feed intake, milk yield and composition, body weight, fate of tDNA	No significant influence on milk yield and composition, no tDNA in milk	Castillo *et al.*, 2004
Insect protected (Bt); cottonseed meal	Broilers	42 days	10%	Feed intake, growth, carcass characteristics, blood constituents	No significant differences to diets with isogenic cottonseed meal	Mandal *et al.*, 2004
Insect protected (Bt); cottonseed meal	Broilers (40)	49 days	10%	Feed intake, growth, carcass characteristics, blood constituents	No significant differences to diets with isogenic cottonseed meal	Elangovan *et al.*, 2006
Insect protected (Bt) and cowpea trypsin inhibitor (CpTI); cottonseed meal	Broilers (108)	42 days	3%	Nutrient digestibility, feed intake, growth, carcass characteristics, Bt + CpTI genes and proteins	No significant differences to cottonseed meal from isogenic cottonseed	Guo *et al.*, 2012
Wheat						
Glyphosate tolerant (RR wheat)	Broilers (90)	40 days	40%	Feed intake, growth, carcass composition	No significant differences in feed intake, growth and slaughtering data	Kan and Hartnell, 2004a
Rice						
Glufosinate tolerant (LL rice)	Growing/fattening pigs (24)	98 days	72.8–85.8%	Feed intake, growth, carcass traits	No significant differences in feed intake, growth and slaughtering data	Cromwell *et al.*, 2005
Potatoes						
2 Bt potatoes (Spunta G2 and G3)	Broilers (9)	14 days	30%	Feed intake, growth, fate of DNA	No significant differences in feed intake and growing, no tDNA in body tissues	Halle *et al.*, 2005

Note: FCM = fat-corrected milk; LL = Liberty Link; RR = Roundup Ready; tDNA = transgenic (recombinant) DNA.

Table 6.4. Effect of the maize event DAS-59122-7 (53–70% maize in the diet) on growth performance and organ weight of broilers in comparison to the near-isogenic control and three non-transgenic conventional hybrids (120 broilers per treatment, 42 days). (From McNaughton *et al.*, 2007.)

Criteria	Control	DAS-59122-7	Confidence interval (95%)
Final weight (g/animal)	1918	1916	1675–2144
Feed: gain (g/g)	1.88	1.87	1.70–2.03
Relative weights of some organs (g/kg body weight)			
Kidney			
♂	20	20	8.5–33.2
♀	20	21	8.2–33.2
Liver			
♂	35	36	20.5–50.6
♀	34[a]	37[b]	19.5–51.0
Post-chill carcass (g/kg body weight)			
♂	708	713	626–792
♀	705	707	622–791

Note: [a,b]Signifies differences between treatments.

Side effects might be expected in GM plants, particularly in GM plants with multiple modifications (multi-stacked events; Cellini *et al.*, 2004), and these should be analysed scientifically in detail. Furthermore, the high biological range for many parameters should be considered.

Such stacked events were also named as first-generation GM plants (e.g. insect-protected and herbicide-tolerant plants). Feeds from stacked events were also fed to animals (Taylor *et al.*, 2005a,b, 2007a,b) and did not show significant differences to feed from non-GM plants (see Table 6.3). Presently, no other foods/feeds have been analysed as extensively and tested in various studies as is the case for first-generation GM plant products. It can be concluded that the safety and nutritional evaluation of GM versus conventionally bred plants is not well balanced (Kok *et al.*, 2008).

Some authors fed diets with feeds from two or more GM plants (Table 6.5). The results did not show any biologically relevant effect of feeds from first-generation GM plants on animal health and welfare, animal yield, quality of food of animal origin (see Chapter 10) or the fate of tDNA or newly expressed proteins (see Chapter 9).

The objective of the efficiency trials as shown in Tables 6.3–6.5 is to measure the effect of feed from GM plants on the performance of food-producing animals, and to compare the results with an isogenic counterpart and some commercial products. Questions concerning the tolerance of some feeds in animals (tolerance studies) may be also included in efficiency trials.

Carman *et al.* (2013) have published results from a field study with 168 pigs (initial weight: 6.8 kg/piglet). Eighty-four pigs (50% males and females each) were each fed with control or GM maize (70.0–81.3%)–soybean meal diets (26.5–16.0% depending on age of pigs) for about 159 days (final weight: about 101 kg/animal). The stacked GM maize contained a combination of NK603, MON863 and MON810 (expressing the CP4 EPSPS, Cry3Bb1 and Cry1Ab proteins) and the soybean was 100% RR soy (expressing the CP4 EPSPS protein). Further control groups (commercial lines) were not included in this study. Chemical analyses of various mixed feeds were not given. Pooled samples of GM feed showed higher total aflatoxins (2.1 ppb) and total fumonisins (3.0 ppm); for non-GM feed, no aflatoxins and 1.2 ppm fumonisins were detected. Mortalities were extremely high with 13 and 14%, respectively, for the non-GM- and the GM-fed groups. There were no differences between pigs of both groups for feed intake, weight gain and routine blood chemistry. Some differences between both groups were

Table 6.5. Examples of animal feeding studies with more than one feed from GM plants in the diet.

GM plant	Traits	Animal species/ category (animals per group)	Duration (days or body weight)	Portion in diet (%)	Parameters	Main results	References
Maize and soybeans	Maize: *gat4621* and *zm-hra* genes; soybean: *gat4601* and *gm-hra* genes	Laying hens (24); 3 × 28 days feeding phases	84	65–71% maize; 13.6–19.6% soybean meal	Feed intake, body weight, egg production, egg quality	All measured values were within the tolerance interval	McNaughton *et al.*, 2011a
Maize and soybeans	Maize: *gat4621* and *zm-hra* genes; soybean: *gat4601* and *gm-hra* genes	Broilers (120)	42	63/66/72% maize; 28/26/21% soybean meal in starter/grower/ finisher	Feed intake, weight gain, carcass composition	All measured values were within the tolerance interval. GM maize and soybeans are nutritionally equivalent to non-transgenic plants	McNaughton *et al.*, 2011b
Maize and soybeans	Insect-protected (Bt) maize and glyphosate-tolerant (RR) soybean meal	Growing/fattening pigs (12)	30–110 kg BW	Maize: not given; 18/14% soybean meal in grower/ finisher	Feed intake, weight gain, carcass composition, meat quality, fate of tDNA	No significant influence on animal performance, slaughtering results, meat quality, no tDNA in tissues	Swiatkiewicz *et al.*, 2011
Maize and soybeans	Insect-protected (Bt) maize and glyphosate-tolerant (RR) soybean meal	Broiler (40)	42	55/60% maize; 39/32% soybean meal in starter/ grower	Feed intake, weight gain, slaughtering yield, meat quality, fate of tDNA	No significant influence on animal performance, slaughtering results, small effects on meat colour, no tDNA in tissues	Swiatkiewicz *et al.*, 2010a,b; Stadnik *et al.*, 2011

Table 6.6. Performance and some metabolic parameters of the first and second lactation of a long-term feeding study with dairy cows ($n = 18$ per treatment, 25 months with Bt maize (MON 810, 63% of roughage, 41% of concentrate from maize).[a] (From Steinke *et al.*, 2010.)

Lactation of experiment	First		P level	Second		P level
	Isogenic	Transgenic		Isogenic	Transgenic	
Dry matter intake						
(kg/day)	18.7	18.9	0.532	21.0	20.4	0.030
Milk yield (kg/day)	23.9	23.7	0.566	29.2	28.8	0.419
Milk fat (%)	3.95	4.03	0.015	3.75	3.86	0.055
Milk protein (%)	3.62	3.71	<0.001	3.59	3.56	0.299
NEFA (µmol/l)	287	281	0.991	292	290	0.988
BHBA (mmol/l)	0.46	0.44	0.107	0.50	0.49	0.304
AST (U/l)	92.6	89.8	0.263	94.3	88.8	0.177
GLDH (U/l)	19.5	19.1	0.922	13.8	16.1	0.178
γ-GT (U/l)	23.2	23.9	0.426	23.5	23.9	0.575

Notes: [a]No fragments of Cry1Ab DNA in blood, milk, faeces and urine of cows; traces of Cry1Ab protein were detected in faeces, but not in blood, milk and urine (Guertler *et al.*, 2008, 2009). AST = aspartate-amino-transferase; BHBA = beta-hydroxy butyric acid; GLDH = glutamate-dehydrogenase; γ-GT = gamma-glutamyl-transferase; NEFA = free fatty acids.

reported for the uterus weight (0.10 and 0.12% of the body weight, respectively, for non-GM- and GM-fed animals) and the rate of severe stomach inflammation. This is the first and only report on changes in organ weight and adverse findings on gross pathology. Further studies should describe materials and methods better from the nutritional point of view and should pay more attention to such end points.

Apart from zootechnical parameters (e.g. feed intake, animal yield, feed conversion rate, composition and quality of food of animal origin), some authors also investigated the metabolic parameters in the animal, as shown, for example, in a 2-year study of dairy cattle fed with a high portion of Bt maize (Table 6.6). The higher milk protein content in the first lactation of cows fed with Bt maize was the only one significant result, but this result should not be overestimated because of other effects in the second lactation and its biological relevance (EFSA, 2011b).

Based on the results mentioned previously, the necessity of animal feeding studies with feed from first-generation GM plants is often questioned with regard to their importance and their scientific yield. According to various guidance documents (e.g. EFSA, 2006, 2008, 2011a), such studies

are not urgently needed. No animal feeding studies are required if the differences in compositional analyses between isogenic and transgenic plants are small or negligible (plants are substantially equivalent; first-generation GM plants) because of the costs of such studies and the reduction in the numbers of experimental animals.

On the other hand, feeding experiments with first-generation GM plants with target animals may contribute to demonstrating the nutritional equivalence and the safety of the feed to the public, and therefore the experiments could improve the public acceptance of GM feed. Furthermore recommendations for optimal amounts in target animal feeding may be deduced.

All animal feeding studies described in peer-reviewed journals and summarized above show that feeds from first-generation GM plants which are assessed by a scientific body with responsibility for regulating food/feed safety (e.g. EFSA, FDA, USDA) are as safe as or safer than crops produced with traditional methods. About 10 years ago, Chassy (2002) came to the same conclusion that 'after extensive safety testing and some five years of experience with such crops in the marketplace, there is not a single report that would lead an expert food scientist to question the safety

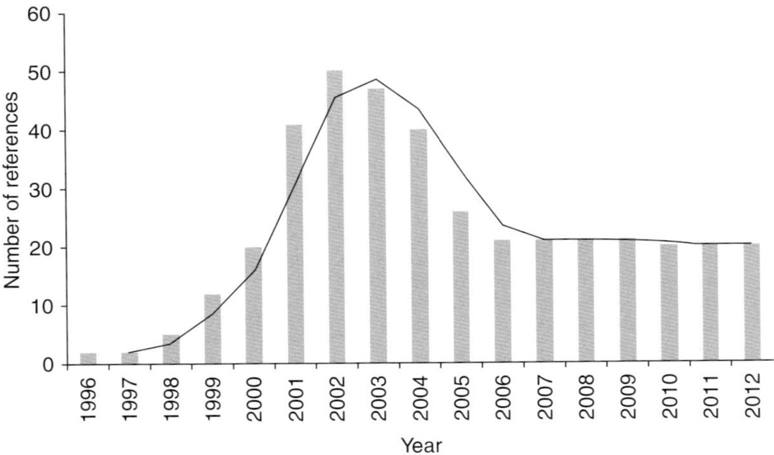

Fig. 6.1. Registered references in the FASS System (2013) on feeding of transgenic crops in livestock.

of such transgenic crops now in use'. Aumaitre *et al.* (2002), Cockburn (2002) and Faust and Glenn (2002) came to the same conclusion.

In summary, from the present perspective, there is no need for further feeding studies with feed from first-generation GM plants (GM plants with input traits) in target animals. Such studies do not contribute substantially to more and better knowledge about the safety and nutritious value of such feed. This statement is also reflected in the published scientific feeding studies (Fig. 6.1). Most feeding studies were done immediately after the cultivation of GM plants under farm conditions (about 50 papers/year). Afterwards, there was a stabilization of between 15 and 30 peer-reviewed papers/year.

The influences of feeds from first-generation GM plants on animal health and welfare, animal yields, body composition, product quality and transfer of transgenic DNA and newly expressed proteins in animal body/tissue are described in Chapters 9 and 10. More references about animal feeding studies with transgenic crops to food-producing animals can be found in FASS (2013).

6.5 Conclusions

Since 1996 (Hammond *et al.*, 1996), about 150 feeding studies with feeds from first-generation GM plants (GM plants with input traits) in food-producing animals have been reported in the scientific literature (see FASS, 2013). Such plants did not show biologically relevant effects on the composition of the feed. Therefore, no biologically relevant effects on animal health and welfare, animal yields and the quality of products of animal origin are expected (see Tables 6.2–6.6 and Chapter 10).

In summary, feeding studies with feeds from first-generation GM plants in food-producing animals do not add substantial knowledge to feed science and animal nutrition because of the substantial equivalence of such plants/feeds to their isogenic counterparts.

From the present perspective, there is no reason to use other feed value tables for such feeds in animal feeding. Feeds from first-generation GM plants can be used as traditional feeds under consideration of their composition to meet the energy and nutrient requirements (NRC, 1994, 1998, 2001; GfE, 1999, 2001, 2008) of food-producing animals.

References

Aeschbacher, K., Messikommer, R., Meile, L. and Wenk, C. (2005) Bt176 corn in poultry nutrition: physiological characteristics and fate of recombinant plant DNA in chickens. *Poultry Science* 84, 385–394.

Al-Deeb, M.A. and Wilde, G.E. (2003) Effects of Bt corn expressing the Cry38b1 toxin for corn rootworm control on aboveground non-target arthropods. *Environmental Entomology* 32, 1164–1170.

Alexander, T.W., Sharma, R., Deng, M.Y., Whetsell, A.J., Jennings. J.C., Wang, Y., *et al.* (2004) Use of quantitative real-time and conventional PCR to assess the stability of the *cp4 epsps* from Roundup Ready canola in the intestinal, ruminal, and fecal contents of sheep. *Journal of Biotechnology* 112, 255–266.

Alexander, T.W., Reuter, T., Aulrich, K., Sharma, R., Okine, E.K., Dixon, W.T., *et al.* (2007) A review of the detection and fate of novel plant molecules derived from biotechnology in livestock production. *Animal Feed Science and Technology* 133, 31–62.

Alston, J., Hyde, J., Marra, M. and Mitchell, P.D. (2002) An *ex ante* analysis of the benefits from the adoption of corn rootworm resistant transgenic corn technology. *AgBioForum* 5, 71–84.

Ash, J., Novak, C. and Scheideler, S.E. (2003) The fate of genetically modified protein from Roundup Ready soybeans in laying hens. *Journal of Applied Poultry Research* 12, 242–245.

Aulrich, K., Böhme, H., Daenicke, R., Halle, I. and Flachowsky, G. (2001) Genetically modified feeds in animal nutrition. 1st communication: *Bacillus thuringiensis* (Bt) corn in poultry, pig and ruminant nutrition. *Archives of Animal Nutrition* 54, 183–195.

Aulrich, K., Pahlow, G. and Flachowsky, G. (2004) Influence of ensiling on the DNA-degradation in isogenic and transgenic corn. *Proceedings of the Society of Nutrition Physiology* 13, 112 (Abstract).

Aumaitre, A., Aulrich, K., Chesson, A., Flachowsky, G. and Piva, G. (2002) New feeds from genetically modified plants: substantial equivalence, nutritional equivalence, digestibility, and safety for animals and the food chain. *Livestock Production Science* 74, 223–238.

Bakan, B., Melcion, D., Richard-Molard, D. and Cahagnier, B. (2002) Fungal growth and fusarium mycotoxin content in isogenic traditional maize and genetically modified maize grown in France and Spain. *Journal of Agricultural and Food Chemistry* 50, 728–7.1.

Barriere, Y., Verite, R., Brunschig, P., Surault, F. and Emile, J.C. (2001) Feeding value of corn silage estimated with sheep and dairy cows is not altered by genetic incorporation of Bt resistance to *Ostrinia nubilalis. Journal of Dairy Science* 84, 1863–1871.

Barros, G., Magnoli, C., Reynoso, M.M., Ramirez, M.L., Farnochi, M.C., Torres, A., *et al.* (2009) Fungal and mycotoxin contamination in Bt maize and non-Bt maize grown in Argentina. *World Mycotoxin Journal* 2, 53–60.

Berger, B., Aulrich, K., Fleck, G. and Flachowsky, G. (2003) Influence of processing of isogenic and transgenic rapeseed on DNA-degradation. *Proceedings of the Society of Nutrition Physiology* 12, 108 (Abstract).

Böhme, H., Aulrich, K., Daenicke, R. and Flachowsky, G. (2001) Genetically modified feeds in animal nutrition. 2nd communication: Glufosinate-tolerant sugar beets (roots and silage) and maize grains for ruminants and pigs. *Archives of Animal Nutrition* 54, 197–207.

Brake, J. and Vlachos, D. (1998) Evaluations of transgenic event 176 'Bt' corn in broiler chickens. *Poultry Science* 77, 648–653.

Brake, J., Faust, M.A. and Stein, J. (2003) Evaluation of transgenic event Bt11 hybrid corn in broiler chickens. *Poultry Science* 82, 551–559.

Brookes, G. and Barfoot, P. (2008) *GM Crop: Global Socio-Economic and Environmental Impact. 1996–2006.* PG Economics Ltd, Dorchester UK, 118 pp.

Brouk, M.J., Cvetkovic, B., Rice, D.W., Smith, B.L. Hinds, M.A., Owens, F.N., *et al.* (2011) Performance of lactating dairy cows fed corn as whole plant silage and grain produced from genetically modified corn containing event DAS-59122-7 compared to a non-transgenic, near-isogenic control. *Journal of Dairy Science* 94, 1991–1996.

Brown, P.B., Wilson, K.A., Jonker, Y. and Nickson, T.E. (2003) Glyphosate-tolerant canola meal is equivalent to the parental line in diets fed to rainbow trout. *Journal of Agriculture, Food and Chemistry* 51, 4268–4272.

Buzoianu, S.G. (2011) Fate of transgenic DNA and protein in pigs fed genetically modified Bt maize and effects on growth and health. PhD thesis, Waterford Institute of Technology, Ireland. 308 pp.

Buzoianu, S.G., Walsh, M.C., Rea, M.C., O'Sullivan, O., Cotter, P.D., Ross, R.P., *et al.* (2012a) High-throughput sequence-based analysis of the intestinal microbiota of weanling pigs fed genetically modified MON810 maize expressing *Bacillus thuringiensis* Cry1AB (Bt maize) for 31 days. *Applied Environmental Microbiology* 78, 4217–4224.

Buzoianu, S.G., Walsh, M.C., Rea, M.C., O'Sullivan, O., Crispie, F., Cotter, P.D., *et al.* (2012b) The effect of feeding Bt MON810 maize to pigs for 110 days on intestinal microbiota. *PLoS ONE* 7(5): e33668. DOI:10.1371/journal.pone. 0033668.

Cahagnier, B. and Melcion, D. (2000) Mycotoxines de Fusarium dans les mais-grains a la recolte: Relation entre la presence d'insectes (pyrale, sesamie) et la teneur en mycotoxines. In: Piva, G. and Masoero, F. (eds) *Proceedings 6th International Feed Production Conference*, Piacenza, Italy, 27–28 November 2000. Faculta UCSC, Piacenza, Italy, pp. 237–249.

Calsamiglia, S., Hernandez, B., Hartnell, G.F. and Phipps, R. (2007) Effect of corn silage derived from genetically modified variety containing two transgenes on feed intake, milk production, and composition, and the absence of detectable transgenic deoxyribonucleic acid in milk of Holstein dairy cows. *Journal of Dairy Science* 90, 4718–4723.

Carman, J.A., Vlieger, H.R., Ver Steeg, L.J., Sneller, V.E., Robinson, G.W., Clinch-Jones, C.A., *et al.* (2013) A long-term toxicology study on pigs fed a combined genetically modified (GM) soy and GM maize diet. *Journal of Organic Systems* 8(1), 38–54.

CAST [Council for Agricultural Science and Technology] (2006) Safety of meat, milk, and eggs from animals fed crops derived from modern biotechnology. Issue Paper 34, 8 pp. CAST, Ames, Iowa.

Castillo, A.R., Gallardo, M.R., Maciel, M., Giordano, J.M., Conti, G.A., Gaggiotti, M.C., *et al.* (2004) Effect of feeding rations with genetically modified whole cottonseed to lactating Holstein cows. *Journal of Dairy Science* 87, 1778–1785.

Cellini, F., Chesson, A., Coquhonn, I., Constable, A., Davies H.V., Engel, K.-H., *et al.* (2004) Unintended effects and their detection in genetically modified crops. *Food and Chemical Toxicology* 42, 1089–1123.

Chassy, B.M.M. (2002) Food safety evaluation of crops produced by biotechnology. *Journal of the American College of Nutrition* 21, 166–173.

Chowdhury, E.H., Kuribara, H., Hino, A., Sultana, P., Mikami, O., Shimada, N., *et al.* (2003a) Detection of corn intrinsic and recombinant DNA fragments and Cry 1Ab protein in the GI contents of pigs fed genetically modified corn Bt11. *Journal of Animal Science* 81, 2546–2551.

Chowdhury, E.H., Shimada, N., Murata, H., Mikami, O., Sultana, P., Miyazaki, S., *et al.* (2003b) Detection of Cry1Ab protein in GI tract but not visceral organs of genetically modified Bt11-fed calves. *Veterinary and Human Toxicology* 45, 72–75.

Clark, J.H. and Ipharraguerre, I.R. (2001) Livestock performance: Feeding biotech crops. *Journal of Dairy Science* 84, E. Suppl. E9–E18, 237–249.

Cockburn, A. (2002) Assuring the safety of genetically modified (GM) foods: The importance of an holistic, integrative approach. *Journal of Biotechnology* 98, 79–106.

Cromwell, G.L., Lindemann, M.D., Randolph, J.H., Parker, G.R., Coffey, R.D., Laurent, K.M., *et al.* (2002) Soybean meal from Roundup Ready or conventional soybeans in diets for growing-finishing swine. *Journal of Animal Science* 80, 708–715.

Cromwell, G.L., Henry, B.J., Scott, A.L., Gerngross, M.F., Dusek, D.L. and Fletcher, D.W. (2005) Glufosinate herbicide-tolerant (Liberty Link) rice vs. conventional rice in diets for growing-finishing swine. *Journal of Animal Science* 83, 1068–1074.

Custodio, M.A., Powers, W.J., Huff-Lonergan, E., Faust, M.A. and Stein, J. (2006) Growth, pork quality, and excretion characteristics of pigs fed Bt corn or non-transgenic corn. *Canadian Journal of Animal Science* 86, 461–469.

Davis, H.V. and Kuiper, H.A. (2011) GM-risk assessment in the EU: Current perspective. *Aspects of Applied Biology* 110, 37–45.

Donkin, S.S., Velez, J.C., Totten, A.K., Stanisiewski, E.P. and Hartnell, G.F. (2003) Effects of feeding silage and grain from glyphosate-tolerant or insect-protected corn hybrids on feed intake, ruminal digestion, and milk production in dairy cattle. *Journal of Dairy Science* 86, 1780–1788.

Dowd, P.F. (2000) Indirect reduction of ear melds and associated mycotoxins in *Bacillus thuringiensis* corn under controlled and open field conditions: utility and limitations. *Journal of Economic Entomology* 93, 1669–1679.

EFSA (2006) Guidance document of the Scientific Panel on Genetically Modified Organisms for the risk assessment of genetically modified plants and derived food and feed. *EFSA Journal* 99, 1–100.

EFSA (2008) Safety and nutritional assessment of GM plant derived food and feed. The role of animal feeding trials. *Food and Chemical Toxicology* 46, 2–70.

EFSA (2010) Scientific opinion on the assessment of allergenicity of GM plants and microorganisms and derived food and feed. *EFSA Journal* 8(7) 1700, 165 pp.

EFSA (2011a) Guidance for risk assessment of food and feed from genetically modified plants. *EFSA Journal* 9(5), 2150, 37 pp.

EFSA (2011b) Statistical significance and biological relevance. *EFSA Journal* 9(10) 2372, 17 pp.

Elangovan, A.V., Praveen, K., Tyagi, K., Shrivastav, A.K., Johri, A.K., Kaur, S., *et al.* (2006) GMO (*Bt-Cry1Ac* gene) cottonseed meal is similar to non-GMO low free gossypol cottonseed meal for growth performance of broiler chicken. *Animal Feed Science and Technology* 129, 252–263.

Erickson, G.E., Robbins, N.D., Simon, J.J., Berger, L.L., Klopfenstein, T.J., Stanisiewski, E.P., *et al.* (2003) Effect of feeding glyphosate-tolerant (Roundup-Ready events GA21 or nk603) corn compared with reference hybrids on feedlot steer performance and carcass characteristics. *Journal of Animal Science* 81, 2600–2608.

FASS [Federation of Animal Science Societies] (2013) References – Feeding transgenic crops to livestock; dated June 2013 (and updated monthly).

Faust, M.A. and Glenn, B.P. (2002) Animal feeds from crops derived through biotechnology: farm animal performance and safety. In: Thomas, J.A. and Fuchs, R.L. (eds) *Biotechnology and Safety Assessment*, 3rd edn. Academic Press, San Diego, California, pp. 143–189.

FDA [Food and Drug Administration] (2007) Guidance for industry and other stakeholders, toxicological principles for the safety assessment of food ingredients. Redbook 2000; July 2000, revised July 2007. US Department of Health and Human Services, Food and Drug Administration, Center for Food Safety and Applied Nutrition (http://www.cfsan.fda.gov/guidance.html, accessed 3 June 2013).

Flachowsky, G. and Aulrich, K. (1999) Animal nutrition and genetically modified organisms (GMOs). *Landbauforschung Völkenrode* 49, 13–20. [In German.]

Flachowsky, G. and Aulrich, K. (2001) Nutritional assessment of GMO in animal nutrition. *Journal of Animal Feed Science* 10, Suppl. 1, 181–194.

Flachowsky, G., Chesson, A. and Aulrich, K. (2005) Animal nutrition with feeds from genetically modified plants. *Archives of Animal Nutrition* 59, 1–40.

Flachowsky, G., Aulrich, K., Böhme, H. and Halle, I. (2007) Studies on feeds from genetically modified plants (GMP) – contributions to nutritional and safety assessment. *Animal Feed Science and Technology* 133, 2–30.

Flachowsky, G., Meyer, U. and Schafft, H. (2012) Animal feeding studies for nutritional and safety assessments of feed from genetically modified plants: a review. *Journal of Consumer Protection and Food Safety* 7, 179–194.

Folmer, J.D., Grant, R.J., Milton, C.T. and Beck, J. (2002) Utilization of Bt corn residues by grazing beef steers and Bt corn silage and grain by growing beef cattle and lactating dairy cows. *Journal of Animal Science* 80, 1352–1361.

GfE [Gesellschaft für Ernährungsphysiologie Society for Nutrition Physiology] (1999) Recommendations for energy and nutrient requirements for laying hens and broilers. DLG-Verlag, Frankfurt, Germany, 185 pp. [In German.]

GfE (2001) Recommendations for energy and nutrient requirements for dairy cows and heifers. DLG-Verlag, Frankfurt, Germany, 157 pp. [In German.]

GfE (2008) Recommendations for the supply of energy and nutrients to pigs. DLG-Verlag, Frankfurt, Germany, 247 pp.

Giddings, L. val, Potrykus, I., Ammann, K. and Fedoroff, N.V. (2012) Opinion: Confronting the Gordian knot. *Nature Biotechnology* 30, 208–209.

Grant, R.J., Fanning, K.C., Kleinschmidt, D., Stanisiewski, E.P. and Hartnell, G.F. (2003) Influence of glyphosate-tolerant (event nk603) and corn rootworm protected (event MON863) corn silage and grain on feed consumption and milk production in Holstein cattle. *Journal of Dairy Science* 86, 1707–1715.

Guertler, P., Lutz, B., Kuehn. R., Meyer, H.-H.D. Einspanier, R., Killermann, B., *et al.* (2008) Fate of recombinant DNA and Cry1Ab protein after ingestion and dispersal of genetically modified maize in comparison to rapeseed by fallow deer (*Dama dama*). *European Journal of Wildlife Research* 54, 36–43.

Guertler, P., Paul, V., Albrecht, C. and Meyer, H.H.D. (2009) Sensitive and highly specific quantitative real-time PCR and ELISA for recording a potential transfer of novel DNA and Cry1Ab protein from feed into bovine milk. *Analytical and Bioanalytical Chemistry* 393, 1629–1638.

Guo, L., Gong, L.M., Zhang, L.Y., Li, J.T. and Zhang, W.J. (2012) Evaluation of transgenic *Bt+CpTI* cotton seed meal fed to broilers. *Poultry Science* 91, (in press).

Halle, I., El Sanhoty, R.M.E.S. and Flachowsky, G. (2005) Nutritional assessment of GM-potatoes in broiler feeding. *Proceedings of the Society Nutrition Physiology* 14, 63 (Abstract).

Hammond, B.G., Vicini, J.L., Hartnell, G.F., Naylor, M.W., Knight, C.D., Robinson, E.H., *et al.* (1996) The feeding value of soybeans fed to rats, chickens, catfish and dairy cattle is not altered by genetic incorporation of glyphosate tolerance. *Journal of Nutrition* 126, 717–727.

Hartnell, G.F., Hvelplund, T. and Weisbjerg, M.R. (2005) Nutrient digestibility in sheep fed diets containing Roundup Ready or conventional fodder beet, sugar beet, and beet pulp. *Journal of Animal Science* 83, 400–407.

Herman, R.A., Chassy, B.M. and Parrott, W. (2009) Compositional assessment of transgenic crops: an idea whose time has passed. *Trends in Biotechnology* 746, 1–3.

Huls, T.J., Erickson, G.E., Klopfenstein, T.J., Luebbe, M.K., Vander Pol, K.J., Rice, D.W., *et al.* (2008) Effect of feeding DAS-59122-7 corn grain and nontransgenic corn grain to individually fed finishing steers. *The Professional Animal Scientist* 24, 572–577.

Hyun, Y., Bressner, G.E., Ellis, M., Lewis, A.J., Fischer, R., Stanisiewski, E.P., *et al.* (2004) Performance of growing-finishing pigs fed diets containing Roundup Ready corn (event nk603), a nontransgenic genetically similar corn, or conventional corn lines. *Journal of Animal Science* 82, 571–580.

Hyun, Y., Bressner, G.E., Fischer, R.L., Miller, P.S., Ellis, M., Peterson, B.A., *et al.* (2005) Performance of growing-finishing pigs fed diets containing YieldGard Rootworm corn (MON 863), a nontransgenic genetically similar corn, or conventional corn hybrids. *Journal of Animal Science* 83, 1581–1590.

ILSI (2003) Best practices for the conduct of animal studies to evaluate crops genetically modified for input traits. International Life Sciences Institute, Washington, DC, 62 pp.

ILSI (2007) Best practices for the conduct of animal studies to evaluate crops genetically modified for output traits. International Life Sciences Institute, Washington, DC, 194 pp.

Ipharraguerre, I.R., Younker, R.S., Clark, J.H., Stanisiewski, E.P. and Hartnell, G.F. (2003) Performance of lactating dairy cows fed corn as whole plant silage and grain produced from a glyphosate-tolerant hybrid (event NK603). *Journal of Dairy Science* 86, 1734–1741.

Jacobs, C.M., Etterback, P.L., Parsons, C.M., Rice, D., Smith, B., Hinds, M., *et al.* (2008) Performance of laying hens fed diets containing DAS-59122-7 maize grain compared with diets containing nontransgenic maize grain. *Poultry Science* 87, 475–479.

Jennings, J.C., Kolwcyk, D.C., Kays, S.B., Whetsell, A.J., Surber, J.B., Cromwell, G.L., *et al.* (2003) Determining whether transgenic and endogenous plant DNA and transgenic protein are detectable in muscle from swine fed Roundup Ready soybean meal. *Journal of Animal Science* 81, 1447–1455.

Kan, C.A. and Hartnell, G. (2004a) Evaluation of broiler performance when fed insect-protected, control, or commercial varieties of dehulled soybean meal. *Poultry Science* 83, 2029–2038.

Kan, C.A. and Hartnell, G.F. (2004b) Evaluation of broiler performance when fed Roundup Ready wheat (event MON 71800), control, and commercial wheat varieties. *Poultry Science* 83, 1325–1334.

Kok, E.J., Keijer, J., Kleter, G.A. and Kuiper, H.A. (2008) Comparative safety assessment of plant-derived feeds. *Regulatory Toxicology and Pharmacology* 50, 98–113.

Kuiper, H.A., Kleter, G.A., Noteborn H.P.J.M. and Kok, E.J. (2001) Assessment of food safety issues related to genetically modified foods. *Plant Journal* 27, 503–528.

Llorente, B., Alonso, G.D., Bravo-Almonacid, F., Rodriguez, V., Lopez, M.G., Carrari, F., *et al.* (2011) Safety assessment of nonbrowning potatoes: opening the discussion about the relevance of substantial equivalence on next generation biotech crops. *Plant Biotechnology Journal* 9, 136–150.

Lutz, B., Wiedemann, S., Einspanier, R., Mayer, J. and Albrecht, C. (2005) Degradation of Cry1AB-protein from genetically modified maize in the bovine gastrointestinal tract. *Journal of Agricultural and Food Chemistry* 53, 1453–1456.

Lutz, B., Wiedemann, S. and Albrecht, C. (2006) Degradation of transgenic Cry1Ab DNA and protein in Bt-176 maize during the ensiling process. *Journal of Animal Physiology and Animal Nutrition* 90, 116–123.

McNaughton, J., Roberts, M., Smith, B., Rice, D., Hinds, M., Schmidt, J., *et al.* (2007) Comparison of broiler performance when fed diets containing event DP-356043-5 (Optimum GAT), non-transgenic near-isoline control, or commercial reference soybean meal, hulls, and oil. *Poultry Science* 86, 2569–2581.

McNaughton, J., Roberts, M., Smith, B., Rice, D., Hinds, M., Rood, T., *et al.* (2008) Comparison of broiler performance and carcass yields when fed diets containing transgenic maize grains from event DP-O98140-6 (Optimum GAT), near-isogenic control maize grain, or commercial reference maize grains. *Poultry Science* 87, 2562–2572.

McNaughton, J., Roberts, M., Rice, D., Smith, B., Hinds, M., Delaney, B., *et al.* (2011a) Nutritional equivalency evaluation of transgenic maize grain from event DP-Ø98140-6 and transgenic soybeans containing event DP-356Ø43-5:

laying hen performance and egg quality measures. *Poultry Science* 90, 377–389.

McNaughton, J., Roberts, M., Rice, D., Smith, B., Hinhs, M., Delaney, B., *et al.* (2011b) Comparison of broiler performance and carcass yields when fed transgenic maize grain containing event DP-Ø9814Ø-6 and processed fractions from transgenic soybeans containing event DP-356043-5. *Poultry Science* 90, 1701–1711.

McNaughton, J., Roberts, M., Rice, D., Smith, B., Hinds, M., Delaney, B., *et al.* (2011c) Evaluation of broiler performance and carcass yields when fed diets containing corn grain from transgenic stacked-trait product DAS-Ø15OO7-1xDAS-59122-7xMON-ØØ81Ø-6xMON-ØØ6Ø3-6. *Journal of Applied Poultry Research* 20, 542–553.

Magg, T., Bohn, M., Klein, D., Meriditaj, V. and Melchinger, A.J. (2003) Concentration of moniliformin produced by *Fusarium* species in grains of transgenic Bt maize hybrids compared to their isogenic counterparts and commercial varieties under European corn borer pressure. *Plant Breeding* 122, 322–327.

Mandal, A.B., Elangovan, A.V., Shrivastav, A.K., Johri, A.K., Kaur, S. and Johri, T.S. (2004) Comparison of broiler chicken performance when fed diets containing meals of Bollgard II hybrid cotton containing *Cry-X* gene (*Cry1Ac* and *Cry2ab* gene), parenteral line or commercial cotton. *British Poultry Science* 45, 657–663.

Matten, S.R., Head, G.P. and Quemada, H.D. (2008) How governmental regulation can help or hinder the integration of *Bt* crops within IMP programs. *Progress in Biological Control* 5, 27–39.

Mazza, R., Soave, M., Morlacchini, M., Piva, G. and Marocco, A. (2005) Assessing the transfer of genetically modified DNA from feed to animal tissues. *Transgenic Research* 14, 775–784.

Mejia, L., Jacobs, C.M., Utterback, P.L., Parsons, C.M., Rice, D., Sanders, C., *et al.* (2010) Evaluation of the nutritional equivalency of soybean meal with the genetically modified trait DP-3 Ø5423-1 when fed to laying hens. *Poultry Science* 89, 2634–2639.

Munkvold, G.P., Hellmich, R.L. and Rice, L.G. (1999) Comparison of fumonisin concentrations in kernels of transgenic Bt-maize hybrids and non-transgenic hybrids. *Plant Diseases* 83, 130–138.

NRC [National Research Council] (1994) *Nutrient Requirements of Poultry*, 9th revised edn. National Academy Press, Washington, DC, 176 pp.

NRC (1998) *Nutrient Requirements of Swine*, 10th revised edn. National Academy of Science, Washington, DC, 210 pp.

NRC (2001) *Nutrient Requirements of Dairy Cattle*, 7th revised edn. National Academy Press, Washington, DC, 408 pp.

OECD (1993) *Safety Evaluation of Foods Derived by Modern Biotechnology. Concepts and Principles*. OECD, Paris.

Papst, C., Utz, H.F., Melchinger, A.E. Eder, J., Magg, T., Klein, D., *et al.* (2005) Mycotoxins produced by *Fusarium* spp. in isogen Bt vs non-Bt maize hybrids under European corn borer pressure. *Agronomy Journal* 97, 219–224.

Pietri, A. and Piva, G. (2000) Occurrence and control of mycotoxins in maize grown in Italy. Proceedings 6th International Feed Production Conference, Piacenza, 27–28 November 2000, pp. 226–236.

Reuter, T. and Aulrich, K. (2003) Investigations on genetically modified maize (Bt-maize) in pig nutrition: fate of feed-ingested foreign DNA in pig bodies. *European Food Research and Technology* 216, 185–192.

Reuter, T., Aulrich, K. and Berk, A. (2002) Investigations on genetically modified maize (Bt-maize) in pig nutrition: fattening performance and slaughtering results. *Archives of Animal Nutrition* 56, 319–326.

Rossi, F., Morlacchini, M., Fusconi, G., Pietri, A., Mazza, R. and Piva, G. (2005) Effect of Bt corn on broiler growth performance and fate of feed-derived DNA in the digestive tract. *Poultry Science* 84, 1022–1030.

Sanden, M., Johannessen, L.E., Berdal, K.G., Sissener, N. and Hemre, G.I. (2011) Uptake and clearance of dietary DNA in the intestine of Atlantic salmon (*Salmo salar* L.) fed conventional or genetically modified soybeans. *Aquaculture Nutrition* 17, E750–E759.

Scheideler, S.E., Hileman, R.E., Weber, T., Robeson, L. and Hartnell, G.F. (2008a) The *in vivo* digestive fate of the Cry3Bb1 protein in laying hens fed diets containing MON 863 corn. *Poultry Science* 87, 1089–1097.

Scheideler, S.E., Rice, D., Smith, B., Dana, C. and Sauber, T. (2008b) Evaluation of nutritional equivalency of corn grain from DAS-Ø15O7-1 (Herculex I) in the diets of laying hens. *Journal of Applied Poultry Research* 17, 383–389.

Sharma, R., Damgaard, D., Alexander, T.W., Dugan, M.E.R., Aalhus, J.L., Stanford, K., *et al.* (2006) Detection of transgenic and endogenous plant DNA in digesta and tissues of sheep and pigs fed Roundup Ready canola meal. *Journal of Agricultural and Food Chemistry* 54, 1699–1709.

Shimada, N., Murata, H., Mikami, O., Yoshioka, M., Guruge, K.S., Yamanaka, N., *et al.* (2006)

Effects of feeding calves genetically modified corn Bt11: a clinico-biochemical study. *Journal of Veterinary Medicine Science* 68, 1113–1115.

Sidhu, R.S., Hammond, B.G., Fuchs, R.L., Mutz, J.-N., Holden, L.R., George, B., *et al.* (2000) Glyphosate-tolerant corn: the composition and feeding value of grain from glyphosate-tolerant corn is equivalent to that of conventional corn (*Zea mays* L.). *Journal of Agricultural and Food Chemistry* 48, 2305–2312.

Siegfried, B., Vaughn, T. and Spencer, T. (2005) Baseline susceptibility of western corn rootworm (Coleoptera: Crysomelidae) to Cry38b1 *Bacillus thuringiensis* toxin. *Journal of Economic Entomology* 98, 1320–1324.

Sissener, N.H., Sanden, M., Bakke, A.M., Krogdahl, A. and Hemre, G.I. (2009a) A long term trial with Atlantic Salmon (*Salmo salar* L.) fed genetically modified soy, focusing general health and performance before, during and after the parr-smolt transformation. *Aquaculture* 294, 108–117.

Sissener, N.H., Bakke, A.M., Gu, J., Penn, M.H., Eie, E., Krogdahl, A., *et al.* (2009b) An assessment of organ and intestinal histo-morphology and cellular stress response in Atlantic salmon (*Salmo salar* L.) fed genetically modified Roundup Ready Soy. *Aquaculture* 298, 101–110.

Spiekers, H., Meyer, H.H.D. and Schwarz, F.J. (2009) Effects of long term utilization of genetically modified maize (MON 810) in dairy cattle feeding on performance and metabolism parameters. *Schriftenreihe, Bayerische Landesanstalt für Landwirtschaft* (LfL), Vol. 18, 105 pp. [In German.]

Stadnik, J., Karwowska, M., Dolatowski, Z.J., Swiatkiewicz, S. and Kwiatek, K. (2011) Effect of genetically modified insect resistant corn (MON 810) and glyphosate tolerant soybean meal (Roundup ready) on physico-chemical properties of broilers breast and thigh muscles. *Bulletin of the Veterinary Institute in Pulawy* 55, 541–546.

Stanford, K., Aalhus, J.L., Dugan, M.E.R., Wallins, G.L., Sharma, R. and McAllister, T.A. (2003) Effects of feeding transgenic canola on apparent digestibility, growth performance and characteristics of lambs. *Canadian Journal of Animal Sciences* 83, 299–305.

Stein, H.H., Rice, D.W., Smith, B.L., Hinds, M.A., Sauber, T.E., Pedersen, C., *et al.* (2009a) Evaluation of corn grain with genetically modified input trait DAS-59122-7 fed to growing-finishing pigs. *Journal of Animal Sciences* 87, 1254–1260.

Stein, H.H., Sauber, T.E., Rice, D.W., Hinds, M.A., Smith, B.L., Dana, G., *et al.* (2009b) Growth performance and carcass composition of pigs fed corn grain from DAS- Ø15Ø7-1 (Herculex I) hybrids. *The Professional Animal Scientist* 25, 689–694.

Steinke, K., Guertler, P., Paul, V., Wiedemann, S., Ettle, T., Albrecht, C., *et al.* (2010) Effects of long-term feeding of genetically modified corn (event MON810) on the performance of lactating dairy cows. *Journal of Animal Physiology and Animal Nutrition* 94, 185–195.

Swiatkiewicz, S., Swiatkiewicz, M., Koreleski, J. and Kwiatek, K. (2010a) Nutritional efficiency of genetically-modified insect resistant corn (MON810) and glyphosate-tolerant soybean meal (Roundup ready) for broilers. *Bulletin of the Veterinary Institute in Pulawy* 54, 43–48.

Swiatkiewicz, S., Twardowska, M., Markowsky, J., Mazur, M., Sieradzki, Z. and Kwiatek, K. (2010b) Fate of transgenic DNA from Bt corn and Roundup Ready soybean meal in broilers fed GMO feed. *Bulletin of the Veterinary Institute in Pulawy* 54, 237–242.

Swiatkiewicz, M., Hanczakowska, E., Twardowska, M., Mazur, M., Kwiatek, K., Kozaczynski, W., *et al.* (2011) Effect of genetically modified feeds on fattening results and transfer of transgenic DNA to swine tissues. *Bulletin of the Veterinary Institute in Pulawy* 55, 121–125.

Taylor, M.L., Hyun, Y., Hartnell, G.F., Riordan, S.G., Nemeth, M.A., Karunanandaa, K., *et al.* (2003) Comparison of broiler performance when fed diets containing grain from YieldGard Rootworm (MON863), YieldGard Plus (MON810 × MON863), nontransgenic control, or commercial reference corn hybrids. *Poultry Science* 82, 1948–1956.

Taylor, M.L., Stanisiewski, E.P., Riordan, S.G., Nemeth, M.A., George, B. and Hartnell, G. (2004) Comparison of broiler performance when fed diets containing Roundup ready (Event RT73), nontransgenic control, or commercial canola meal. *Poultry Science* 83, 456–461.

Taylor, M.L., Hartnell, G., Nemeth, M., Karunanandaa, K. and George, B. (2005a) Comparison of broiler performance when fed diets containing corn grain with insect-protected (corn rootworm and European corn borer) and herbicide-tolerant (glyphosate) traits, control corn, or commercial reference corn. *Poultry Science* 84, 587–593.

Taylor, M.L., Hartnell, G., Nemeth, M., Karunanandaa, K. and George, B. (2005b) Comparison of broiler performance when fed diets containing corn grain with insect-protected (corn rootworm and European corn borer) and herbicide-tolerant (glyphosate) traits, control corn, or commercial reference corn – revisited. *Poultry Science* 84, 1893–1899.

Taylor, M.L., Hartnell, G., Lucas, D., Davis, S. and Nemeth, M. (2007a) Comparison of broiler performance and carcass parameters when fed diets containing soybean meal produced from glyphosate-tolerant (MON 89788), control, or conventional reference soybeans. *Poultry Science* 86, 2608–2614.

Taylor, M.L., Hartnell, G., Nemeth, M., Lucas, D. and Davis, S. (2007b) Comparison of broiler performance when fed diets containing grain from second-generation insect protected and glyphosate-tolerant, conventional control and commercial reference corn. *Poultry Science* 86, 1972–1979.

Taylor, M., Lucas, D., Nemeth, M., Davis, S. and Hartnell, G. (2007c) Comparison of broiler performance and carcass parameters when fed diets containing combined trait insect-protected and glyphosate-tolerant corn (MON 89034 × NK603), control, or conventional reference corn. *Poultry Science* 86, 1988–1994.

Tony, M.A., Butschke, A., Broll, A., Zagon, J., Halle, I., Dänicke, S., *et al.* (2003) Safety assessment of Bt 176 maize on broiler nutrition: degradation of maize-DNA and its metabolic fate. *Archives of Animal Nutrition* 57, 235–252.

Tudisco, R., Mastellone, V., Cutrignelli, M.I., Lombardi, P., Bovera, F., Mirabella, N., *et al.* (2010) Fate of transgenic DNA and evaluation of metabolic effects in goats fed genetically modified soybean and in their offsprings. *Animal* DOI:10.1017/S1751731110000728.

Valenta, H., Dänicke, S., Flachowsky, G. and Böhme, T. (2001) Comparative study on concentrations of the *Fusarium* mycotoxins deoxynivalenol and zearalenone in kernels of transgenic Bt maize hybrids and nontransgenic hybrids. *Proceedings of the Society of Nutrition Physiology* 10, 182 (Abstract).

Vander Pol, K.J., Erickson, G.E., Robbins, N.B., Berger, L.L., Wilson, C.B., Klopfenstein, T.J., *et al.* (2005) Effects of grazing residues or feeding corn from a corn rootworm-protected hybrid (MON 865) compared with reference hybrids on animal performances and carcass characteristics. *Journal of Animal Sciences* 83, 2826–2834.

Walsh, M.C., Buzoianu, S.G., Gardiner, G.E., Rea, M.C., Gelencser, E., Janosi, A., *et al.* (2011) Fate of transgenic DNA from orally administered Bt MON810 maize and effects on immune response and growth in pigs. *PLoS ONE* 6(11): e27177. DOI:10.1371/journal.pone.0027177.

Walsh, M.C., Buzoianu, S.G., Gardiner, G.E., Rea, M.C., Ross, R.P., Cassidy, J.P., *et al.* (2012) Effects of short term feeding of Bt MON810 maize on growth performance, organ morphology and function in pigs. *British Journal of Nutrition* 107, 364–371.

Wiedemann, S., Lutz, B., Kurtz, H., Schwarz, F.J. and Albrecht, C. (2006) *In situ* studies on the time dependent degradation of recombinant corn DNA and protein in the bovine rumen. *Journal of Animal Science* 84, 135–144.

Wu, F. (2006a) Bt corn's reduction of mycotoxins: regulatory decisions and public opinion. In Just, R.E., Alston, J.M. and Zilberman, D. (eds) *Regulating Agricultural Biotechnology: Economics and Policy*. Springer, New York, pp. 179–200.

Wu, F. (2006b) Mycotoxin reduction in Bt corn: potential economic, health, and regulatory impacts. *Transgenic Research* 15, 277–289.

7 Feeding Studies with Second-generation GM Plants (Output Traits) with Food-producing Animals

Gerhard Flachowsky*

Institute of Animal Nutrition, Friedrich-Loeffler-Institute (FLI), Federal Research Institute for Animal Health, Braunschweig, Germany

7.1 Introduction

Apart from water and energy, humans and animals require many nutrients to meet their metabolic need (so-called essential nutrients). Table 7.1 reviews such nutrients known to be essential for sustaining human and animal life.

Inadequate consumption of one of these nutrients will result in metabolic disturbances, leading not only to lower feed intake, weaker performance of animals and lower feed efficiency but also to sickness, poor health, impaired development of juveniles and higher costs for humans and animals (Welch and Graham, 2004; Mayer *et al.*, 2008). Therefore, nutritional supplementation of diets is common for humans and animals in deficit situations.

In some cases, plants enriched with adequate nutrients could be more sustainable not only for human but also for animal nutrition. A large number of genetically modified (GM) plants (crops and vegetables) of the so-called 'second generation' (plants with output traits or with substantial changes in composition) with specific benefits for the consumer and animals are being developed or are in development (see Chapters 2 and 12). These plants with increased nutritional content are also called biofortified plants or crops (Bouis, 2002; Welch, 2002; White and Broadley, 2005; Hirschi, 2009). They can have a great impact on improving the already existing food and feed supply (Nestel *et al.*, 2006).

Specific advantages are higher content(s) of important nutrients and substances with nutritional values, such as (mentioned with some references):

- protein and/or amino acids (Sevenier *et al.*, 2002; Lucas *et al.*, 2007; Ufaz and Galili, 2008; see Section 7.2);
- fat or specific fatty acids (Cahoon *et al.*, 2007; Napier, 2007; see Section 7.3);
- starch or special carbohydrates (Sevenier *et al.*, 2002; see Section 7.4);
- specific minerals (Goto *et al.*, 1999; Welch *et al.*, 2000; Lucca *et al.*, 2001; Gregorio, 2002; Holm *et al.*, 2002; Welch and Graham, 2004; White and Broadley, 2005; Broadley *et al.*, 2006; see Section 7.5);
- vitamins or vitamin precursors (Ye *et al.*, 2000; Potrykus, 2001; Beyer *et al.*, 2002; Rocheford *et al.*, 2002; Diaz de la Garza *et al.*, 2004; van Eenennaam *et al.*, 2004; van Jaarsveld *et al.*, 2005; DellaPenna, 2007; see Section 7.6);

*E-mail: gerhard.flachowsky@t-online.de

Table 7.1. Nutrient groups and nutrients essential for humans and animals.

Groups of nutrients	Essential nutrients
Amino acids	Histidine, isoleucine, leucine, lysine,[b] methionine,[b] phenylalanine, threonine[b] tryptophane,[b] valine (semi-essential: arginine, cystine)
Fatty acids	Linoleic acid, linolenic acid
Major elements	Ca,[b] Mg,[b] P,[b] Na, K, S, Cl
Trace elements	Fe,[b] Zn,[b] Cu, Mn, I,[b] Se,[b] Co (cobalamin; vitamin B_{12})
Ultra-trace elements[a]	F, B, Mo, Ni, Cr, V, Si, As, Cd, Pb, Li, Sn
Vitamins (fat soluble)	A[b] (precursor β-carotene), D, E,[b] K
(water soluble)	B_1 (thiamin), B_2 (riboflavin), B_6 (pyridoxin), B_{12}[b] (cobalamin), pantothenic acid, niacin, folate,[b] biotin, C (ascorbic acid)

Notes: [a]Essentiality of some elements is unclear (McDowell, 2003); occasionally beneficial elements (Suttle, 2010) [b]best limiting nutrients.

- enzymes (Zhang *et al.*, 2000; Nyannor *et al.*, 2007; Gao *et al.*, 2012; see Section 7.7);
- antioxidative substances (Sevenier *et al.*, 2002);

and lower contents of undesirable substances, such as:

- glucosinolates (Vageeshbabu and Chopra, 1997);
- gluten (Vasil and Anderson, 1997);
- mycotoxins (Munkvold *et al.*, 1999; Duvick, 2001, see also Table 6.1); and
- phytate (see Section 7.8).

A similar structure is used by Hirschi (2009) to characterize biofortified crops for human nutrition. He distinguishes between protein and amino acids, carbohydrates, micro-nutrients and functional metabolites on the one hand and plant components with suggested functionality such as dietary fibre, carotinoids, fatty acids, flavonoids, gluco-sinolates, phenolics, plant sterols, phyto-oestrogens, sulfides and tannins on the other hand (see also ILSI, 2008; Newell-McGloughlin, 2008).

Such biofortification may be more important for human nutrition than for animal nutrition (see Table 7.2). Food from biofortified crops can reach rural populations for reducing levels of micronutrient mal-nutrition, as has been discussed and demonstrated by many authors during the past few years (DellaPenna, 1999; Dawe *et al.*, 2002; King, 2002; Bouis *et al.*, 2003; Johns, 2003; McKeon, 2003; Zimmermann *et al.*, 2004; White and Broadley, 2005; Sautter *et al.*, 2006; Mayer *et al.*, 2008; Gilligan, 2012; see Chapter 12 for further details).

Many feed additives are available for animals (see the present EU feed law). Such additives are mostly cheaper and their development/production is faster than via plant bioengineering. Many additives are produced by GM microorganisms (see Chapter 11).

Table 7.3 shows some examples of GM plants with altered composition. Some fundamentals are described in Chapter 7 and future developments in the field of GM plants are shown in Chapter 12.

Plants with output traits are not substantially equivalent to their isogenic counterparts because of substantial changes in composition and nutritive value (Cloren *et al.*, 2011), and the paradigm of substantial equivalence (OECD, 1993) cannot be used for the safety and nutritional assessment of food/feed from such plants. New and changed procedures are necessary for the safety and nutritional assessment of food/feed from the second generation of GM plants (ILSI, 2007; EFSA, 2008, 2011). The higher content of some nutrients in food/feed is one side of genetic modification; their bioavailability in humans and animals is the other side.

Apart from the safety assessment of tDNA and newly expressed protein(s), nutritional assessments and investigation into the consequences of changes in nutrient content should be undertaken (see Bouis

Table 7.2. Pros and cons of substantial changes in plant compositions (second-generation plants; plants with output traits, or biofortified plants).

Pros	Cons
More advantages for human nutrition (meet requirements; e.g. fatty acids, minerals, vitamins, etc.) than for animal nutrition	Plant breeding takes a long time (longer than the development of food/feed additives)
Lower content of undesirable substances	Many feed additives are available for animal nutrition
Improvement of properties of food/feed	High amounts of food/feed may be necessary to meet requirements

Table 7.3. Examples of GM plants with improved characteristics intended to provide nutritional benefits (biofortified plants). (From EFSA, 2008.)

Plant/species	Altered characteristic	Transgene/mechanism
Maize	Improved amino acid profile ↑	Various enzymes
	Vitamin C ↑	Dehydroascorbate reductase
	Bioavailable iron ↑	Ferritin and phytase
	Fumonisin ↓	De-esterase and de-aminase
Potatoes	Starch ↑	ADP glucose pyrophosphorylase
	Solanine ↓	Antisensesterol glycotransferase
Rapeseed	Vitamin E ↑	Gamma-Tocopheryltransferase
	β-Carotene ↑	Phytoene-synthase
	Linoleic and/or linolenic acids ↑	Various desaturases
Rice	β-Carotene ↑	Phytoene-synthetase and -desaturase, lycopene cyclase
	Iron ↑	Ferritin, metallothionein, phytase
Soybean	Oleic acid ↑	Suppression of desaturase
	Stearidonic acid ↑	Various desaturases

et al., 2003; ILSI, 2007; EFSA, 2008, 2011; see Chapter 5) by:

- *in vitro* studies;
- studies with animal models to determine the bioavailability;
- efficacy trial with animals; and
- effectivity studies with target animals/humans.

King (2002) proposed a three-step process for the nutritional assessment of biofortified food from such plants in human nutrition:

- Test the bioavailability of newly expressed nutrient(s) or nutrient(s) expressed in higher amounts.
- Feeding trial(s) to test the efficacy of the biofortified food for improving the nutrition and health of the target population.

- Final trial for evaluating the nutritional, health, agricultural, societal, environmental and economic aspects of biofortified food in the community.

Similar steps seem to be necessary in animal nutrition. Experimental designs for such studies are discussed in Chapter 5 and are described in detail by Flachowsky and Böhme (2005), ILSI (2007), EFSA (2008) and Llorente *et al.* (2011).

Crops can also be genetically modified to produce oils, starch, fibre, protein or other substances useful for food/feed and industrial processes (McKeon, 2003). Generally, such substances are mainly extracted from the crops and so-called co-products (e.g. soybean meal, rapeseed meal, cottonseed meal) could be available for animal nutrition. In general, extracted co-products do not vary strongly from their isogenic counterparts in the composition of main nutrients (McNaughton *et al.*, 2008; Mejia *et al.*, 2010), but analyses are necessary of the composition of such co-products (e.g. stearidonic acids in soybean meal; see Section 7.3). If GM crops are used for industrial purposes and they are not suitable for animal and human nutrition, they should not enter the food chain or contaminate feed and food or other crops with their transgenes (McKeon, 2003). Some examples of the biofortification of various crops will be described in the following subsections.

In addition, there also exist GM plants that express specific proteins for the prevention of diseases in humans and animals (Pribylova *et al.*, 2006), which are not considered in the following text.

7.2 Protein and Amino Acids

Essential amino acids such as lysine, methionine, threonine and tryptophane (see Table 7.1) play an important role in the protein metabolism of humans and animals (Wu *et al.*, 2010; see nutrient requirements of food-producing animals, e.g. NRC, 1994, 1998, 2001; GfE, 1999, 2001, 2008).

Normally, such amino acids lacking in non-ruminant feeding are supplemented by adequate crystalline amino acids (Nelson *et al.*, 1986; Kidd *et al.*, 1998; Heger *et al.*, 2002, 2003; Susenbeth, 2006); so-called rumen-protected amino acids are used in ruminant nutrition (Bertrand *et al.*, 1998; Robinson, 2010; Robinson *et al.*, 2010; Chen *et al.*, 2011).

Nowadays, new technologies such as genetic engineering allow improvement of the protein composition of plants (Galili, 2002; Christou and Twyman, 2004; Galili *et al.*, 2005; Beauregard and Hefford, 2006; Ufaz and Galili, 2008; Maruyama *et al.*, 2011; see Chapter 12). Under such conditions, there is no longer a need for separate amino acid supplementation as the lower amounts now provided by the GM plant are sufficient to meet the requirements of animals.

Such studies have been done with typical protein sources such as soybeans, either to increase the protein content (Edwards *et al.* 2000) or to change the protein compo tion (Falco *et al.*, 1995; Parsons and Zhang 1997), and also with other legumes, for example lupins (Molvig *et al.*, 1997; Muntz *et al.*, 1998; White *et al.*, 2000; Ravindar *et al.*, 2002), oilseeds (Falco *et al.*, 1995), cereals (Maruyama *et al.*, 2001; Lee *et al.*, 2003; Wu *et al.*, 2003; Huang *et al.*, 2005, 2006; Glenn, 2007; Houmard *et al.*, 2007; Lucas *et al.*, 2007; Wakasa *et al.*, 2006), potatoes or sweet potatoes (Sevenier *et al.*, 2002) and forage (Avraham *et al.*, 2005).

For example, lysine maize (*Zea mays* LY038) was developed to accumulate free lysine (Newell-McGloughlin, 2008) in the germ portion of maize grain (see Table 7.4) and to provide an alternative to a direct supplementation of lysine or to the feeding of lysine-rich feeds in non-ruminant nutrition. GM maize LY038 × MON810 was produced from two GM strains by conventional breeding of LY038 with MON810, which provided the maize plant with protection against feeding damage from the European corn borer. Both maize varieties contained significantly more lysine than the control maize (Table 7.4). The crude protein content, and also most of the other essential amino acids, were increased in GM maize. Lucas *et al.* (2007) fed the GM maize (59.5% in starter and 66.1% in grower/finisher diets) to broilers and compared these maize varieties with unsupplemented and L-lysine-supplemented diets. Broiler performance and carcass data demonstrated that the bioefficacy of the incremental lysine in GM maize was not different from that of lysine in conventional maize diets supplemented with L-lysine HCl (see Table 10.12). This type of technology of free amino acids was also used for increased lysine content in canola and soybean and produced a significant increase in tryptophan levels in grain (Hirschi, 2009).

Apart from higher protein or amino acid concentration, there are also activities to develop transgenic crops producing seed storage proteins with bioactive peptides (Maruyama *et al.*, 2011).

Table 7.4. Protein (%) and amino acid content (g/kg as-fed basis) of parenteral maize line and GM maize (LY038 and LY038 × MON810). (From Lucas *et al.*, 2007.)

Item	Control	LY038	LY038 × MON810
Crude protein	8.9	9.5	9.8
Lysine	2.55	3.70	3.49
Free lysine	0.05	0.96	0.78
Arginine	3.83	3.73	3.75
Histidine	2.84	3.08	3.00
Isoleucine	3.02	3.24	3.36
Leucine	11.1	11.8	12.4
Methionine	1.69	1.95	1.87
Phenylalanine	4.15	4.46	4.57
Threonine	2.91	3.05	3.03
Tryptophane	0.61	0.62	0.69
Valine	4.38	4.53	4.65

7.3 Fat and Fatty Acids

Genetically modified oilseeds are able to express modified fatty acid patterns (McKeon, 2003; Hirschi, 2009; see also Chapter 12). The expression of the *gm-fad2-1* gene in soybeans results in a higher concentration of oleic acid (C18:1) by suppressing the expression of endogenous *FAD2-1* gene, which encodes an n-6 fatty acid desaturase enzyme that catalyses desaturation of linoleic acid (C18:2; Okuley *et al.*, 1994; Heppard *et al.*, 1996; Small, 2007) to C18:1. More oleic acid in oil instead of linoleic acid confers a higher oxidative stability to the oil. Feeding studies were done with full fat soybeans (Delaney *et al.*, 2008), with soybean oil or with soybean meal (McNaughton *et al.*, 2008; Mejia *et al.*, 2010).

Conjugated linoleic acid (CLA) may influence the metabolism as well as body composition and milk composition of animals (Bauman and Lock, 2006; Pappritz *et al.*, 2011; von Soosten *et al.*, 2012). Hornung *et al.* (2002) and Iwabuchi *et al.* (2003) introduced a conjugate gene isolated from *Tricosanthes kirilowii* into *Brassica napus* by using an *Agrobacterium*-mediated transformation method to produce a genetically modified rapeseed with CLA. The result was a rapeseed oil with a low content of 2.5% punicic acid, a C18:3 9cis, 11trans, 13cis fatty acid (Table 7.5).

Plant oils are very important energy sources for humans and animals. Some unsaturated fatty acids are characterized by specific health effects (Lunn and Theobald, 2006; Dyer *et al.*, 2008). Therefore, many activities exist to increase the content of specific unsaturated fatty acids in oilseeds. The introduction of two new genes affects the expression of Δ6 and higher expression of Δ15-desaturases (Fig. 7.1) and the biosynthesis of a highly unsaturated fatty acid with four double bonds (stearidonic acid).

Stearidonic soybean oil contains between 20 and 30% stearidonic acid (SDA; C18:4 n-3), but the contents of oleic (C18:1) and

Table 7.5. Fatty acid composition (% of oil) of rapeseed oil and punicic acid (PA) GM rapeseed oil. (From Koba *et al.*, 2007.)

Fatty acids	Rapeseed oil	GM PA oil
C16:0	4.1	5.0
C18:0	1.5	1.7
C18:1	62.5	68.4
C18:2	19.3	15.3
C18:3	9.8	3.8
PA C18:3	0.0	2.5

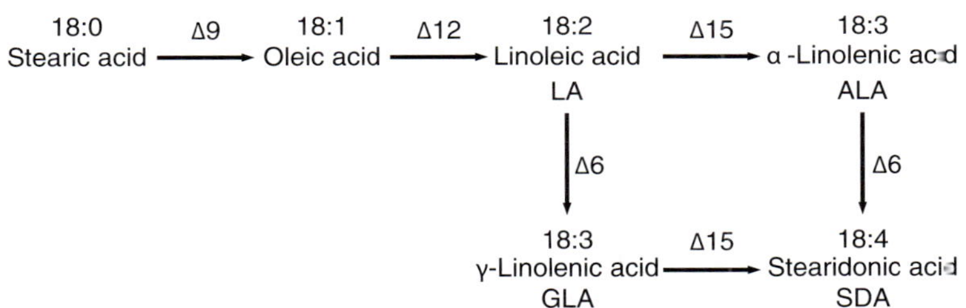

Fig. 7.1. Fatty acid biosynthesis in plants and the new introduced changes to produce stearidonic acid (C18:4 n-3) and the effects of various desaturases (Δ6 and Δ15). (From Ursin, 2003; Whelan 2009.)

linoleic acid (C18:2) are adequately decreased (Table 7.6).

Rymer *et al.* (2011) added 45 (grower) and 50 g SDA oil containing 24% of stearidonic acid per kg finisher broiler diet and compared this with conventional soybean oil. The authors did not observe any significant influence of SDA oil on feed intake, weight gain and feed conversion rate in the animals, but they found some influence on fatty acids in various body fats (see Chapter 10). Similar results are described by Bernal-Santos *et al.* (2010) in lactating cows after duodenal infusion of SDA soybean oil; by Kitessa and Young (2011) after feeding of ruminally protected SDA oil to dairy cows; and by Mejia *et al.* (2010) in laying hens. Such long-chain omega-3 polyunsaturated fatty acids may also be important for fish nutrition and to replace fish oils in human nutrition as a precursor of eicosapentaenoic acid (EPA, C20:5 n-3) and docosahexaenoic acid (DHA, C22:6 n-3; Cahoon *et al.*, 2007).

Amaro *et al.* (2012) tested the influence of increasing levels of stearidonic acid (up to 50 mg/l incubation media) on methane production in a rumen *in vitro* system. The SDA supplementation was not associated with a reduction in methanogenesis. Higher levels or an alternative form might be needed.

Apart from stearidonic acid, the ability to genetically engineer plants has facilitated the generation of oilseeds also synthesizing fatty acids that are seldom in oils or are non-native (Napier, 2007), but which are used for various technical purposes. For example, Böhme *et al.* (2007) tested oil from rapeseed in which the acyl-thioesterase gene of *Cuphea lanceolata* was inserted. The portion of mid-chain fatty acids (C_{12}–C_{14}) increased from 5.2 to 33.3% in the transgenic rape; those of oleic acid reduced from 68.6 to 42.6%. The glucosinolate content increases from 12.4 in the parenteral plant to 19.0 μmol/g DM in the GM plant and requires a critical analysis (Böhme *et al.*, 2007).

7.4 Carbohydrates

Carbohydrates in plants may also be a subject of genetic modification; for example inulin, as polymers of fructans, was intended for use as a prebiotic functional food in human nutrition. Heyer *et al.* (1999) and Sevenier *et al.* (1998) introduced genes of globe artichoke (*Cynara scolymus*) into potatoes or sugarbeets and found an ability to synthesize high molecular weight fructan as inulin. The inulin concentration in the dry matter of the transgenic tubers amounted to about 5% (Hellwege *et al.*, 2000), and starch

Table 7.6. Fatty acid composition of soybean oil and 'SDA oil' (% of oil) from soybeans used in the study by Rymer *et al.* (2011).

Fatty acids	Soybean oil	SDA soybean oil
C16:0	7.3	8.4
C18:0	3.1	3.0
C18:1	14.0	9.8
C18:2	34.0	12.4
C18:3	5.0	12.3
C18:4 (SDA)	0	24.1

content was reduced. The content of total alkaloids in transgenic potatoes has been increased and needs further research (Böhme *et al.*, 2005).

Lignin does not belong to the group of carbohydrates, but it is an important ingredient of most fibre fractions of plant cell walls. Lignin is also considered to be an undesirable constituent of feed. Lignin and fibre fractions (e.g. NDF, ADF, crude fibre) are very important for the digestibility and feed value of the vegetative parts of plants, as well as the performance of ruminants. A reduction of lignin or a less intensive connection between lignin and cellulose/ hemicellulose may contribute substantially to a higher digestibility and a higher dry matter intake of many roughages. So-called brown midrib (bm) hybrids (e.g. maize, sorghum), as a result of traditional mutation breeding (Cherney *et al.*, 1991; Grant *et al.*, 1995; Taylor and Allen, 2005a,b), demonstrate the influence on digestibility (Rook *et al.*, 1977; Koehler *et al.*, 1989; Ivan *et al.*, 2005; Gorniak *et al.*, 2012) and rumen fermentation, feed intake and the performance of ruminants (Oba and Allen, 1999, 2000a,b; Barriere *et al.*, 2004; Ivan *et al.*, 2005).

Based on these data, genetic engineering may also be helpful to increase feed value and feed intake of low-quality roughages. Such studies may also contribute to increasing the feed value of some co-products, e.g. straw from grain production. On the other hand, lignin is a very important ingredient for cell wall stability, and therefore for the steadiness of the stalks of cereals and maize. The propensity for layers of such plants including bm hybrids is higher than for plants with higher lignin content.

7.5 Minerals

Humans and animals require about 20 major and trace elements (see Table 7.1). The normal animal diet consisting of forage/ roughage and concentrates or co-products from concentrates does not meet the nutritional requirements of animals. About three billion people are malnourished in terms of micronutrients (minerals and vitamins; Welch and Graham, 2004). Therefore, the supplementation of nutrition with minerals is necessary and usual.

In some regions, it is difficult to supplement human diets with minerals, and often mineral supplements such as lick stones or other mixtures are not available in adequate composition and amounts for animals. Therefore, biofortifying of crops with essential mineral elements (e.g. Fe, Zn, Ca, Zn, Se, etc.) and increasing the bioavailability of minerals may contribute towards overcoming the gap between feed/ food content and requirements for animals/ humans (Welch and Graham, 2004; White and Broadley, 2005). During the past few years, plant breeding has made progress by using biotechnological tools to increase the pace and prospects for success of the biofortification of many plants with minerals, especially for staple food crops such as rice, cassava, wheat, maize and beans (Gregorio, 2002; Holm *et al.*, 2002). Special attention has been paid to iron and zinc, because iron deficiency is estimated to affect about 30% of the world population (WHO, 2008; Lynch, 2011). Plant breeding might provide a sustainable and cost-effective solution in the long run, delivering minerals to the entire population (White and Broadley, 2005; see also Chapter 12). Two approaches have been used to improve the mineral content in feed/food (Frossard *et al.*, 2000; Colangelo and Guerinot, 2006):

1. To increase the efficiency of uptake and transport of minerals into edible plant tissues.
2. To increase the amount of bioavailable mineral accumulation in the plant.

There are various studies to test the mineral bioavailability in model animals as shown for iron (see Tako *et al.*, 2010):

- Fe and Zn in genetically enriched beans and rice (Welch *et al.*, 2000; Welch, 2002; Tako *et al.*, 2009).
- Fe in genetically modified grains expressing a microbial phytase or reduced level of phytate (Holm *et al.*, 2002; Sautter *et al.*, 2006).

- Fe in genetically modified rice grains (ferritin gene from *Phaseolus vulgaris*, phytase from *Aspergillus fumigatus*) into the rice endosperm and cysteine peptides as enhancers of iron absorption (Lucca *et al.*, 2002; Sautter *et al.*, 2006).

Most bioavailability studies have been carried out with rats, but such studies are considered to be of little use in predicting the bioavailability of trace elements in humans and food-producing animals (Hurrell, 1997; Sandstorm, 1997). Human studies (Lucca *et al.*, 2002; Petry *et al.*, 2012) or adequate studies with food-producing animals (e.g. broilers; Tako *et al.*, 2010) are considered necessary, but sufficient test material must be available for adequate studies to be carried out.

Some data of Fe and Zn radio-labelled bean seeds and rice grains are shown in Table 7.7. These data demonstrate that increasing amounts of iron or zinc in enriched bean seeds or rice grain increase significantly the amount of iron or zinc bioavailable to rats, but not the percentage of bioavailability. Iron absorption inhibitors such as phytate and polyphenols may influence iron absorption. An efficacious iron biofortification may be difficult to achieve in plants rich in phytate and polyphenols (Petry *et al.*, 2012). More details about the mineral biofortification of plants are described by Hirschi (2009).

7.6 Vitamin Precursors and Vitamins

Vitamins are deficient in humans and animals in many regions. Therefore, core activities deal with an increase of vitamins in plants or specific plant parts. For example, much attention has been paid to the enhancement of β-carotene as a vitamin A precursor (Potrykus, 2001, 2003; Beyer *et al.*, 2002; Ha *et al.*, 2010) and vitamin E (Shintani and DellaPenna, 1998; Schledz *et al.*, 2001; Rocheford *et al.*, 2002; Cahoon *et al.*, 2003; van Eenennaam *et al.*, 2004; DellaPenna and Pogson, 2006) in some cereals or folate (DellaPenna, 2007; Storozhenko *et al.*, 2007) in tomatoes (up to 25 times more than controls; Diaz de la Garza, 2007).

In the case of substantial changes in plant composition (GM plants with output traits, or second-generation GM plants), studies are necessary to measure the digestibility/ availability of some nutrients or nutrient precursors (see Flachowsky and Böhme, 2005; ILSI, 2007; EFSA, 2008). Such studies have been done mostly with model animals.

Table 7.7.a Concentrations of iron in bean seeds (radio-labelled with ^{59}Fe) on the iron bioavailability by iron-depleted rats (5 examples out of 24 bean genotypes). (From Welch *et al.*, 2000.)

Bean genotype	Fe (µg/g)	Phytate (µmol/g)	Bioavailable Fe (µg/g meal)	Bioavailable Fe (%
G12610	51.6	19.6	32.4	62.8
G2774	75.3	20.7	48.8	64.8
G23063	88.9	21.9	51.7	58.7
G2572	103.6	23.1	66.9	64.6
G734	156.9	24.1	88.4	53.2

Table 7.7.b Concentration of zinc in rice grain (radio-labelled with [65]Zn) on the zinc bioavailability by zinc-depleted rats (5 examples out of 10 genotypes). (From Welch *et al.*, 2000.)

Rice genotype	Zn (µg/g)	Phytate (µmol/g)	Bioavailable Zn (µg/g meal)	Bioavailable Zn (%)
IR74	44.1	19.7	35.8	81.2
Heibao	46.7	17.3	35.5	76.1
IR58	48.1	19.2	37.4	77.8
Madhukar	51.1	13.4	39.6	77.6
IR101198-66-2	60.5	19.4	49.7	82.1

(mice, rats, rabbits) or small target animals (chicks, quails, piglets), because of the high costs and the limited feed amounts available in some cases, especially in early breeding stages. A model for such studies is proposed in Chapter 5.

Table 7.8 shows the results of measuring the bioconversion of β-carotene into vitamin A. The retinol concentration in the liver of Mongolian gerbils as a model animal was used as an end point. The liver is considered to be the most suitable indicator organ for vitamin A. After a depletion period, four different diets were fed to the gerbils (see Table 7.8).

The results showed that the retinol concentration in the liver of gerbils fed with carotene-rich maize was similar to animals fed with maize poor in carotene and supplemented with adequate amounts of β-carotene. This means that, in this case, β-carotene from maize is converted into vitamin A almost as identically as supplementary β-carotene. In the case of 'Golden Rice' containing between 1.6 (Ye *et al.*, 2000) and up to 37 mg β-carotene/kg dry rice (Paine *et al.*, 2005), the first studies to determine the vitamin A value of β-carotene were done using deuterium-labelled rice with five humans (Tang *et al.*, 2009) and not with laboratory or target animals. Recently, Tang *et al.* (2012) compared β-carotene in 'Golden Rice' with pure β-carotene and that in green spinach in providing vitamin A to children ($n = 68$; 6–8 years). The conversion of β-carotene into vitamin A is influenced by many factors (Tanumihardjo, 2002). β-carotene in 'Golden Rice' was as effective as pure β-carotene converted in vitamin A (2.0:1 and 2.3:1 by weight), but much better than that in spinach (7.5:1).

7.7 Enzymes

The use of exogenous enzymes such as phytase (see Section 7.8.), xylanase or β-glucanase as feed additives in feeding regimes of non-ruminants has led to significant improvements in feed efficiency and has increased the ability of animals to use a wide range of feed ingredients. Therefore, one objective of plant breeding is also to express various enzymes in plants by genetic modifications (see also Chapter 12).

Some studies have been done with potatoes expressing a β-glucanase gene from *Fibrobacter succinogenes* (Armstrong *et al.*, 2002; Baah *et al.*, 2002) or with maize expressing phytase to increase phosphorus utilization.

Armstrong *et al.* (2002) transferred an *F. succinogenes* 1,3-1,4 β-glucanase (1,3-1,4 β-D-glucan 4-glucanohydrolase) gene into potatoes and measured a specific activity in the leaves (1693 units/mg) and tubers (2978 units/mg β-glucanase), but the tuber yield in this study was reduced significantly by 28–72%. In some cases, 0.6 g GM potatoes/kg barley-based diets for broilers improved

Table 7.8. Experimental design to assess the conversion of β-carotene from maize into vitamin A in Mongolian gerbils (60% maize in diets; $n = 10$; depletion phase: 4 weeks; feeding: 8 weeks). (From Howe and Tanumihardjo, 2006.)

	Unsupplemented control (maize poor in carotene)	Maize rich in carotene	Control + β-carotene	Control + vitamin A
β-Carotene (nmol/g)	0	8.8	8.8	4.4
Theoretical retinol intake (nmol/day)	0	106	106	106
Retinol in serum (μmol/l)	1.23 ±0.20	1.25 ±0.22	1.23 ±0.20	1.22 ±0.16
Retinol in liver (μmol/g)	0.10[a] ±0.04	0.25[b] ±0.15	0.25[b] ±0.08	0.56[c] ±0.15

Notes: [a,b,c]Means with different letters differ ($p < 0.05$).

feed conversion and reduced ileal digesta viscosity (Baah *et al.*, 2002). The authors concluded that there might be potential for using transgenic potatoes to deliver enzyme(s) to improve poultry production, but that it might be necessary to improve the specific activity and/or the level of expression of the enzyme in the potato tuber in order to achieve consistent results.

7.8 Phytate and Phytase

Balance and feeding studies are necessary to demonstrate the efficacy of enzymes expressed in plants or to show the higher phosphorus (P) availability in plants with lower phytate content. Phytate (phytic acid) is one of the most important inhibitors of P availability in various plants. Phytic acid is a hexa-phosphorus-acid ester of the cycle alcohol inositol and it is called D-myo-inositol-(1,2,3,4,5,6)-hexakis dihydrogen phosphate. Six P atoms and some bivalent major and trace elements (e.g. Mg, Ca, Fe, Zn, Cu) can be found in phytic acid (Johnson and Tate, 1969). About 50–80% of the total P content is present in many seeds as phytate P (Eeckhout and De Paepe, 1994; Rodehutscord *et al.*, 1996). Most cereal seeds contain only low concentrations of the specific enzyme phytase and non-ruminants (pigs, poultry, humans) do not express phytase in their digestive tract, or only in very small amounts (Yang *et al.*, 1991; Angel

et al., 2002). As a consequence, most of the organically bound P passes through the digestive tract and can be found in considerable amounts in the faeces of those animals, and may contribute to environmental pollution in regions with high animal concentrations.

Apart from the supplementation of animal diets with inorganic P sources to meet the P requirements of animals, or to supplement microbially synthesized phytase as feed additive (Wodzinski and Ullah, 1996; see Chapter 11) to the diets, there are also some opportunities in plant and animal breeding to improve P availability, such as:

- Reduction of phytate synthesis in plant via plant breeding to create low phytate hybrids such as maize, barley, rice or soybeans (Spencer *et al.*, 2000a,b; Raboy 2002).
- Expression of phytase in plants and a higher bioavailability of phytate P (Chen *et al.*, 2008; Gao *et al.*, 2012).
- Expression of phytase in digestive tract of non-ruminants (Golovan *et al.*, 2001a,b; Cho *et al.*, 2006).

In a study with pigs (Spencer *et al.*, 2000a, b), low phytate maize showed the same results as traditional maize supplemented with 2 or 1.5 g inorganic P/kg feed, but a significantly lower P excretion (Table 7.9).

Gao *et al.* (2012) tested a phytase transgenic maize, an *Aspergillus niger*-derived phytase expressed in the endosperm of the

Table 7.9. Conventional and low-phytate maize (78.5% of the mixture) in the feed of fattening pigs. (From Spencer *et al.*, 2000a,b.)

Parameter	Control (0.3 g of available P/kg)		Low-phytate maize (1.7 g of available P/kg)	
Inorganic P supplement	−	+	−	+
P content (g/kg)				
29–73 kg live weight	3.4	5.4[d]	3.4	5.4[d]
73–112 kg live weight	3.2	4.7	3.2	4.7[e]
Feed intake (kg/day)	2.23[a]	2.50[b]	2.53[b]	2.51[b]
Live weight gain (g/day)	730	870[b]	900[b]	880[b]
Feed per gain (kg/kg)	3.05[a]	2.87[b]	2.81[b]	2.85[b]
P excreted (g/kg)	4.6[a]	8.9[c]	3.8[b]	8.8[c]
Strength (4th metacarpal bone, kg)	79.3[a]	138.5[b,c]	132.2[b]	153.9[c]
Ash content (% in 4th metacarpal bone)	53.5[a]	60.1[b,c]	59.3[b]	61.2[c]

Notes: [a,b,c]Different letters in one line indicate significant differences ($p < 0.05$); [d]+2.0 g P/kg; [e]+1.5 g P/kg.

maize (Chen *et al.*, 2008). In three experiments with brown roosters, phytase transgenic maize (phytase activity: 8037 FTU/kg) was compared with isogenic conventional maize (37 FTU/kg) and extraneous microbial phytase for enhancing the utilization of phytate phosphorus (Table 7.10).

The chemical composition of phytase transgenic maize and isogenic maize was not different, nor were differences observed in the energy content and the true amino acid availability of control maize and transgenic maize. The true P utilization of transgenic maize was significantly greater (55.8%) than for isogenic maize (37.9%). There were no differences in P utilization between animals fed phytase transgenic maize or supplemented with equivalent amounts of extraneous microbial phytase (see Table 7.10). Similar results have been reported in weanling pigs (Nyannor *et al.*, 2007) and in growing broilers (Zhang *et al.*, 2000; Nyannor and Adeola, 2008; Nyannor *et al.*, 2009). In addition, phytase expression in plants may also improve the bioavailability of trace elements such as iron, zinc, etc. (Lucca *et al.*, 2002).

Some recent studies show that genetically modified pigs (Golovan *et al.*, 2001a,b) and poultry (Cho *et al.*, 2006) are also able to express phytase via saliva. The so-called 'environmentally friendly pig' (Enviropig[TM]) carries a bacterial phytase gene under the transcriptional control of a gland-specific promoter, which allows the animals to digest certain amounts of plant phytate. However, Health Canada was unable to assess the safety of the GM pigs for human consumption; therefore, the University of Guelph, Canada, where the studies were done, lost its funding for Enviropigs[TM] and the last remaining animals from the 10th generation were euthanized on 24 May 2012 (Anon., 2012).

7.9 Conclusions

Biofortified plants may contribute to reducing micronutrient undernutrition in many parts of the world. Presently, nutritionally improved transgenic plants have not been fully developed (see Stein and Rodriguez-Cerezo, 2009) and tested for their potential to improve the micronutrient status of humans and animals. Many biofortified plants are still in the pipeline (see Chapter 12) and must be tested in animal feeding studies during the forthcoming years. More *in vitro* studies and animal experiments are necessary to assess the bioavailability of micronutrients in biofortified plants and to demonstrate the effects of further desirable ingredients such as enzymes. Furthermore, there is a need for better communication between plant breeders and human and animal nutritionists about the potentials of plant breeding for future improvements in nutrition and health.

More references on the feeding of biofortified transgenic crops to humans and livestock can be found at FASS (2012 and updated monthly).

Table 7.10. Influence of phytase transgenic maize on the true P utilization in maize compared with control maize and supplemented microbial phytase on the P availability in maize–soybean diets of roosters. (From Gao *et al.*, 2012.)

Parameter	Control maize	Control + microbial phytase[a]	Phytase transgenic maize[a]
True P utilization (%)	37.9		55.8
Available P (g/kg)	1.3		1.9
P utilization in maize–soybean diet (2.5:1; %)	49.9	70.2	72.8

Note: [a]5000 FTU/kg diet of microbial or maize-based phytase.

References

Amaro, P., Maja, M.R.G., Dewhurst, R.J., Fonseca, A.J.M. and Cabrita, A.R.J. (2012) Effects of increasing levels of stearidonic acid on methane production in a rumen *in vitro* system. *Animal Feed Science and Technology* 173, 252–260.

Angel, R., Tamin, N.M., Applegate, T., Dhandu, A.S. and Ellestad, L.E. (2002) Phytic acid chemistry: influence on phytin-phosphorus availability and phytase efficiency. *Journal of Applied Poultry Research* 11, 471–480.

Anon. (2012) Canada's 'enviropigs' killed for failing to bring profit. *Journal of Humanitarian Affairs* 6 June 2012, 1.

Armstrong, J.D., Inglis, G.D., Lawrence, M., Kawchuk, L.M., McAllister, T.A., Leggett, F., *et al.* (2002) Expression of a *Fibrobacter succinogenes* 1,3-1,4-β-glucanase in potato (*Solanum tuberosum*). *American Journal of Potato Research* 79, 39–48.

Avraham, T., Badani, H., Galili, S. and Amir, R. (2005) Enhanced levels of methionine and cysteine in transgenic alfalfa (*Medicago sativa* L.) plants over-expressing the *Arabidopsis crystathionine* gamma-synthetase gene. *Plant Biotechnology Journal* 3, 71–79.

Baah, J., Scott, T.A., Kawchuk, L.M., Armstrong, J.D., Selinger, L.B., Cheng, K.-L., *et al.* (2002) Feeding value in broiler chicken diets of a potato expressing a β-glucanase gene from *Fibrobacter succinogenes*. *Canadian Journal of Animal Science* 82, 111–113.

Barriere, Y., Dias Goncalves, J.D.G., Emile, J.C. and Lefevre, B. (2004) Higher intake of DK265 corn silage by dairy cattle. *Journal of Dairy Science* 87, 1439–1445.

Bauman, D.E. and Lock, A.L. (2006) Conjugated linoleic acid: biosynthesis and nutritional significance. In: Fox, P.F. and McSweeney, P.L.H. (eds) *Advanced Dairy Chemistry, Vol. 2: Lipids.* Springer, New York, pp. 95–135.

Beauregard, M. and Hefford, M.A. (2006) Enhancement of essential amino acid content in crops by genetic engineering and protein design. *Plant Biotechnology Journal* 4, 561–574.

Bernal-Santos, B., O'Donell, S.M., Vicine, J.L., Hartnell, G.F. and Bauman, D.E. (2010) Hot topic: Enhancing omega-3 fatty acids in milk fat of dairy cows by using stearidonic acid-enriched soybean oil from genetically modified soybeans. *Journal of Dairy Science* 93, 32–37.

Bertrand, J.A., Pardue, F.E. and Jenkins, T.C. (1998) Effects of ruminally protected amino acids on milk yield and composition of Jersey cows fed whole cottonseed. *Journal of Dairy Science* 81, 2215–2220.

Beyer, P., Al-Babili, S., Ye, X., Lucca, P. Schaub, Welsch, R., *et al.* (2002) Golden rice introducing the β-carotene biosynthesis pathway into rice endosperm by genetic engineering to defeat vitamin A deficiency. *Journal of Nutrition* 132, 506S–510S.

Böhme, H., Hommel, B. and Flachowsky, G. (2005) Nutritional assessment of silage from transgenic inulin synthesizing potatoes for pigs *Journal of Animal and Feed Sciences* 14, Suppl. 1, 333–336.

Böhme, H., Rudloff., E., Schöne, F., Schumann, W., Hüther, L. and Flachowsky, G. (2007) Nutritional assessment of genetically modified rapeseed synthesizing high amounts of mid-chain fatty acids including production responses of growing-finishing pigs. *Archives of Animal Nutrition* 61, 308–316.

Bouis, H.E. (2002) Plant breeding: a new tool for fighting micronutrient malnutrition. *Journal of Nutrition* 132, 491S–494S.

Bouis, H.E., Chassy, B.M. and Ochanda, J.O. (2003) Genetically modified food crops and their contribution to human nutrition and food quality. *Trends in Food Science and Technology* 14, 191–209.

Broadley, M.R., White, P.J., Bryson, R.J., Meacham, M.C., Bowen, H.C., Johnson, S.E., *et al.* (2006) Biofortification of UK food crops with selenium. *Proceedings of the Nutrition Society* 65, 169–181.

Cahoon, E.B., Hall, S.E., Ripp, K.G., Ganzke, T.S., Hitz, W.D. and Coughlan, S.J. (2003) Metabolic redesign of vitamin E synthesis in plants for tocotrienol production and increased antioxidant content. *Nature and Biotechnology* 21, 1082–1087.

Cahoon, E.B., Shockey, J.M., Dietrich, C.R., Gidda, S.K., Mullen, R.T. and Dyer, J.M. (2007) Engineering oilseeds for sustainable production of industrial and nutritional feedstocks: solving bottlenecks in fatty acid flux. *Current Opinion in Plant Biology* 10, 236–244.

Chen, R., Xue, G.X., Chen, P., Yao, B., Yang, W.Z., Ma, Q.L., *et al.* (2008) Transgenic maize plants expressing a fungal phytase gene. *Transgenic Research* 17, 633–643.

Chen, Z.H., Broderick, G.A., Luchini, N.D., Sloan, B.K. and Devillard, E. (2011) Effect of different sources of rumen-protected methionine on milk production and N-utilisation in lactating dairy cows. *Journal of Dairy Science* 94, 1978–1988.

Cherney, J.H., Cherney, D.J.R., Akin, D.E. and Axtell, J.D. (1991) Potential of brown-midrib, low-lignin mutants for improving forage quality. *Advanced Agronomy* 46, 157–198.

Cho, J., Choi, K., Darden, T., Reynolds, P.R., Petitte, J.N. and Shears, S.B. (2006) Avian multiple inositol polyphosphate phosphatase is an active phytase that can be engineered to help ameliorate the planets 'phosphate crisis'. *Journal of Biotechnology* 126, 248–259.

Christou, P. and Twyman, R.M. (2004) The potential of genetically enhanced plants to address food insecurity. *Nutrition Research Revue* 17, 23–42.

Colangelo, E.P. and Guerinot, M.I. (2006) Put the metal to the petal: metal uptake and transport throughout plants. *Current Opinion of Plant Biology* 9, 322–330.

Dawe, D., Robertson, R. and Unnevehr, L. (2002) Golden rice: What role could it play in alleviation of vitamin A deficiency? *Food Policy* 27, 541–560.

Delaney, B., Appenzeller, L.M., Munley, S.M., Hoban, D., Sykes, G.P., Malley, L., *et al.* (2008) Subchronic feeding study with high oleic acid soybeans (Event DP-3O5423-1) in Sprague-Dawlay rats. *Food and Chemical Toxicology*, DOI:10.1016/j.fct.2008.10.003.

DellaPenna, D. (1999) Nutritional genomics: manipulating plant micronutrients to improve human health. *Science* 285, 375–379.

DellaPenna, D. (2007) Biofortification of plant-based food enhancing folate levels by metabolic engineering. *Proceedings of National Academy of Sciences, USA* 104, 3675–3676.

DellaPenna, D. and Pogson, B.J. (2006) Vitamin synthesis in plants: tocopherols and carotenoids. *Annual Review of Plant Biology* 57, 711–738.

Diaz de la Garza, R.I., Quinlivan, E.P., Klaus, S.M., Basset, G.J., Gregory, J.F. and Hanson, A.D. (2004) Folate biofortification in tomatoes by engineering the pteridine branch of folate synthesis. *Proceedings of the National Academy of Sciences, USA* 101, 13720–13725.

Diaz de la Garza, R.I., Gregory, J.F. and Hanson, A.D. (2007) Folate biofortification of tomato fruit. *Proceedings of the National Academy of Sciences, USA* 104, 4218–4222.

Duvick, J. (2001) Prospects for reducing fumonisin contamination of maize through genetic modification. *Environmental Health Perspectives* 109 (Suppl. 2), 337–342.

Dyer, M.J., Stymme, S., Green, A.G. and Carlsson, A.S. (2008) High-value oils from plants. *The Plant Journal* 54, 640–655.

Edwards, H.M., Douglas, M.W., Parsons, C.M. and Baker, D.H. (2000) Protein and energy evaluation of soybean meals processed from genetically modified high-protein soybeans. *Poultry Science* 79, 525–527.

Eeckhout, W. and De Paepe, M. (1994) Total phosphorus, phytate phosphorus and phytase activity in plant feedstuffs. *Animal Feed Science and Technology* 47, 19–29.

EFSA (2008) Safety and nutritional assessment of GM plant derived food and feed. The role of animal feeding trials. *Food and Chemical Toxicology* 46, 2–70.

EFSA (2011) Scientific opinion on guidance for risk assessment of food and feed from genetically modified plants. *EFSA Journal* 9(5), 2150, 1–37.

Falco, S.C., Guida, T., Locke, J., Mauvals, J., Sanders, C., Ward, R.T., *et al.* (1995) Transgenic canola and soybean seeds with increased lysine. *Biotechnology* (NY) 13, 577–582.

FASS [Federation of Animal Science Societies] (2012) References – Feeding transgenic crops to livestock; dated May 2012.

Flachowsky, G. and Böhme, H. (2005) Proposals for nutritional assessments of feed from genetically modified plants. *Journal of Animal Feed Science* 14, 49–70.

Frossard, E., Bucher, M., Mächler, F.M., Mozafar, A. and Hurrell, R. (2000) Potential for increasing the content and bioavailability of Fe, Zn, and Ca in plants for human nutrition. *Journal of Science in Food and Agriculture* 80, 861–879.

Galili, G. (2002) New insights into the regulation and functional significance of lysine metabolism in plants. *Annual Review of Plant Biology* 53, 27–43.

Galili, G., Amir, R., Hoefgen, R. and Hesse, H. (2005) Improving the levels of essential amino acids and sulfur metabolites in plants. *Biological Chemistry* 386, 817–831.

Gao, C.Q., Ma, Q.C., Ji, C., Luo, X.G., Tang, H.F. and Wei, Y.M. (2012) Evaluation of the compositional and nutritional values of phytase transgenic corn to conventional corn in roosters. *Poultry Science* 91, 1142–1148.

GfE [Gesellschaft für Ernährungsphysiologie; Society for Nutrition Physiology] (1999) *Recommendations for Energy and Nutrient Requirements for Laying Hens and Broilers.* DLG-Verlag, Frankfurt, Germany, 185 pp. [In German.]

GfE (2001) *Recommendations for Energy and Nutrient Requirements for Dairy Cows.* DLG-Verlag, Frankfurt, Germany, 85 pp. [In German.]

GfE (2008) *Recommendations for Energy and Nutrient Requirements for Pigs.* DLG-Verlag. Frankfurt, Germany, 247 pp.

Gilligan, D.O. (2012) Biofortification, agricultural technology adoption, and nutrition policy: some lessons and emerging challenges. *CESifo Economic Studies* 58, 405–421.

Glenn, K.C. (2007) Nutritional and safety assessments of foods and feeds nutritionally improved through biotechnology lysine maize as a case study. *Journal of AOAC International* 90, 1470–1479.

Golovan, S.P., Hayes, M.A., Phillips, J.P. and Forsberg, C.W. (2001a) Transgenic mice expressing bacterial phytase as a model for phosphorus pollution control. *Nature Biotechnology* 19, 429–433.

Golovan, S.P., Meidinger R.G., Ajakaiye, A., Cottrill, M., Wiederkehr, M.Z., Barney, D.J., *et al.* (2001b) Pigs expressing salivary phytase produce low-phosphorus manure. *Nature Biotechnology* 19, 741–745.

Gorniak, T., Meyer, U. and Dänicke, S. (2012) Effects of corn silage of a Brown-midrib hybrid on dry matter intake, milk yield and milk composition in German Holstein cows compared to a common hybrid. *Proceedings of the Society of Nutrition Physiology* 21, 159 (Abstract).

Goto, F.T., Yoshihara, T., Shigemoto, N., Toki, S. and Tahaiwa, F. (1999) Iron fortification of rice seed by the soybean ferritin gene. *Nature Biotechnology* 17, 282–286.

Grant, R.J., Haddad, S.G., Moore, K.J. and Pedersen, J.F. (1995) Brown midrib sorghum silage for midlactation dairy cows. *Journal of Dairy Science* 78, 1970–1980.

Gregorio, G.B. (2002) Progress in breeding trace minerals in staple crops. *Journal of Nutrition* 132, 500S–502S.

Ha, S.H., Liang, Y.S., Jung, H., Ahn, M.J., Suh, S.C., Kweon, S.J., *et al.* (2010) Application of two bicistronic systems involving 2A and IRES sequences to the biosynthesis of carotenoids in rice endosperm. *Plant Biotechnology Journal* 8, 928–938.

Heger, J., Van Phung, T. and Krizowa, L. (2002) Efficiency of amino acid utilization in the growing pig at suboptimal levels of intake: lysine, threonine, sulphur amino acids and tryptophan. *Journal of Animal Physiology and Animal Nutrition* 86, 153–165.

Heger, J., van Phung, T., Krizowa, L., Sustala, M. and Simecek, K. (2003) Efficiency of amino acid utilization in the growing pig at suboptimal levels of intake: branched-chain amino acids, histidine and phenylalanine + tyrosine. *Journal of Animal Physiology and Animal Nutrition* 87, 62–65.

Hellwege, E.M., Czapla, S., Jahnke, A., Willmitzer, L. and Heyer, A.G. (2000) Transgenic potato (*Solanum tuberosum*) tubers synthesize the full spectrum of inulin molecules naturally occurring in globe artichoke (*Cynara scolymus*) roots. *Proceedings of the National Academy of Sciences (PNAS)* 97, 8699–8704.

Heppard, E.P., Kinney, A.J., Stecca, K.L. and Miao, G.-H. (1996) Developmental and growth temperature regulation of two different microsomal omega-6-desaturase genes in soybeans. *Plant Physiology* 110, 311–319.

Heyer, A.G., Lloyd, J.R. and Kossmann, J. (1999) Production of modified polymeric carbohydrates. *Current Opinion in Biotechnology* 10, 169–174.

Hirschi, K.D. (2009) Nutrient biofortification of food crops. *Annual Review of Nutrition* 29, 401–421.

Holm, P.B., Kristiansen, K.N. and Pedersen, B. (2002) Transgenic approaches in commonly consumed cereals to improve iron and zinc content and bioavailability. *Journal of Nutrition* 132, 514S–516S.

Hornung, E., Pernstich, C. and Feussner, I. (2002) Formation of conjugated Δ11, Δ13-double bounds by Δ12-linoleic acid (1-4)-acyl-lipid-desaturase in pomegranate seeds. *European Journal of Biochemistry* 269, 4852–4859.

Houmard, N.M., Mainville, I.I., Bonin, C.P., Huang, S., Luethy, M.H. and Malvar, T.M. (2007) High-lysine corn generated by endosperm-specific suppression by lysine catabolism using RNAi. *Plant Biotechnology Journal* 5, 605–614.

Howe, J.A. and Tanumihardjo, S.A. (2006) Cartenoid-biofortified maize maintains adequate vitamin A status in Mongolian gerbils. *Journal of Nutrition* 136, 2562–2567.

Huang, S., Kruger, D.F., Frizzi, A., D'Ordine, R.L., Florida, C.A., Adams, W.R., *et al.* (2005) High-lysine corn produced by the combination of enhanced lysine biosynthesis and reduced zein accumulation. *Plant Biotechnology Journal* 3, 555–569.

Huang, S., Frizzi, A., Florida, C.A., Kruger, D.F. and Luethy, M.H. (2006) High lysine and high tryptophan transgenic maize resulting from the reduction of both 19- and 22-kD alpha-zeins. *Plant Molecular Biology* 61, 525–535.

Hurrell, R.F. (1997) Bioavailability of iron. *European Journal of Clinical Nutrition* 51, S4–S8.

ILSI (2007) *Best Practices for the Conduct of Animal Studies to Evaluate Crops Genetically Modified for Output Traits.* International Life Sciences Institute, Washington, DC, 194 pp.

ILSI (2008) Nutritional and safety assessment of foods and feeds nutritionally improved through biotechnology: case studies. *Comprehensive Review of Food Science and Food Safety* 7, 50–99.

Ivan, S.K., Grant, R.J., Weakley, D. and Beck, J. (2005). Comparison of a corn silage hybrid with high cell-wall content and digestibility with a hybrid of lower cell-wall content on performance of Holstein cows. *Journal of Dairy Science* 88, 244–254.

Iwabuchi, M., Kohno-Murase, J. and Imamura, J. (2003) Δ12-oleate desaturase-related enzymes associated with formation of conjugated trans Δ11, cis-Δ13double bonds. *Journal of Biology and Chemistry* 278, 4603–4610.

Johns, T. (2003) Plant biodiversity and malnutrition: simple solutions to complex problems. *African Journal of Food, Agriculture, Nutrition and Development* 3, 45–52.

Johnson, L.F. and Tate, M.E. (1969) Structure of phytic acid. *Canadian Journal of Biochemistry* 47, 63–73.

Kidd, M.T., Kerr, B.J., Halpin, K.M., McWard, G.W. and Quarles, C.L. (1998) Lysine levels in starter and grower-finisher diets affect broiler performance and carcass traits. *Journal of Applied Poultry Research* 7, 351–358.

King, J.C. (2002) Evaluating the impact of plant fortification on human nutrition. *Journal of Nutrition* 132, 511S–513S.

Kitessa, S.M. and Young, P. (2011) Enriching milk fat with n-3 polyunsaturated fatty acids by supplementing grazing dairy cows with ruminally protected echium oil. *Animal Feed Science and Technology* 170, 35–44.

Koba, K., Imamura, J., Akashoshi, A., Kohno-Murase, J., Nishizono, S., Iwabuchi, M., *et al.* (2007) Genetically modified rapeseed oil containing cis-9,trans-11,cis-13-octadecatrienoic acid affects body fat mass and lipid metabolism in mice. *Journal of Agricultural and Food Chemistry* 55, 3741–3748.

Koehler, R., Leuoth, A., Jeroch, H., Flachowsky, G., Gebhardt, G., Hielscher, H., *et al.* (1989) Composition and digestibility of fibre fractions of various maize hybrids. *Archives and Animal Nutrition* 39, 187–192. [In German.]

Lee, T.T.T., Wang, M.M.C., Hou, R.C.W., Chen, L.I., Su, R.C., Wang, C.S., *et al.* (2003) Enhanced methionine and cysteine levels in transgenic rice seeds by the accumulation of sesame 25 albumin. *Bioscience, Biotechnology and Biochemistry* 67, 1299–1705.

Llorente, B., Alonso, G.D., Bravo-Almonacid, F., Rodriguez, V., Lopez, M.G., Carrari, F., *et al.* (2011) Safety assessment of non-browning potatoes: opening the discussion about the relevance of substantial equivalence on next generation biotech crops. *Plant Biotechnology Journal* 9, 136–150.

Lucas, D.M., Taylor, M.L., Hartnell, G.F., Nemeth, M.A., Glenn, K.C. and Davis, S.W. (2007) Broiler performance and carcass characteristics when fed diets containing lysine maize (Ly038 or Ly038 × Mon 810), control, or conventional reference maize. *Poultry Science* 86, 2152–2161.

Lucca, P., Hurrell, R. and Potrykus, I. (2001) Genetic engineering approaches to improve the bioavailability and the level of iron in rice grains. *Theoretical and Applied Genetics* 102, 392–397.

Lucca, P., Hurrell, R. and Potrykus, I. (2002) Fighting iron deficiency anaemia with iron-rich rice. *Journal of the American College of Nutrition* 21, 184S–190S.

Lunn, J. and Theobald, H.E. (2006) The health effects of dietary unsaturated fatty acids. *Nutrition Bulletin* 31,178–224.

Lynch, S.R. (2011) Why nutritional iron deficiency persists as a worldwide problem. *Journal of Nutrition* 141, 763S–768S.

McDowell, L.R. (2003) *Minerals in Animal and Human Nutrition*, 2nd edn. Elsevier, Amsterdam, 644 pp.

McKeon, T.A. (2003) Genetically modified crops for industrial products and processed and their effects on human health. *Trends in Food Science and Technology* 14, 229–241.

McNaughton, J., Roberts, M., Smith, B., Rice, D., Hinds, M., Sanders, C., *et al.* (2008) Comparison of broiler performance when fed diets containing event DP-3O5423-1, nontransgenic near-isoline control, or commercial reference soybean meal, hulls and oil. *Poultry Science* 87, 2549–2561.

Maruyama, N., Mikami, B. and Utsumi, S. (2011) The development of transgenic crops to improve human health by advanced utilization of seed storage proteins. *Bioscience, Biotechnology and Biochemistry* 75(5), 823–828.

Mayer, J.E., Pfeiffer, W.H. and Beyer, P. (2008) Biofortified crops to alleviate micronutrient malnutrition. *Current Opinion in Plant Biology* 11, 166–170.

Mejia, L., Jacobs, C.M., Utterback, P.L., Parsons, C.M., Rice, D., Sanders, C., *et al.* (2010) Evaluation of the nutritional equivalency of soybean meal with the genetically modified trait DP-3O5423-1 when fed to laying hens. *Poultry Science* 89, 2634–2639.

Molvig, L., Tabe, L.M., Eggum, B.O., Moore, A.E., Craig, S., Spencer, D., *et al.* (1997) Enhanced methionine levels and increased nutritive value of seeds of transgenic lupins (*Lupinus angustifolius*) expressing a sunflower seed albumin gene. *Proceedings of the National Academy of Sciences, USA* 94, 8389–8398.

Munkvold, G., Hellmich, R. and Rice, L.G. (1999) Comparison of fumonisin concentration in kernels of transgenic Bt maize hybrids and nontransgenic hybrids. *Plant Disease* 83, 130–138.

Muntz, K., Christov, V., Saalbach, G., Saalbach, I., Waddell, D., Pickardt, T., *et al.* (1998) Genetic engineering for high methionine grain legumes. *Nahrung* 42, 125–127.

Napier, J.A. (2007) The production of unusual fatty acids in transgenic plants. *Annual Review of Plant Biology* 58, 295–319.

Nelson, T.S., Kirby, L.K. and Halley, J.T. (1986) Digestibility of crystalline amino acids and the amino acids in corn and poultry blend. *Nutrition Report International* 34, 903–906.

Nestel, P., Bouis, H.E., Meenakshi, J.V. and Pfeiffer, W. (2006) Biofortification of staple food crops. *Journal of Nutrition* 136, 1064–1067.

Newell-McGloughlin, M. (2008) Nutritionally improved agricultural crops. *Plant Physiology* 147, 939–953.

NRC [National Research Council] (1994) *Nutrient Requirements of Poultry*, 9th revised edn. National Academy Press, Washington, DC, 176 pp.

NRC (1998) *Nutrient Requirements of Swine*, 10th revised edn. National Academy of Sciences, Washington, DC, 210 pp.

NRC (2001) *Nutrient Requirements of Dairy Cattle*, 7th revised edn. National Academy Press, Washington, DC, 408 pp.

Nyannor, E.K.D. and Adeola, O. (2008) Corn expressing an *Escherichia coli*-derived phytase gene: comparative evaluation study in broiler chicks. *Poultry Science* 87, 2015–2022.

Nyannor, E.K.D., Williams, P., Bedford, M.R. and Adeola, O. (2007) Corn expressing an *Escherichia coli*-derived phytase gene: a proof-of-concept nutritional study in pigs. *Journal of Animal Science* 85, 1946–1952.

Nyannor, E.K.D., Bedford, M.R. and Adeola, O. (2009) Corn expressing an *Escherichia coli*-derived phytase gene: residual phytase activity and microstructure of digesta in broiler chicks. *Poultry Science* 88, 1413–1420.

Oba, M. and Allen, M.S. (1999) Effects of brown midrib 3 mutation in corn silage on dry matter intake and productivity of high yielding cows. *Journal of Dairy Science* 82, 135–142.

Oba, M. and Allen, M.S. (2000a) Effects of brown midrib 3 mutation on corn silage on productivity of dairy cows fed two concentrations of dietary neutral detergent fiber: 1. Feeding behavior and nutrient utilization. *Journal of Dairy Science* 83, 1333–1341.

Oba, M. and Allen, M.S. (2000b) Effects of brown midrib 3 mutation in corn silage on productivity of dairy cows fed two concentrations of dietary neutral detergent fiber. 3. Digestibility and microbial efficiency. *Journal of Dairy Science* 83, 1350–1356.

OECD (1993) *Safety Evaluation of Foods Derived by Modern Biotechnology. Concepts and Principles.* OECD, Paris, France.

Okuley, J., Lightner, J., Feldmann, K., Yadav, N., Lark, E. and Browse, J. (1994) Arabidopsis FAD2 gene encodes the enzyme that is essential for polyunsaturated lipid synthesis. *Plant Cell* 6, 147–158.

Paine, J.A., Shipton, C.A., Chaggar, S., Howell, R.M., Kennedy, M.J., Vernon, G., *et al.* (2005) Improving the nutritional value of Golden Rice through increased pro-vitamin A content. *National Biotechnology* 23, 482–487.

Pappritz, J., Meyer, U., Kramer, R., Weber, E.-M., Jahreis, G., Rehage, J., *et al.* (2011) Effects of long-term supplementation of dairy cow diets with rumen-protected conjugated linoleic acids (CLA) on performance, metabolic parameters and fatty acid profile in milk fat. *Archives of Animal Nutrition* 65, 89–107.

Parsons, C.M. and Zhang, Y. (1997) Digestibility of amino acids in high-lysine soybean meal. *Poultry Science* 76 (Suppl. 1), 85 (Abstract).

Petry, N., Egli, I., Gahutu, J.B., Tugirimana, P.L., Boy, E. and Hurrell, R. (2012) Stable iron isotope studies in Rwandese women indicate that the common bean has limited potential as a vehicle for iron biofortification. *Journal of Nutrition* 42, 492–497.

Potrykus, I. (2001) Golden rice and beyond. *Plant Physiology* 125, 1157–1161.

Potrykus, I. (2003) Nutritionally enhanced rice to combat malnutrition disorders of the poor. *Nutrition Review* 61(6Pt.2), S101–S104.

Pribylova, R., Pavlik, I. and Bartos, A. (2006) Genetically modified potato plants in nutrition and prevention of diseases in humans and animals: a review. *Veterinarni Medicina* 51, 212–223.

Raboy, V. (2002) Progress in breeding low phytate crops. *Journal of Nutrition* 132, 503S–505S.

Ravindran, V., Tabe, I.M., Molvig, L., Higgins, T.J.V. and Bryden, W.I. (2002) Nutritional evaluation of transgenic high-methionine lupins (*Lupinus angustifolius* L.) with broiler chickens. *Journal of the Science of Food and Agriculture* 82, 280–285.

Robinson, P.H. (2010) Impacts of manipulating ration metabolizable lysine and methionine levels on performance of lactating dairy cows: a systematic review of the literature. *Journal of Livestock Science* 127, 115–126.

Robinson, P.H., Swanepoel, N. and Evans, E. (2010) Effects of ruminally protected lysine product with or without isoleucine, valine and histidine to dairy cows on their productive performance and plasma amino acid profiles. *Animal Feed Science and Technology* 161, 75–84.

Rocheford, T.R., Wong, J.C.W., Egesel, C.O. and Lambert, R.J. (2002) Enhancement of vitamin E levels in corn. *Journal of American College of Nutrition* 21, 191S–198S.

Rodehutscord, M., Faust, M. and Lorenz, H. (1996) Digestibility of phosphorus contained in soybean meal, barley, and different varieties of wheat, without and with supplemental phytase fed to pigs and additivity of digestibility in a wheat–soybean-meal diet. *Journal of Animal Physiology and Animal Nutrition* 75, 40–48.

Rook, J.A., Muller, L.D. and Shank, D.B. (1977) Intake and digestibility of brown midrib corn silage by lactating dairy cows. *Journal of Dairy Science* 60, 1894–1904.

Rymer, C., Hartnell, G.F. and Givens, D.I. (2011) The effect of feeding modified soyabean oil enriched with C18:4n-3 to broilers on the deposition of n-3 fatty acids in chicken meat. *British Journal of Nutrition* 105, 866–878.

Sandstorm, B. (1997) Bioavailability of zinc. *European Journal of Clinical Nutrition* 51, S17–S19.

Sautter, C., Poletti, S., Zhang, P. and Gruissem, W. (2006) Biofortification of essential nutritional compounds and trace elements in rice and cassava. *Proceedings of the Nutrition Society* 65, 153–159.

Schledz, M., Seidler, A., Beyer, P. and Neuhaus, G. (2001) A novel phytyltransferase from Synechocystis sp PCC 6803 involved in tocopherol biosynthesis. *FEBS Letter* 499, 15–20.

Sevenier, R., Hall, R.D., van der Meer, I.M., Hakkert, J.C., van Tunen, A.J. and Koops, A.J. (1998) High level fructan accumulation in a transgenic sugar beet. *Nature Biotechnology* 16, 843–846.

Sevenier, R., van der Meer, I.M., Bino, R. and Koops, A.J. (2002) Increased production of nutriments by genetically engineered crops. *Journal of the American College of Nutrition* 21, 199S–204S.

Shintani, D. and DellaPenna, D. (1998) Elevating the vitamin E content of plants through metabolic engineering. *Science* 282, 2098–2100.

Small, I. (2007) RNAi for revealing and engineering plant gene functions. *Current Opinion in Biotechnology* 18, 148–153.

Spencer, J.D., Allee, G.L. and Sauber, T.E. (2000a) Phosphorus bioavailability and digestibility of normal and genetically modified low-phytate corn for pigs. *Journal of Animal Science* 78, 675–681.

Spencer, J.D., Allee, G.L. and Sauber, T.E. (2000b) Growing-finishing performance and carcass characteristics of pigs fed normal or genetically modified low-phytate corn. *Journal of Animal Science* 78, 1529–1536.

Stein, A.J. and Rodriguez-Cerezo, E. (2009) The global pipeline of new GM crops: implications of asynchronous approval for international trade. JRC Scientific and Technical Reports. Office for Official Publications of the European Communities, Luxembourg, 102 pp.

Storozhenko, S., De Brouwer, V., Navarrete, M.V.O., Blancquaert, D. and Zhang, G.-F. (2007) Folate fortification of rice by metabolic engineering. *Nature Biotechnology* 25, 1277–1279.

Susenbeth, A. (2006) Optimum threonine:lysine ratio in diets for growing pigs: analysis of literature data. *Livestock Science* 101, 32–45.

Suttle, N. (2010) *Mineral Nutrition of Livestock*, 4th edn. CAB International, Wallingford, UK, and Cambridge, Massachusetts, 587 pp.

Tako, E., Laparra, M., Glahn, R.P., Welch, R.M., Lei, X.G., Beebe, S., et al. (2009) Biofortified black beans in a maize and bean diet provide more bioavailable iron to piglets than standard black beans. *Journal of Nutrition* 139, 305–309.

Tako, E., Rutzke, M.A. and Glahn, R.P. (2010) Using the domestic chicken (*Gallus gallus*) as an *in vivo* model for iron bioavailability. *Poultry Science* 89, 514–521.

Tang, G., Qin, J., Dolnikowski, G.G., Russell, R.M. and Grusak, M.A. (2009) Golden rice is an effective source of vitamin A. *The American Journal of Clinical Nutrition* 89, 1776–1783.

Tang, G., Hu, Y., Yin, S., Wang, Y., Dallal, G.E., Grusak, M.A. and Russell, R.M. (2012) β-carotene in golden rice is as good as β-carotene in oil providing vitamin A to children. *American Journal of Clinical Nutrition* 96, 658–664.

Tanumihardjo, S.A. (2002) Factors influencing the conversion of carotinoids to retinol: bioavailability to bioconversion to bioefficacy. *International Journal of Vitamin Nutrition Research* 72, 40–45.

Taylor, C.C. and Allen, M.S. (2005a) Corn grain endosperm type of brown midrib 3 corn silage: feeding behaviour and milk yield of lactating cows. *Journal of Dairy Science* 88, 1425–1433.

Taylor, C.C. and Allen, M.S. (2005b) Corn grain endosperm type and brown midrib 3 corn silage: ruminal fermentation and N partitioning in lactating cows. *Journal of Dairy Science* 88, 1434–1442.

Ufaz, S. and Galili, G. (2008) Improving the content of essential amino acids in crop plants: goals and opportunities. *Plant Physiology* 147, 954–961.

Ursin, V.M. (2003) Modification of plant lipids for human health: development of functional land-based omega-3 fatty acids. *Journal of Nutrition* 133, 4271–4274.

Vageeshbabu, H. and Chopra, V. (1997) Genetic and biotechnological approaches for reducing glucosinolates from rapeseed-mustard meal. *Journal of Plant Biochemistry and Biotechnology* 6, 53–62.

Van Eenennaam, A.L., Li, G., Venkatramesh, M., Levering, C., Gong, X., Jamieson, A.C., *et al.* (2004) Elevation of seed α-tocopherol levels using plant based transcription factors targeted to an endogenous locus. *Metabolic Engineering* 6, 101–108.

Van Jaarsveld, P.J., Faber, M., Tanumihardjo, S.A., Nestel, P., Lombard, C.J. and Spinnler-Benade, A.J. (2005) β-carotene-rich orange-fleshed potato improves the vitamin A status of primary school children assessed with the modified-relative-dose–response test. *American Journal of Clinical Nutrition* 81, 1080–1087.

Vasil, I. and Anderson, O. (1997) Genetic engineering of wheat gluten. *Trends in Plant Science* 2, 292–297.

Von Soosten, D., Meyer, U., Piechotta, M., Flachowsky, G. and Dänicke, S. (2012) Effect of conjugated linoleic acid supplementation on body composition, body fat mobilization, protein accretion, and energy utilization in early lactating dairy cows. *Journal of Dairy Science* 95, 1222–1230.

Wakasa, K., Hasenawa, H., Nemoto, H., Matsuda, F., Miyazawa, H., Tozawa, Y., *et al.* (2007) High-level tryptophan accumulation in seeds of transgenic rice and its limited effects on agronomic traits and seed metabolite profile. *Journal of Experimental Botany* 57, 3069–3078.

Welch, R.M. (2002) Breeding strategies for biofortified staple plant foods to reduce micronutrient malnutrition globally. *Journal of Nutrition* 132, 495S–499S.

Welch, R.M. and Graham, R.D. (2004) Breeding for micronutrients in staple food crops from human nutrition perspectives. *Journal of Experimental Botany* 55, 353–364.

Welch, R.M., House, W.A., Beebe, A., Senadhira, D., Gregorio, G.B. and Cheng, Z. (2000) Testing iron and zinc bioavailability in genetically enriched beans (*Phaseolus vulgaris* L.) and rice (*Oryza sativa* L.) in a rat model. *Food Nutrition Bulletin* 21, 428–433.

Whelan, J. (2009) Dietary stearidonic acid is a long chain (n-3) polyunsaturated fatty acid with potential health benefits. *Journal of Nutrition* 139, 5–10.

White, C.L., Tabe, L.M., Dove, H., Hamblin, J., Young, P., Phillips, N., *et al.* (2000) Increased efficiency of wool growth and live weight gain in Merino sheep fed transgenic lupin seed containing sunflower albumin. *Journal of the Science of Food and Agriculture* 81, 147–154.

White, P.J. and Broadley, M.R. (2005) Biofortifying crops with essential mineral elements. *Trends in Plant Science* 10, 586–593.

WHO [World Health Organization] (2008) *Vitamin and Mineral Requirements in Human Nutrition* 2nd edn. WHO, Geneva.

Wodzinski, R.J. and Ullah, A.H.J. (1996) Phytase. *Advances in Applied Microbiology* 42, 263–302.

Wu, G., Bazer, F.W., Burghardt, R.C., Johnson, G.A., Kim, S.W., Knabe, D.A., *et al.* (2010) Functional amino acids in swine nutrition and production. In: Doppenberg, J. and van der Aar, P. (eds) *Dynamics in Animal Nutrition*. Wageningen Academic Publishers, Wageningen, the Netherlands, pp. 69–98.

Wu, X.R., Chen, Z.H. and Folk, M.R. (2003) Enrichment of cereal protein lysine content by altered tRNA(lys) coding during protein synthesis. *Plant Biotechnology Journal* 1, 187–194.

Yang, W.J., Matsuda, Y., Sano, S., Masatani, H. and Nagagawa, H. (1991) Purification and characterization of phytase from rat intestinal mucosa. *Biochimica et Biophysica Acta* 1075, 75–82.

Ye, X.D., Al-Babili, S., Kloti, A., Zhang, J., Lucca, P., Beyer, P., *et al.* (2000) Engineering the provitamin A (beta-carotene) biosynthetic pathway into (carotenoid-free) rice endosperm. *Science* 287, 303–305.

Zhang, Z.B., Kornegay, E.-T., Radcliffe, J.S., Denbow, D.M., Veit, H.P. and Larsen, C.T. (2000) Comparison of genetically engineered microbial and plant phytase for young broilers. *Poultry Science* 79, 709–717.

Zimmermann, R., Stein, A. and Qaim, M. (2004) Agricultural technology to fight micronutrient malnutrition? A health economics approach. *Agrarwirtschaft* 53, 67–76. [In German.]

8 Long-term and Multi-generational Animal Feeding Studies

Agnes E. Ricroch,[1]* Aude Berheim,[1] Chelsea Snell,[1,2] Gérard Pascal,[3] Alain Paris[4] and Marcel Kuntz[5]

[1]*Université Paris-Sud, CNRS, AgroParisTech, Paris, France;* [2]*University of Nottingham, Loughborough, UK;* [3]*INRA, Saint Alyre d'Arlanc, France;* [4]*INRA, met@risk, AgroParisTech, Paris, France;* [5]*CNRS/CEA/Inra/Univ. Joseph Fourier, Grenoble, France*

8.1 Introduction

In the European Union, the European Food Safety Authority (EFSA) holds the responsibility to assess the safety of genetically modified (GM) food and feed. It recommends that:

> The safety assessment of GM plants and derived food and feed follows a comparative approach, i.e. the food and feed are compared with their non-GM counterparts in order to identify intended and unintended (unexpected) differences which subsequently are assessed with respect to their potential impact on the environment, safety for humans and animals, and nutritional quality.
>
> (EFSA, 2008)

The GM line is compared to its near-isogenic counterpart, according to specific determinants such as molecular characteristics and agronomic and phenotypic traits (see Chapters 3 and 4). When 'molecular, compositional, phenotypic, agronomic and other analyses have demonstrated equivalence of the GM food/feed, animal feeding trials do not add to the safety assessment' (EFSA, 2009; updated in EFSA, 2011; see Chapters 5 and 6). However, valuable information can be added to the assessment of GM food and feed safety by animal feeding studies, especially if elements are found that lead to the suspicion of possible undetected effects.

In the EFSA report, it is therefore suggested that:

> The use of 90-day studies in rodents should be considered for the detection of possible unintended effects in food and feed derived from GM plants which have been more extensively modified in order to cope with environmental stress conditions like drought or high salt conditions, or GM plants with quality or output traits with the purpose to improve human or animal nutrition and/or health.
>
> (EFSA, 2008; see Chapter 7)

The protocols for *in vivo* toxicological studies are adapted from the 90-day rodent study depicted in OECD Test Guideline No 408 (OECD, 1998). The guidelines describing the 90-day rodent study serve as a basis to define the experimental material; to test the practical conditions (target animal species, housing, number of doses administered, gender and number of animals, etc.); and to find the appropriate methods used to measure phenotypic responses (body weight, food consumption, clinical biochemistry, etc.) for *in vivo* toxicological studies. The last few decades have seen a growing presence of low molecular weight xenobiotics (drugs, pesticides, additives). This has led to a refinement and an improvement of toxicological assessments that constitute a solid foundation for the

**E-mail: agnes.ricroch@u-psud.fr*

evaluation of GM-based food or feed. One of the major problems concerning the adaptation of the test is the fact that feeds (diets) are comparable between animal groups (control or treated) in 90-day rodent studies, whereas in food safety assessment studies, the feeds themselves are tested. In most GM-based food or feed studies, 33% of GM feed is included in animal diets (see recommendations of the French Agency for Food, Environmental and Occupational Health and Safety, ANSES, 2011); the remaining part provides a balanced diet. Nevertheless, it should be kept in mind that 90-day animal feeding studies are designed to expose any change (e.g. a compositional change) in the GM feed, whether linked directly to the genetic modification (i.e. the transgene) or not (e.g. due to other genetic differences between plant varieties). Such a broad method leads to a drawback in which these studies may not actually be able to detect weak effects, as stated by the expert panel, 'It is unlikely that substances present in small amounts and/or with a low toxic potential will result in any observable unintended effects' (EFSA, 2008). For the testing of a specific molecule, it is possible to increase the dose in order to observe the biological effect; however, this is impossible for feed tests, as it would compromise the balance of the diet (see Chapter 5). Therefore, it is beneficial to examine whether tests beyond a 90-day rodent feeding study are needed.

Moreover, the EFSA states that:

> The subchronic, 90-day rodent feeding study is not designed to detect effects on reproduction or development, other than effects on adult reproductive organ weights and histopathology. Thus, in some cases, testing of the whole food and feed beyond a 90-day rodent feeding study may be recommended. In cases where structural alerts, indications from the subchronic study or other information on the whole GM plant derived food and feed are available that suggest the potential for reproductive, developmental or chronic toxicity, the performance of such testing should be considered.
>
> (EFSA, 2008)

In these specific cases, such late effects may not be detected by 90-day rodent feeding studies. Therefore, it is interesting to examine whether long-term studies performed over a period longer than 90 days) and multi-generational studies can detect unintended effects and whether the findings between such studies and 90-day feeding studies differ.

In this chapter, we assess critically recently published studies on the long-term effects of GM plants, i.e. studies significantly longer than the 90-day subchronic tests (17 studies), as well as multi-generational studies (16 studies). The GM feed were derived mainly from marketed insect-resistant (Bt) and herbicide-tolerant varieties, while some studies tested experimental GM lines. We examine whether these publications reveal adverse effects. The long-term studies and multi-generational studies are compared to 90-day studies that have already been performed. The possible need to update the current regulatory framework is discussed (see Chapter 5 for more of studies).

8.2 Results

8.2.1 Long-term studies

Long-term studies (longer than the classical 90- to 96-day feeding trials using rodents) are listed in Table 8.1 and discussed below. The duration of GM-based diet feeding varies between 110 days (16 weeks) and 775 days (110 weeks). Long-term studies that also involved mating/reproduction/offspring will be examined in the following section on multi-generational studies. Various animal models have been used such as mice, pigs, cows, quail, macaques and fish; however, rodents were the predominant models. Several criteria have been evaluated (body and organ weights, haematology analyses, enzymatic activities, histopathological observations of organs and detection of transgenic DNA). More details can be found in Snell et al. (2012). The studies presented in Table 8.1 concern

Table 8.1. Compilation of long-term animal feeding studies (>90 days).

Plant species/trait	Animal species/category (animals per treatment and duration of study)	Parameters measured	Main results	Authors' interpretation of results	References
Maize					
Bt 176 (*cry1Ab* gene)	Beef cattle (20) 246 days (35 weeks)	Feed intake, growth, slaughtering results.	No significant influence on feed intake, fattening and slaughtering results.	Silage from Bt 176 is similar to isogenic control maize.	Aulrich *et al.*, 2001
Bt-MON810 (*cry1Ab* gene)	Simmental dairy cows (9 primiparous, 9 multiparous) 25 months (765 days)	Feed intake, milk production and composition, body condition.	No significant influence on all parameters measured. All changes fall within normal ranges.	Safe, no long-term effects. Bt-MON810 and its isogenic control are equivalent.	Steinke *et al.*, 2010
Bt-MON810 (*cry1Ab* gene)	Pigs (cross-bred Large White × Landrace male pigs) (10–15) 110 days (16 weeks)	Feed intake, growth, characteristics and body composition. Heart, kidneys, spleen and liver weight and histological analysis. Blood and urine analysis.	No difference in overall growth, body composition, organ weight and histology and serum and urine biochemistry. A significant treatment × time interaction observed for serum urea, creatinine and aspartate aminotransferase.	Safe, no long-term effects. Differences observed in serum biochemistry within normal reference intervals, probably resulting from the lower enzyme-resistant starch content of GM.	Buzoianu *et al.*, 2012a
Bt (*cry1Ab* gene)	Japanese quail (60 or 30) 22 weeks	Test the effect of an active immunization using BSA as antigen, ELISA for total IgY and for BSA-specific IgY titres in eggs, serum zinc, percentage of eggs per hen.	Higher serum zinc concentrations in GM group than in the isogenic group and probably ranges within the normal variation of zinc serum concentrations in quails.	Possible causes for higher zinc serum concentrations in GM maize-fed group attributable to random differences in feed intake or zinc content. No effect of GM maize on laying of quails.	Scholtz *et al.*, 2010

Bt-MON810 (*cry1Ab* gene)	Atlantic salmon parr (20) 8 months (32 weeks)	Enzyme activities of maltase, leucine aminopeptidase (LAP) and acid phosphatase (AcP). Transport activity and protein expression in intestinal brush border membrane vesicles. Endocrine pancreatic response. Lysozyme activity. Immunoglobulin M.	No differences between GM and non-GM feed. GM and non-GM diets result in higher LAP activity compared to standard diet. Activity of maltase and AcP highest in standard diet.	The two GM varieties at inclusion levels of up to 6% (ingredient) appear to be as safe as the two counterparts.	Bakke-McKellep *et al.*, 2008
Rice					
7Crp#10 (*7Crp* gene derived from cedar pollen *Cryj I* and *Cryj II* allergen protein genes)	Macaques (11/sex) 182 days (26 weeks)	Gross necropsy, histopathology and absolute and relative organ weights. Blood composition (haematology; blood biochemistry).	No significant differences in haematological or biochemical values between treatment groups (a high dose of GM rice, a low dose of GM rice and a high dose of the parental rice strain).	No adverse effects on behaviour or body weight, haematological and biochemical variables. Neither pathological symptoms nor histopathological abnormalities observed.	Domon *et al.*, 2009
Soybean					
Glyphosate tolerant (*CP4-EPSPS* gene)	Wistar rats (10 males) 455 days (65 weeks)	Growth. Blood composition.	No significant differences between GM group and organic group.	Safe, no long-term effects.	Daleprane *et al.*, 2009a
Glyphosate tolerant (*CP4-EPSPS* gene)	Wistar rats (10) 455 days (65 weeks)	Aorta wall tissue. Cholesterol, triaclyglycerol, insulin, glucose and testosterone.	Lower body weight and fat mass in control. Cholesterol, triaclyglycerol, glucose and aortic tunics reduced in non-GM and GM.	Safe, no long-term effects.	Daleprane *et al.*, 2010
Glyphosate tolerant (*CP4-EPSPS* gene) Cv. 90B72 (near-isogenic line)	F344/Du Crj (Fischer) rats (26-week diet: 20/sex; 52-week diet: 10/sex) Two durations: 26 and 52 weeks	Feed intake. Growth. Organ weight. Haematology, serum. Histology. Eosinophils and goblet cells in jejunal mucosa.	Minor differences between GM- and non-GM-fed males in some blood biochemical parameters (26-week groups). Minor differences between GM- and non-GM-fed males in heart and spleen weight (52-week groups).	Safe, no long-term effects.	Sakamoto *et al.*, 2007

Table 8.1. Continued

Plant species/trait	Animal species/category (animals per treatment and duration of study)	Parameters measured	Main results	Authors' interpretation of results	References
Glyphosate tolerant (*CP4-EPSPS* gene) Cv. 90B72 (near-isogenic line)	F344/Du Crj (Fischer) rats (GM and non-GM-diet: 50/sex; standard CE-2 diet: 35/sex) 104 weeks	Feed intake. Growth. Organ weight. Haematology, serum. Neoplastic and non-neoplastic lesions. Bile duct proliferation. Basophilic, clear and eosinophilic altered hepatocellular foci. Chronic nephropathy.	Minor differences between GM- and non-GM-fed males and between GM- and non-GM-fed females in some haematological parameters.	Safe, no long-term effects.	Sakamoto *et al.*, 2008
Glyphosate tolerant (*CP4-EPSPS* gene)	Salmon (24) 7 months (28 weeks)	Growth. Body weight. Organ development. Histology. Haematology, plasma enzymes and nutrients. mRNA transcription. Differential white blood cell count.	No growth differences. Glycogen deposits in liver decreased in the GM-fed fish. Minor differences observed between the diet groups.	Safe, no long-term effects.	Sissener *et al.*, 2009
Glyphosate tolerant (*CP4-EPSPS* gene) (No isogenic line compared to GM soybean)	Atlantic salmon (20) 8 months (32 weeks)	Enzyme activities of maltase, leucine aminopeptidase, acid phosphatase (AcP) and fructose-1,6-bisphosphatase activity. Transport activity and protein expression in intestinal brush border membrane vesicles. Endocrine pancreatic response. Lysozyme levels. Total amount of immunoglobulin M.	The increased AcP activity reported in GM soybean-fed fish most likely due to increased macrophage infiltration. Highest values in intestinal Na$^+$-dependent D-glucose uptake and SGLT1 protein level in the region pyloric caeca found in GM, intermediate in the non-GM and lowest in the standard fed groups.	Atlantic salmon parr appear to tolerate a 12.5% inclusion level of full-fat soybean meal. GM soybeans at inclusion levels of up to 6% appear to be as safe as counterparts. Differences due to origin as well as heat treatment of the non-GM soybean varieties.	Bakke-McKellep *et al.*, 2008
Glyphosate tolerant (*CP4-EPSPS* gene) (No isogenic line compared to GM soybean; 'wild type')	Swiss mice (12 females) 1, 2, 5 or 8 months (4 weeks to 30 weeks)	Ultrastructural morphetrical and immunocytochemical analyses of hepatocyte nuclei.	Irregularly shaped nuclei, higher number of nuclear pores, numerous small fibrilla centres and abundant dense fibrillar component, nucleoplasmic and nuclear splicing factor more abundant in GM-fed mice.	Higher metabolic rate and molecular trafficking. Influence of GM soybean intake on hepatocyte nuclear features in young and adult mice (mechanisms unknown).	Malatesta *et al.*, 2002a

Glyphosate tolerant (*CP4-EPSPS* gene) (No isogenic line compared to GM soybean; 'wild type')	Swiss mice (12 females) 240 days (30 weeks)	Histocytochemistry pancreatic acinar cells.	No differences in body weight and no macroscopic changes in the pancreas. No structural modifications but quantitative changes in some cellular constituents. Reduction of α-amylase synthesis.	A diet containing significant amounts of GM food seems to influence the zymogen synthesis and processing in pancreatics acinar cells (reasons remain unknown).	Malatesta *et al.*, 2002b
Glyphosate tolerant (*CP4-EPSPS* gene) (No isogenic line compared to GM soybean; 'wild type')	Swiss mice (12 females) 1, 2, 5 or 8 months (4 weeks to 30 weeks)	Ultrastructural morphetrical and immunocytochemical analyses of pancreatic acinar cell nuclei.	Decrease of the shape index and the fibrillar centre density and increase of the pored density, the perichromatin granule density, the percentage of fibrillar centres in GM-fed mice. Lower labelling for the nucleoplasmic splicing factors.	A diet containing significant amounts of GM food seems to influence the pancreatic metabolism (reasons remain unclear).	Malatesta *et al.*, 2003
Glyphosate tolerant (*CP4-EPSPS* gene) (No isogenic line compared to GM soybean; 'wild type')	Swiss mice (12 females) 1, 2, 5 or 8 months (4 weeks to 30 weeks)	Enzyme chemistry of serum, liver and pancreas.	Enlarged vesicles of the smooth endoplasmic reticulum. Decrease in the number of nuclear pores. Reduced labelling during the 2–8 month interval. Increase in perichromatin granules in Sertoli cells and in spermatocytes of GM-fed mice.	A transient transcriptional decrease during the 2–8 month interval. Most of the effects reversible. Causes of the alteration not established, especially because glyphosate residues might influence transcriptional process.	Vecchio et al., 2004
Glyphosate tolerant (*CP4-EPSPS* gene) (No isogenic line compared with GM soybean; 'wild type')	Swiss mice (10 females) 2 years (104 weeks)	Histocytochemistry of hepatocytes. Total protein content of the liver (2-DE).	Different expression of proteins related to hepatocyte metabolism, stress response, calcium signalling and mitochondria in GM-fed mice. Indications of reduced metabolic rate in GM-fed mice.	GM-soybean intake can influence some liver features during ageing (mechanisms remain unknown).	Malatesta *et al.*, 2008

feeding studies with first-generation GM plants with agronomic traits (see Chapter 6); one study uses a plant line with an output trait (see Chapter 7).

It should be noted that none of the five papers published by Malatesta and colleagues (including Vecchio *et al.*, 2004) explicitly states the exact identity of the soybean lines used. The control/non-GM plant materials used in these studies are unlikely to come from isogenic lines or grown in the same location, as already discussed (Snell *et al.*, 2012). Therefore, the differences claimed in these studies cannot be interpreted meaningfully as resulting from the genetic modification. Unfortunately, in 3 of the remaining 11 long-term studies summarized in Table 8.1 (using soybean), which show no or little long-term effects, it is unclear whether near-isogenic lines were used (most of the time the transformation event was not specified), making it possible to argue that they did not comply with the required standards to compare GM and non-GM soybean soundly (Bakke-McKellep *et al.*, 2008; Daleprane *et al.*, 2009a, 2010).

Accompanied by a high-profile media campaign, a publication by Séralini *et al.* (2012) claimed that the glyphosate-tolerant GM maize, NK603, treated or not by a herbicide formulation, caused organ damage, tumours and an earlier death among rats fed this maize variety for 2 years. This publication is not included in Table 8.1, since it has been refuted by nine food safety agencies from 11 countries (the German agencies, BVL and BfR, the Food Standards Australia and New Zealand, the Danish agency, DTU, the Netherlands agency, NVWA, the National Biosafety Technical Commission of Brazil, Health Canada (Federal department) and the Canadian food inspection agency, the Belgian Biosafety Advisory Council, the French agencies, ANSES and the High Council of Biotechnologies), and by the European authority, the EFSA, six national academies of France, the European Society of Toxicologic Pathology and the French Society of Toxicologic Pathology, as well as by many scientists.

What can be learned from long-term studies?

In 2009, the BEETLE report on the long-term effects of GM crops on health and the environment, subtitled 'Prioritisation of potential risk and delimitation of uncertainties', analysed the scientific literature and collected via an online survey the contributions of a wide range of experts. The literature review did not find evidence of long-term health effects, and the expert survey confirmed this view while recommending methodical improvements of the risk assessment procedure, including a higher number of replications and additional control groups to demonstrate the biological range of measured parameters (see also Chapter 5 and Flachowsky *et al.*, 2012).

The general conclusion drawn from the present compilation was that no biologically significant differences or adverse health effects were reported. No new safety concerns were raised by the authors. These studies are in line with the previously demonstrated nutritional equivalence between the studied GM varieties (most of them being commercial products subjected to a pre-marketing safety assessment) and their non-GM conventional counterparts (see Chapter 6). It is important to draw attention to the diversity of the animal models used (rat, mouse, cattle, pig, macaque, quail and salmon), as well as the varying feeding durations. These long-term feeding studies cover the whole lifespan or a very long life period of some animals. Other studies (e.g. in the case of laying hens or dairy cows) cover a longer period of time than normally used in classical nutritional studies and are expected to be able to reveal the presence of toxic compounds in feed (and not only assess their nutritional quality).

8.2.2 Multi-generational studies

Multi-generational studies were performed on animals that were fed GM-based diets throughout their whole lives or only on a short-term (less than 90 days), or long-term (more than 90 days), basis, but in both cases these animals were bred to produce future generations (studies performed on two to ten generations). Further details can be found in Snell *et al.* (2012). Farm animals (dairy cows, bulls, goats, pigs, sheep, hens and quails) or rodents were used (see Chapter 5). Parameters measured included body weight, feed intake, detection of DNA from the GM plant in animal organs, enzyme concentrations or activities and some reproductive parameters. The main goal of these studies was to assess whether feeding a generation (n) with a GM-based diet had adverse effects on subsequent generations ($n + x$). The studies presented in Table 8.2 concern feeding studies with first-generation GM plants with agronomic traits (see Chapter 6); one study used a plant line with an output trait (see Chapter 7).

What can be learned from multi-generational studies?

Multi-generational (or trans-generational) studies (most of them over several generations) were carried out to test the influence of GM feed on reproduction, long-term health and metabolic effects in laboratory and target animals. In laboratory animals, no negative effects were described for growth, testicular cells or reproductive traits in mice fed Bt maize, a glyphosate-tolerant soybean or a transgenic triticale grain tolerant to the herbicide, glufosinate, when compared with conventional maize, soybean or triticale (Brake and Evenson, 2004; Brake *et al.*, 2004; Baranowski *et al.*, 2006). Rats and their offspring were not influenced significantly in a five-generation study if fed 5% GM glufosinate-tolerant potatoes or conventional potatoes (Rhee *et al.*, 2005). Kiliç and Akay (2008) found no differences in the organ weights of the offspring and no differences in the reproduction rates of rats fed up to 20% Bt maize or conventional maize. Krzyzowska

et al. (2010) fed pellets containing 20% control triticale or 20% glufosinate-tolerant triticale to mice for five consecutive generations and found some changes in lymph nodes and in immune response (increased IL-2 levels and decreased IL-1 levels) in the fifth generation. Trabalza Marinucci *et al.* (2008) and Tudisco *et al.* (2010) reported some minor metabolic changes when comparing glyphosate-tolerant soybean and Bt maize, respectively, to their control. The results of the latter study did not demonstrate any health hazards, but the authors suggested that these changes should be investigated further. It would be particularly interesting to know whether or not these changes were reproducible.

In target animals, Buzoianu *et al.* (2012b) specifically examined pig offspring at birth, while the publication of the same team (2012c) examined pigs for 115 days post-weaning. Interestingly, the longest multi-generational study consisted of feeding laying quail with a diet containing up to 50% Bt maize over ten generations. Bt maize did not influence significantly the production and reproductive performances of animals compared with animals fed a diet containing 50% isogenic maize. Unfortunately, further multi-generational studies using food-producing animals are missing (Flachowsky *et al.*, 2012).

It should be highlighted that some of these studies suffer from serious weaknesses such as lack of an appropriate control group (see Table 8.2), which could be the main reason for the observed differences. Statistical criticisms of these studies can also be raised, especially as the EFSA (2010) have underlined the necessity for an improved methodology when statistics are involved: poor definition of a control (or group control), weak definition of factor levels, lack of a complete combination of factors inside experimental designs, no evaluation of the statistical power, as well as too few multivariate approaches, are weaknesses that have often been observed in these studies.

Table 8.2. Compilation of multi-generational animal feeding studies.

Plant species/trait	Animal species (animal number per group and duration)	Parameters	Main results	Authors' interpretation of results	Reference
Maize					
Bt 11 (38PO6), non-transgenic control (38PO5)	Mice (10 females per diet), (3 males chosen in progeny for each of six time points), 8, 16, 26, 32, 63 and 87 days after birth (from approximately 1 week to approximately 12 weeks) 4 generations	Testicular development, litter size, body weight.	No differences in fetal, postnatal, pubertal, or adult testicular development with the GM diet.	Safe, no multi-generational effects.	Brake et al., 2004
Bt 176 (cry1Ab gene)	Quails (weeks 1–6: 70 males, 75 females; weeks 7–12: 32 hens) 10 × 12 weeks; 840 days 10 generations	Growing (6 weeks), laying performance, reproduction, hatchability (210 eggs per hatch).	No biologically relevant influence on growth, laying performance and hatchability.	Safe, no multi-generational effects.	Flachowsky et al., 2005; Flachowsky et al., 2007
Bt 176 (cry1Ab gene)	Laying hens (weeks 1–18: 36; weeks 19–31: 18) 4 × 31 weeks; 868 days 4 generations	Growing (18 weeks), laying performance, hatchability.	No significant differences in growth, laying performance and hatchability.	Safe, no multi-generational effects.	Halle et al., 2006
Bt (event not specified) (Isogenic line used)	Wistar albino rats (F0: 18 females, 6 rats/each group, mated with 9 males, one male for two females; F1, F2 and F3 generations obtained by same procedure) (19–37 individuals in progeny were examined according to generation and diet composition)	Histological and biochemical parameters characterizing stomach, duodenum, liver, kidney.	No differences in organ weights. Some minor histological changes in liver and kidney (change in creatinine, total protein and globulin level).	Changes are minor and do not threaten the health of rats. Suggestion that long-term feeding studies be performed on other species. Collaboration with new improving technologies is needed to assure their safety.	Kiliç and Akay, 2008

	Duration not specified but at least 3.5 months (14 weeks) Dams and their offsprings fed with the diets during the periods of mating, gestation, lactation, offspring care and pubescence. F3 rats fed until they reached 3.5 months in age. F0 + 3 generations				
Bt 176 (*cry1Ab* gene)	Sheep (53 Bergamasca × Appenninica ewes and their progeny for 3 years) 44 months (188 weeks) F0 + 3 generations	Immune response, ruminal metabolism, microbial population, meat quality, microscopy, transgene detection.	Differences only observed in some cytosolic observations (liver and pancreas cell nuclei) and immune response to Salmonella vaccination. Significance and reproducibility of these phenomena unclear.	Small effects. Suggestion for more metabolic research, in particular in gastrointestinal organs and the immune response mechanisms.	Trabalza-Marinucci *et al.*, 2008
Bt 11 (*cry1Ab*) gene	Mice (F0: 99 females, 117 males fed non-GM maize; 108 females and 108 males fed GM maize. F1: 21 females and 17 males fed non-GM maize; 52 females and 55 males fed GM maize) Organ weight: 16 GM-fed mice and 22 non-GM; Lifespan: 45 GM-fed mice and 44 non-GM mice 1072 days (approximately 153 weeks) F0 + 4 generations	Growth. Gestation, milking periods, reproduction, lifespan. Breeding performance in 4 generations.	No difference in any parameters.	Safe, no multi-generational effects.	Haryu *et al.*, 2009

Continued

Table 8.2. Continued

Plant species/trait	Animal species (animal number per group and duration)	Parameters	Main results	Authors' interpretation of results	Reference
Bt-MON810 (*cry1Ab* gene)	Pigs (cross-bred Large White × Landrace) (12) Fed for ~143 days (20 weeks) throughout gestation and lactation F0 + 1 generation (offspring at birth)	Haematological and immune functions to detect possible inflammatory and allergenic responses at various times. Attempts to detect Cry1Ab protein in blood and faeces at various times.	Cytokine production similar between treatments. Some differences in monocyte, granulocyte or lymphocyte subpopulations counts at some times, but no significant patterns of changes.	No indication for inflammation or allergy due to GM maize feeding.	Buzoianu *et al.*, 2012b
Bt-MON810 (*cry1Ab* gene)	Pigs (cross-bred Large White × Landrace) (10) Maize dietary inclusion rate identical between treatments (isogenic parent line maize from service to weaning and GM maize from service to weaning (Bt)) and ranged from 86.6% during gestation to 74.4% during lactation) Offspring fed for 115 days in 4 dietary treatments (20 weeks) F0 + 1 generation	Pig growth performance body weight and feed disappearance recorded at the time of each dietary change (weaning (day 0), on day 30, 70, and 100) and at harvest (day 115). At harvest, organ weight, histological observations, cold carcass weight. Serum biochemistry.	No pathology observed in the organs. Offspring of sows fed Bt maize had improved growth throughout their productive life compared to offspring of sows fed non-GM maize, regardless of the maize line fed between weaning and harvest. Some minor differences in average daily gain, carcass and spleen weights, dressing percentage, duodenal crypt depths for offspring from GM fed or in average daily feed intake for offspring from sows fed GM and for GM-fed pigs or in liver weight for pigs in the Bt/Bt.	Trans-generational consumption of GM maize diets not detrimental to pig growth and health.	Buzoianu *et al.*, 2012c
Bt-MON810 (*cry1Ab* gene)	Zebrafish (*Danio rerio*) 2 generations, 3–5 tanks per treatment (14 fish per tank), 19% maize in feed; switch over of feed in offspring	Feed intake, growth, weight of organs, gene and enzyme expressions.	No significant effects on growth for both generations. Bt maize is as safe and nutritious as non-Bt control when fed to zebrafish for 2 generations.	Effects on gene biomarkers for oxidative stress and cell cycle (apoptosis) may be related to mycotoxins in non-Bt maize.	Sanden *et al.*, 2013

Glufosinate ammonium tolerant T25 event	Wistar rats (10) Total of 630 adults and 2837 pups. Five groups fed with GM, control near-isogenic or three other conventional maize lines. 25 days F0 and 2 generations	Feed intake, body and organ weights, macroscopy and histopathology. Haematology. Blood biochemistry. Immunology. Permeability of the intestinal barrier. Locomotor behaviour.	No difference in any parameters.	No impact of GM maize on reproductive function of rats and on progeny development.	Tyshko *et al.*, 2011
Potato Glufosinate ammonium tolerant	Sprague-Dawley rats (25 males and 25 females) 10 weeks 5 generations (F0–F4)	Presence of DNA. Feed consumption. Developmental and reproductive performance.	No difference in any parameters.	Safe, no multi-generational effects.	Rhee *et al.*, 2005
Soybean Glyphosate tolerant (*CP4-EPSPS* gene; event GTS 40-3-2)	8–87 days (from approximately 1 week to approximately 12 weeks; with pregnant mice and male mice) 4 generations	Testicular development in mouse model.	No differences in fetal, post-natal, pubertal or adult testicular development.	Safe, no multi-generational effects.	Brake and Evenson, 2004
Glyphosate tolerant (*CP4-EPSPS* gene; event GTS 40-3-2)	Wistar rats (10 rats) Fed throughout life (duration not precise) in both generations (F0 and F1) 2 generations	Weight gain, ration intake, protein intake and efficacy ratio.	Some differences between experimental and control.		Daleprane *et al.*, 2009b
Glyphosate tolerant (*CP4-FPSPS* gene; event GTS 40-3-2)	Dairy goats (10) 60–67 days (approximately 8–9 weeks) F0 + 1 generation	DNA in milk and blood, and other parameters.	No weight differences. No pathological manifestations. Some possible effects on metabolism. Presence of transgenic DNA in milk (parents) and blood (parents and offspring) reported but doubtful.	LDH[a] modifications suggest a rise of the cell metabolism. No health issue but further studies should be undertaken.	Tudisco *et al.*, 2010

Continued

Table 8.2. Continued

Plant species/trait	Animal species (animal number per group and duration)	Parameters	Main results	Authors' interpretation of results	Reference
Triticale					
Glufosinate ammonium tolerant	C57Bl/6J mice (41–60 according to treatment and generation) 91-day-old mated/killed (at each generation) The next generation, up to F5, over all periods of life (including gestation and lactation), fed only with control or treatment pellets 5 generations	Body weight and growth. Presence of transgenic DNA. Pathological manifestations.	No presence of transgenic DNA. No weight differences. No pathological manifestations.	Safe, no multi-generational effects.	Baranowski *et al.*, 2006
Glufosinate ammonium tolerant	C57BL/6J mice (20) 120 days then mated/killed (at each generation) (approximately 17) 5 generations	Immune system.	In F5 enlarged inguinal and axillary lymph nodes detected. Decrease in T cells in spleen and lymph nodes and decrease in B cells in lymph nodes and blood.	Changes not caused by an allergy or a malignant process but further studies should be undertaken.	Krzyowska *et al.*, 2010
Rice					
The gene encoding lysine-rich protein (LR) from winged bean inserted into the glutelin (Gt1)-encoding region of rice	Sprague-Dawley rats (20 females and 10 males) (20 weanlings/female/group and 10 weanlings/male/group as F1 generation. F2 generation the same procedure. F3 generation: 10 rats/sex/group) 17 weeks (females) and 9 weeks (males, then mated/killed) of diet for F1 and F2 generations; 13 weeks of diet for F3 generation (males and females) 3 generations	Body weight, food consumption, reproductive data, organ weight, ratio and pathology, haematology, serum chemistry, serum sex hormone levels. Gross and anatomic pathology.	No adverse behaviour or clinical effects on F0, F1, F2, F3 generation animals. Mean corpuscular volume values 2% higher in F3 female rats in the GM-rice diet compared with control diet: no indication of toxicology. In serum chemistry, slight differences not considered biologically significant or related to exposure to GM rice.	GM rice as safe as near-isogenic non-GM rice.	Zhou *et al.*, 2012

Note: [a]LDH = lactate dehydrogenase.

8.3 Discussion

8.3.1 Funding of long-term feeding studies

We compiled a total of 33 publications (17 long-term feeding trials and 16 multi-generational feeding trials; Tables 8.1 and 8.2). All studies were conducted by public research laboratories. Nine out of the 17 long-term studies performed (Aulrich *et al.*, 2001; Malatesta *et al.*, 2002a,b, 2003; Vecchio *et al.*, 2004; Daleprane *et al.*, 2009a, 2010; Domon *et al.*, 2009; Steinke *et al.*, 2010) and 8 out of the 15 multi-generational studies (Brake *et al.*, 2004; Rhee *et al.*, 2005; Halle *et al.*, 2006; Flachowsky *et al.*, 2007; Kiliç and Akay, 2008; Haryu *et al.*, 2009; Krzyzowska *et al.*, 2010; Tudisco *et al.*, 2010) did not mention any specific funding. It was therefore assumed that all of the studies compiled here were publicly funded.

8.3.2 Short-term versus long-term feeding studies

The evaluation of GM-based diets relies on the general principles depicted in the OECD Test Guideline (1998) or as discussed by EFSA (2008):

1. The principle of substantial equivalence, in which the goal is to make possible the comparison of chemical composition in macro- and micronutrients and known anti-nutrients and natural toxicants between GM lines and unmodified near-isogenic lines.
2. The toxico-nutritional response in subchronic toxicity tests of animals fed either a GM-based diet or a control diet, and if necessary, long-term or multi-generational studies.

Usually, the comparison of GM lines (with no deliberate metabolic modification) and their comparator shows their nutritional equivalence. The whole process of production of GM commercial lines, including selection in laboratory and field trials by comparison with known non-GM lines on various phenotypic traits, contributes strongly to

nutritional equivalence and to the food safety of such lines (see Chapter 5 for guidance documents). However, if the nutritional equivalence is still in doubt, it is advised by some experts to perform subchronic toxicity 90-day tests to assess this uncertainty. Considering the step-by-step framework for risk assessment, long-term and multi-generational studies would only be performed if doubt remained after a subchronic toxicity 90-day study.

Some of the above-mentioned GM events have also been subjected to 90-day studies on rats (to the best of our knowledge, such studies have not been published on 'old' events such as Bt 11 and Bt 176; however, short-term studies using various farm animals are available for Bt 11; see Chapter 6). Notably, no biologically significant differences were found between a GM diet and a non-GM diet in 90-day studies using glyphosate-tolerant maize (CP4-EPSPS gene, event NK603; Hammond *et al.* 2004), glufosinate-tolerant maize (Mackenzie *et al.* 2007; Malley *et al.*, 2007) or insect-resistant maize (*cry1Ab* gene, event MON810, Hammond *et al.*, 2006). Similar results were obtained using glyphosate-tolerant soybean but which incorporated another gene (DP356043 lines; Appenzeller *et al.*, 2008) in addition to CP4-EPSPS, which has been subjected to several long-term studies. The latter was assessed in short-term animal feeding studies, including two independent 4-week studies in rats (one with unprocessed and one with processed soybeans), a 4-week dairy cow study, a 6-week chicken study, a 10-week catfish study and a 5-day quail study (CERA GM crop database).

Thus, no evidence is currently available indicating that long-term feeding studies of marketed GM crops could detect adverse effects that remained undetected by short-term studies.

8.3.3 Exploratory studies in the context of a step-by-step approach

Most of the studies compiled here were not performed in a step-by-step approach as part of the pre-marketing regulatory process, but had exploratory goals. Taken together, these results validate the step-by-step approach, i.e. there is no indication that long-term studies should be performed mandatorily. Further support showing that the concepts used up to now to assess the safety of GM food and feed are sound has also been provided recently by the use of technologies such as metabolomics, proteomics and transcriptomics (see Ricroch *et al.*, 2011, for a review).

8.3.4 Standard protocols, quality of the studies and harmonization of protocols

The standard procedures outlined in the OECD Test (1998) recommend the use of at least ten animals per sex and per group, with three doses of the test substance and a control group. Two multi-generational studies used a reduced size sample of three animals (Brake and Evenson, 2004; Brake *et al.*, 2004). In some studies, the number of animals is correct per treatment, while the number per sex is not clearly mentioned. Inadequate experimental design in these studies has disabling effects on the statistical analysis (see internationally agreed statistical methods: EFSA, 2011). A balance should be found between an experimental design allowing robust toxicological inter-pretations and a reasonable cost.

The plant material and its description constitute another major problem. Out of the 33 studies examined, 15 did not state the use of isogenic lines as a control. This has been the case in the studies by Malatesta *et al.* (see above), which have been severely criticized (Williams and DeSesso, 2010) due to six methodological errors. In addition, it is a general weakness of toxicological studies where the feed being tested and compared to a near-isogenic line (i.e. the best comparator available) may not provide feed with fully identical composition. Therefore, if changes are observed, they can be caused by the differences between cultivars and not specifically by the transgene. Inclusion of commercial cultivars can help to establish whether the observed values fall within the range of variation observed for different

parameters, but should not replace the use of a well-characterized isogenic line.

Other weaknesses in the studies examined here are the absence of repetitions (see below), over-interpretation of dif-ferences, which are often within the normal range of variation, and poor toxicological interpretation of the data.

The major flaws in some papers underline the need to improve the reviewing process before publication of papers addressing this subject, in order to avoid confusion in the general press. Many unfounded allegations in the media regarding the health hazard of GM food and feed could have been avoided if non-specialized scientific reviews had not published the results of experiments which did not meet the internationally recognized criteria. Thus, qualified scientists (e.g. from the EFSA, etc.) would not lose valuable time in analysing and refuting these publications.

Very few published long-term feeding studies of genetically modified organisms (GMOs) use the same animal model and the same plant model, and do not consider the same parameters. Hence, no trial has been carried out twice in the same conditions by different research teams. This wide diversity of models makes it hard to perform analysis of the results on a large scale. Therefore, improvement in the protocols should be made, particularly focusing on repro-ducibility.

As discussed before (Snell *et al.*, 2012), although long-term and multi-generational studies would rarely be used in a step-by-step assessment of the safety of GM whole food, they could play an important role for its future improvement (e.g. to assess the effects of a particular diet, such as a specific amount of crop, or to find out which amount of GM material per diet is the most appropriate) and to validate new method-ologies.

8.4 Conclusions

Up to now, long-term and multi-generational studies have been performed as exploratory fundamental research projects. To date, none of these studies has proven the need to perform additional studies to the 90-day rodent feeding study as defined by Guideline No 408 (OECD, 1998) in order to assess effectively the risks associated with the use of new GM traits. Therefore, in the context of GM food and feed risk assessment, long-term and multi-generational studies should be conducted only in a case-by-case approach, after reasonable doubt still exists following a 90-day feeding trial.

Complementary fundamental studies should be performed but with a strong need for harmonization between studies, as well as with a broader spectrum of animal models. This type of research would help to choose the most efficient experimental design to assess the risks associated with new GM traits by revealing the physiological differences arising between short-, mid- and long-term tests.

Acknowledgements

The French Academy of Agriculture has awarded Dr Agnes Ricroch the Limagrain Prize 2012.

References

ANSES [Agence nationale de sécurité sanitaire de l'alimentation, de l'environnement et du travail] (2011) Opinion of the French Agency for Food, Environmental and Occupational Health and Safety. Recommendations for carrying out statistical analyses of data from 90-day rat feeding studies in the context of marketing authorisation applications for GM organisms. Request No 2009-SA-0285 (http://www.anses.fr/Documents/BIOT2009sa0285EN.pdf, accessed 6 June 2013).

Appenzeller, L.M., Munley, S.M., Hoban, D., Sykes, G.P., Malley, L.A. and Delaney, B. (2008) Subchronic feeding study of herbicide-tolerant soybean DP-356043-5 in Sprague-Dawley rats. *Food and Chemical Toxicology* 46(6), 2201–2213.

Aulrich, K., Boehme, H., Daenicke, R., Halle, I. and Flachowsky, G. (2001) Genetically modified feeds in animal nutrition: 1st communication: *Bacillus thuringiensis* (Bt) corn in poultry, pig and ruminant nutrition. *Archives of Animal Nutrition* 54, 183–195.

Bakke-McKellep, A.M., Sanden, M., Danieli, A., Acierno, R., Hemre, G.I., Maffia, M. *et al.* (2008) Atlantic salmon (*Salmo salar* L.) parr fed genetically modified soybeans and maize: histological, digestive, metabolic, and immunological investigations. *Research in Veterinary Science* 84(3), 395–408.

Baranowski, A., Rosochacki, S., Parada, F., Jaszczak, K., Zimny, J. and Połoszynowicz, J. (2006) The effect of diet containing genetically modified triticale on growth and transgenic DNA fate in selected tissues of mice. *Animal Science Papers and Reports* 24, 129–142.

BEETLE (2009) Long-term effects of genetically modified (GM) crops on health and the environment (including biodiversity). Executive Summary and Main Report. Federal Office of Consumer Protection and Food Safety (EVL), Berlin, 132 pp. + 6 Annexes (http://ec.europa.eu/food/food/biotechnology/reports_studies/docs/lt_effects_report_en.pdf, accessed 6 November 2012).

Brake, D.G. and Evenson, D.P. (2004) A generational study of glyphosate-tolerant soybeans on mouse fetal, postnatal, pubertal and adult testicular development. *Food and Chemical Toxicology* 42, 29–36.

Brake, D.G., Thaler, R. and Evenson, D.P. (2004) Evaluation of Bt (*Bacillus thuringiensis*) corn on mouse testicular development by dual parameter flow cytometry. *Journal of Agricultural and Food Chemistry* 52, 2097–2102.

Buzoianu, S.G., Walsh, M.C., Rea, M.C., Cassidy, J.P., Cotter, P.D., Ross, R.P., *et al.* (2012a) Effect of feeding genetically modified Bt MON810 maize to 40-day-old pigs for 110 days on growth and health indicators. *Animal* 6(10), 1609–1619. DOI:10.1017/S1751731112000249.

Buzoianu, S.G., Walsh, M.C., Rea, M.C., O'Donovan, O., Gelencser, E., Ujhelyi, G., *et al.* (2012b) Effect of feeding Bt maize to sows during gestation and lactation on maternal and offspring immunity and fate of transgenic material. *PLoS One* 7, 10:e47851.

Buzoianu, S.G., Walsh, M.C., Rea, M.C., Cassidy, J.P., Ryan, T.P., Ross, R.P., *et al.* (2012c) Transgenerational effects of feeding genetically modified maize to nulliparous sows and offspring on offspring growth and health. *Journal of Animal Science* 91(1), 318–330. DOI:10.2527/jas.2012-5360.

CERA (2012) GM Crop Database. Center for Environmental Risk Assessment (CERA), ILSI Research Foundation, Washington, DC (http://cera-gmc.org/index.php?action=gm_crop_database, accessed 6 November 2012).

Daleprane, J.B., Feijó, T.S. and Boaventura, G.T. (2009a) Organic and genetically modified soybean diets: consequences in growth and in hematological indicators of aged rats. *Plant Foods for Human Nutrition* 64, 1–5.

Daleprane, J.B., Pacheco, J.T. and Boaventura, G.T. (2009b) Evaluation of protein quality from genetically modified and organic soybean in two consecutive generations of Wistar rats. *Brazil Archives of Biology and Technology* 52, 841–847.

Daleprane, J.B., Chagas, M.A., Vellarde, G.C., Ramos, C.F. and Boaventura, G.T. (2010) The impact of non- and genetically modified soybean diets in aorta wall remodeling. *Journal of Food Science* 75, 126–131.

Domon, E., Takagi, H., Hirose, S., Sugita, K., Kasahara, S., Ebinuma, H., *et al.* (2009) 26-Week oral safety study in macaques for transgenic rice containing major human T-cell epitope peptides from Japanese cedar pollen allergens. *Journal of Agricultural and Food Chemistry* 57, 5633–5638.

EFSA [European Food Safety Agency] (2008) Safety and nutritional assessment of GM plants and derived food and feed: the role of animal feeding trials. *Food and Chemical Toxicology* 46, S2–S70.

EFSA (2009) EFSA and GMO risk assessment for human and animal health and the environment. EFSA Meeting Summary Report, 210 pp.

EFSA (2010) Scientific Opinion on statistical considerations for the safety evaluation of GMOs. EFSA Panel on Genetically Modified Organisms (GMO). *EFSA Journal* 8(1), 1250.

EFSA (2011) Scientific Opinion. Guidance for risk assessment of food and feed from genetically modified plants. EFSA Panel on Genetically Modified Organisms (GMO). *EFSA Journal* 9 (5), 2150.

Flachowsky, G., Halle, I. and Aulrich, K. (2005) Long term feeding of Bt-corn – a 10 generation study with quails. *Archives of Animal Nutrition* 59, 449–451.

Flachowsky, G., Aulrich, K., Böhme, H. and Halle, I. (2007) Studies on feeds from genetically modified plants (GMP) – contributions to nutritional and safety assessment. *Animal Feed Science and Technology* 133, 2–30.

Flachowsky, G., Schafft, H. and Meyer, U. (2012) Animal feeding studies for nutritional and safety assessments of feeds from genetically modified plants: a review. *Journal of Consumer Protection and Food Safety* 7, 179–194.

Halle, I., Aulrich, K. and Flachowsky, G. (2006) Four generations feeding GMO-corn to laying hens. *Proceedings of the Society of Nutrition Physiology* 15, 114.

Hammond, B., Dudek, R., Lemen, J. and Nemeth, M. (2004) Results of a 13 week safety assurance study with rats fed grain from glyphosate tolerant corn. *Food and Chemical Toxicology* 42, 1003–1014.

Hammond, B., Lemen, J., Dudek, R., Ward, D., Jiang, C., Nemeth, M., *et al.* (2006) Results of a 90-day safety assurance study with rats fed grain from corn rootworm-protected corn. *Food and Chemical Toxicology* 44, 147–160.

Haryu, Y., Taguchi, Y., Itakura, E., Mikami, O., Miura, K., Saeki, T., *et al.* (2009) Long term biosafety assessment of a genetically modified (GM) plant: the genetically modified (GM) insect-resistant Bt11 corn does not affect the performance of multi-generations or life span of mice. *Plant Science Journal* 3, 49–53.

Kiliç, A. and Akay, M.T. (2008) A three-generation study with genetically modified Bt corn in rats: biochemical and histopathological investigation. *Food and Chemical Toxicology* 46, 1164–1170.

Krzyzowska, M., Wincenciak, M., Winnicka, A., Baranowski, A., Jaszczak, K., Zimny, J., *et al.* (2010) The effect of multigenerational diet containing genetically modified triticale on immune system in mice. *Polish Journal of Veterinary Sciences* 13, 423–430.

Mackenzie, S.A., Lamb, I., Schmidt, J., Deege, L., Morrisey, M.J., Harper, M., *et al.* (2007) Thirteen week feeding study with transgenic maize grain containing event DAS015O7-1 in Sprague–Dawley rats. *Food and Chemical Toxicology* 45, 551–562.

Malatesta, M., Caporaloni, C., Gavaudan, S., Rocchi, M.B.L., Serafini, S., Tiberi, C., *et al.* (2002a) Ultrastructural morphometrical and immunocyto-chemical analyses of hepatocyte nuclei from mice fed on genetically modified soybean. *Cell Structure Function* 27, 173–180 (erratum 399).

Malatesta, M., Caporaloni, C., Rossi, L., Battistelli, S., Rocchi, M.B.L., *et al.* (2002b) Ultrastructural analysis of pancreatic acinar cells from mice fed on genetically modified soybean. *Journal of Anatomy* 201, 409–415.

Malatesta, M., Biggiogera, M., Manuali, E., Rocchi, M.B.L., Baldelli, B. and Gazzanelli, G. (2003) Fine structural analyses of pancreatic acinar cell nuclei from mice fed on genetically modified soybean. *European Journal of Histochemistry* 47, 385–388.

Malatesta, M., Boraldi, F., Annovi, G., Baldelli, B., Battistelli, S., Biggiogera, M., *et al.* (2008) A long-term study on female mice fed on a genetically modified soybean: effects on liver ageing. *Histochemistry and Cell Biology* 130, 967–977.

Malley, L.A., Everds, N.E., Reynolds, J., Mann, P.C., Lamb, I., Rood, T., *et al.* (2007) Subchronic feeding study of DAS-59122-7 maize grain in Sprague–Dawley rats. *Food and Chemical Toxicology* 45, 1277–1292.

OECD [Organisation for Economic Co-operation and Development] (1998) Section 4 (Part 408), Health effects: repeated dose 90-day oral toxicity study in rodents, *Guideline for the Testing of Chemicals*. OECD, Paris.

Rhee, G.S., Cho, D.H., Won, Y.H., Seok, J.H., Kim, S.S., Kwack, S.J., *et al.* (2005) Multigeneration reproductive and developmental toxicity study of bar gene inserted into genetically modified potato on rats. *Journal of Toxicology and Environmental Health* Part A, 68, 2263–2276.

Ricroch, A., Bergé, J.B. and Kuntz, M. (2011) Evaluation of genetically engineered crops using transcriptomic, proteomic and metabolomic profiling techniques. *Plant Physiology* 155(4), 1752–1761.

Sakamoto, Y., Tada, Y., Fukumori, N., Tayama, K., Ando, H., Takahashi, H., *et al.* (2007) A 52-week feeding study of genetically modified soybeans in F344 rats. *Journal of the Food Hygienic Society of Japan (Shokuhin Eiseigaku Zasshi)* 48, 41–50. [In Japanese.]

Sakamoto, Y., Tada, Y., Fukumori, N., Tayama, K., Ando, H., Takahashi, H., *et al.* (2008) A 104-week feeding study of genetically modified soybeans in F344 rats. *Journal of the Food Hygienic Society of Japan (Shokuhin Eiseigaku Zasshi)* 49, 272–282. [In Japanese.]

Sanden, M., Ornsrud, R., Sissener, N.H., Jorgensen, S., Gu, J., Bakke, A.-M., *et al.* (2013) Cross-generational feeding of Bt (*Bacillus thuringiensis*) maize to zebrafish (*Danio rerio*) has no adverse effects on the parenteral or offspring generations. *British Journal of Nutrition* 17 June, 1–12.

Scholtz, N.D., Halle, I., Daenicke, S., Hartmann, G., Zur, B. and Sauerwein, H. (2010) Effects of an active immunization on the immune response of laying Japanese quail (*Coturnix coturnix japonica*) fed with or without genetically modified *Bacillus thuringiensis*-maize. *Poultry Science* 89(6), 1122–1128.

Séralini, G.E., Clair, E., Mesnage, R., Gress, S., Defarge, N., Malatesta, M., *et al.* (2012) Long term toxicity of a Roundup herbicide and a Roundup-tolerant genetically modified maize. *Food and Chemical Toxicology* 50(11), 4221–4231.

Sissener, N.H., Sanden, M., Bakke, A.M., Krogdahl, A. and Hemre, G.-I. (2009) A long term trial with Atlantic salmon (*Salmo salar* L.) fed genetically modified soy; focusing general health and performance before, during and after the parr–smolt transformation. *Aquaculture* 294, 108–117.

Snell, C., Berheim, A., Bergé, J.B., Kunz, K., Pascal, G., Paris, A., *et al.* (2012) Assessment of the health impact of GE plant diets in long term and multigenerational animal feeding trials: a literature review. *Food and Chemical Toxicology* 50(3–4), 1134–1148.

Steinke, K., Guertler, P., Paul, V., Wiedemann, S., Ettle, T., Albrecht, C., *et al.* (2010) Effects of long-term feeding of genetically modified corn (event MON810) on the performance of lactating dairy cows. *The Journal of Animal Physiology and Animal Nutrition* 94, 185–193.

Trabalza-Marinucci, M., Brandi, G., Rondini, C., Avellini, L., Giammarini, C., Costarelli, S. *et al.* (2008) A three-year longitudinal study on the effects of a diet containing genetically modified Bt176 maize on the health status and performance of sheep. *Livestock Science* 13, 178–190.

Tudisco, R., Mastellone, V., Cutrignelli, M.I., Lombardi, P., Bovera, F., Mirabella, N., *et al.* (2010) Fate of transgenic DNA and evaluation of metabolic effects in goats fed genetically modified soybean and in their offsprings. *Animal* 1–10. DOI:10.1017/S1751731110000728.

Tyshko, N.V., Zhminchenko, V.M., Pashorina, V.A., Selyaskin, K.E., Saprykin, V.P., Utembaeva, N.T., *et al.* (2011) Assessment of the impact of GMO of plant origin on rat progeny development in 3 generations. *Problems of Nutrition* 80(1), 14–28. [In Russian.]

Vecchio, L., Cisterna, B., Malatesta, M., Martin, T.E. and Biggiogera, M. (2004) Ultrastructural analysis of testes from mice fed on genetically modified soybean. *European Journal of Histochemistry* 48, 449–454.

Williams, A.L. and DeSesso, J.M (2010) Genetically-modified soybeans. A critical evaluation of studies addressing potential changes associated with ingestion. Annual Meeting of the Society of Toxicology Salt Lake City, Utah, 7–11 March 2010. *The Toxicologist* 114(1), 1154.

Zhou, X.H., Dong, Y., Wang, Y., Xiao, X., Xu, Y., Xu, B., *et al.* (2012) A three generation study with high-lysine transgenic rice in Sprague–Dawley rats. *Food and Chemical Toxicology* 50, 1902–1910.

9

The Fate of Transgenic DNA and Newly Expressed Proteins

Ralf Einspanier*

Freie Universität Berlin, Institute of Veterinary Biochemistry, Berlin, Germany

9.1 Introduction

Knowledge of the fate of ingested commercialized genetically modified (GM) feed is a major scope of recent research projects. Concerning the widespread distribution of a growing number of different GM forage plants, the vast amount of such crops is nowadays fed to animals. When tracking the fate of such GM forage and its transgenic components, the specific focus is laid on traceable transgenic biopolymers like DNA and proteins newly present in the genetically engineered plants. Current detection technologies enable a very effective analysis of DNA traces, whereas the presence of newly expressed proteins is, due to degradation, not so easy detectable, even after ingestion. Different mandatory rules deal with the risk assessment of such new biotech materials, resulting in distinct regulations; for example, for the European Union (EC Regulation No 1829/2003). Very often, 90-day rodent feeding studies are mandatory and have been widely introduced as one component of risk assessment concerning consumer safety before market release of such GM products, mainly based on an OECD regulation (OECD, 1998). As another alternative, feeding experiments using farm animals have been introduced to calculate possible direct effects on domestic animal species like cattle, pigs or poultry (Phipps *et al.*, 2006; see Chapters 5 and 6). However, long-term feeding studies extending 14 weeks have seldom been performed (see Chapter 8).

Besides monitoring growth and physiological performance, the fate of specific recombinant biopolymers (recDNA/recProtein) is frequently used to determine possible contamination of animal-derived secondary products like meat, milk and eggs with transgenes (Beever *et al.*, 2003). Earlier, a general transfer of fragmented feed DNA was detected in selected domestic animal species (Einspanier and Flachowsky, 2009), leading to the assumption that an uptake of feed DNA might reflect a natural process potentially useful for GM monitoring in GM-fed animals.

9.2 General Aspects of GM Feed, Transgenic DNA and Newly Expressed Proteins

Modern biotechnology enables plant breeders to introduce foreign genes (recDNA) and corresponding newly expressed recombinant proteins (recProtein) of interest into crop plants, generating novel traits with newly wanted characteristics (see Chapter 2). The first generation of GM crop plants was constructed with emphasis on herbicide tolerance and insect resistance. Global production of such GM plants has increased significantly during the last 10 years. As reported in 2012, a total area of approximately 170 million hectares (Mha) of GM plants has been planted in more than 25 countries worldwide (ISAAA, 2013). Accordingly, the total GM crop area increased

*E-mail: ralf.einspanier@zedat.fu-berlin.de

threefold between 2001 and 2012, probably due to increased agronomic productivity and decreased use of pesticides. Beside the more than 28 commercialized transgenic traits, herbicide-tolerant (EPSPS) and/or insect-resistant (Cry-toxin) maize and soybeans are currently the dominant GM crops. Interestingly, developing countries are at present in the majority and have shown obviously higher increase rates within the last years in planting GM crop compared with the industrial countries (ISAAA, 2011). For example, in 2012 more than 70% of the world production of soybeans contained GM varieties routinely used as the major protein source in farm animal feed. Furthermore, it is expected that a new generation of transgenic plants will be commercialized and will further increase the spectrum and global abundance of GM forage plants in the future (see Chapter 12). As a practical consequence, the continuous increase of GM crop production makes these crops an important source for farm animal feeding. For example, in Germany more than 90% of feedstuff for pigs contains GM material (Bendiek and Grohmann, 2006), indicating that, based on European regulations, feed containing >0.9% GM components must be labelled. The ubiquitous monitoring approach is to trace distinct recDNA sequences within feed or animal substrates (different organs and secondary products like meat, milk and eggs) after GM plant feeding. As has been reviewed before (Phipps *et al.*, 2006; Alexander *et al.*, 2007; Einspanier and Flachowsky, 2009), a broad scientific knowledge regarding the uptake, effect and disposition of GM feed in animals is nowadays available. Despite the fact that nearly all studies were unable to discover a specific health problem for animals fed commercialized GM plants (see Chapters 6–8), some concerns about the fate and impact of consumed transgenic biopolymers (recDNA, recProteins) and derived food products thereof are still being publicly debated (Fagerstroem *et al.*, 2012).

In general, DNA and proteins are common components of feed, and after ingestion, a rapid intestinal degradation into short fragments has been suggested. Based on the fact that a complete destruction of feed DNA and proteins during digestion will not occur, adopted international safety assessments have been liberated (EFSA, 2008, 2011). This chapter summarizes the animal feeding studies published to date concerning the fate of GM feed DNA and proteins, together with the resulting significance and consequence of a potential GM transfer into animals and their secondary products, and the correlating safety aspects.

9.2.1 The fate of ingested feed DNA

Initially, it has to be acknowledged that the integrity of dietary DNA remains stable enough during feed processing to become the most suitable molecule for forensic approaches. One of the first studies introducing the tracing of feed DNA within farm animals fed GM crops provides essential results about the presence and distribution of ingested DNA in selected tissues (Einspanier *et al.*, 2001). By use of specifically developed, highly sensitive amplification techniques (polymerase chain reaction, PCR), it is nowadays possible to trace single DNA molecules quantitatively in most complex samples (Einspanier, 2006). However, due to the variety of different methods, severe specificity problems arise (sample preparation, normalization, cross-contamination), directly interacting with the reliability of the generated results. Therefore, confounded data interpretations are unavoidable when inadvertent DNA contaminations distort such extremely sensitive PCR assays; for example, when detecting highly abundant plant genes versus single copy genes (Klaften *et al.*, 2004). In this context, it has to be assessed that nearly all feeding studies use sensitive PCR techniques searching for foreign DNA in animal tissues or secondary products. Only in conjunction with optimized and professional sample selection and processing, contamination-free DNA extraction and suitable (real-time) PCR methods will reliable data interpretation be possible. Therefore, all studies should cope carefully with a minimum of harmonization and standardization of used DNA-based detection methods avoiding irrelevant data sets.

Since the earliest studies investigating DNA degradation in farm animals fed conventional versus GM plants, it is now generally accepted that nucleic acids are not completely degraded during digestion. Fragments of highly abundant plant DNA (e.g. chloroplast genomes) are routinely found in the digestive tract and/or specific organs of distinct farm animals (cattle, chicken); Fig. 9.1 depicts a theoretical deduced route of feed DNA in animals. In contrast, no transgenic or conventional plant DNA fragments have been found in pig or fish organs (Sanden *et al.*, 2004; Walsh *et al.*, 2011). One exception is bee honey, where native pollen particles collected from GM plants are, in principle, present, causing positive GM signals. For example, the presence of recDNA pollen in honey initiates massive public concerns and subsequently might provoke specific legal decisions. But recent studies have documented that even the adverse effects on the vitality of honeybees directly fed Bt pollen can be neglected (Duan *et al.*, 2008; Hendriksma *et al.*, 2011).

In conclusion, small fragments of feed DNA will pass through the intestinal tract and appear in some body tissues of some farm animal species. No transfer of recDNA from commercialized GM crops has been reliably detected in animal organs or secondary animal products like meat, milk or eggs, as supported by the majority of

publications investigating the fate of ingested feed DNA in animals (Alexander *et al.*, 2007; Einspanier and Flachowsky, 2009). No data have ever indicated that transgenic DNA and native plant DNA are degraded differently during feed processing and digestion in animals.

9.2.2 The fate of ingested feed proteins

Due to the fact that dietary proteins are mostly denatured during feed preparation and the subsequent digestion process, a reliable measurement of degraded feed proteins is challenging. Therefore, the minority of published studies deal with the detection of GM proteins, because trace amounts and fragmented proteins are hardly detectable due to missing the exponential amplification technology of DNA. However, by using highly sophisticated immunoassays, it is nowadays possible to trace degraded recProteins satisfactorily during GM feed processing and intestinal digestion. When summarizing all the current publications, it was found that a rapid degradation of Cry-toxin from Bt maize had been measured within the bovine gastrointestinal tract (GIT) (Lutz *et al.*, 2005; Wiedemann *et al.*, 2006; Paul *et al.*, 2010), the pig gastrointestinal tract (GIT) (Walsh *et al.*, 2012) and in the chicken (Scheideler *et al.*, 2008). All published data affirm a rapid degradation and fragmentation of Bt proteins, starting with ensiling and continuing during the digestion process in all animals investigated. A specific allergic potential of

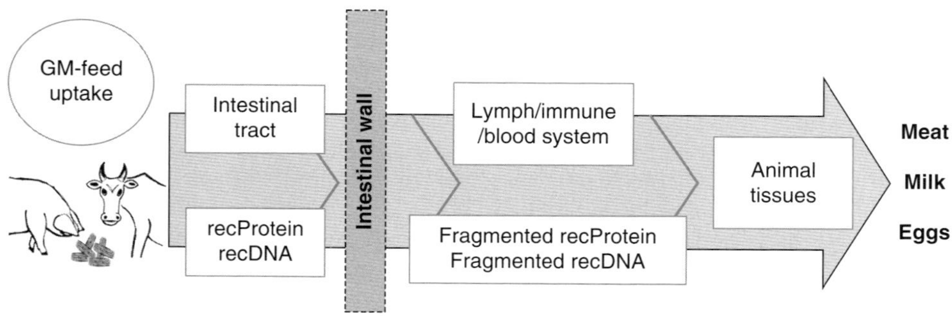

Fig. 9.1. Potential transfer of recombinant feed components in animals.

complete or degraded recProteins cannot be excluded and has been discussed in detail in a recent guidance document (EFSA, 2011), though such effects have never been reported in animals fed commercialized GM plants. As a major result of all these studies, no transfer of recProteins or their fragments into animal organs or secondary products has ever been measured.

To conclude, feed proteins are mainly degraded in the animal GIT to peptides and amino acids. Deduced from a few publications, there is no indication that recProteins and conventional plant proteins are degraded differently. Intact recProteins have never been detected in tissues of animals fed GM crops.

9.2.3 Resumé

A conclusive safety evaluation of a potential transfer of recDNA and recProteins derived from GM forage into animal organs can nowadays be compiled based on feeding studies observing transfer events, together with unexpected endogenous reactions comparing conventional versus GM feed. When reviewing the principal transfer of highly abundant feed polymers, reliable results can only be found when tracing ingested feed DNA (Table 9.1). Besides the majority of reliable studies, several published investigations introducing different feeding experiments, statistical methods and DNA/protein-tracing technologies may need

Table 9.1. Notable feeding experiments investigating a potential transfer of feed-derived recDNA or recProtein into farm animals. (Adapted from Einspanier and Flachowsky, 2009.)

GM trait	Animal species	recDNA or recProtein found in animal tissues	Reference
Bt maize			
	Chicken, cattle	None	Einspanier et al., 2001
	Pig	None	Klotz et al., 2002
	Pig	None	Reuter and Aulrich, 2003
	Chicken	None	Tony et al., 2003
	Pig	None	Chowdhury et al., 2003a
	Cattle	None	Yonemochi et al., 2003
	Calf	None	Chowdhury et al., 2003b
	Quail	None	Flachowsky et al., 2005
	Chicken	None	Aeschbacher et al., 2005
	Cattle	None	Calsamiglia et al., 2007
	Chicken	None	Scheideler et al., 2008
	Cattle	None	Guertler et al., 2010
	Pig	None	Walsh et al., 2011
EPSPS soybean			
	Pig	None	Aulrich et al., 2002
	Salmon	None	Sanden et al., 2004
	Rabbit	rDNA	Tudisco et al., 2006
	Goat	rDNA	Tudisco et al., 2010
	Salmon	None	Sanden et al., 2011
EPSPS rapeseed			
	Pig, sheep	rDNA	Sharma et al., 2006
GM potato			
	Chicken	None	El-Sanhoty et al., 2004
	Pigs	None	Broll et al., 2005
Bt cottonseed			
	Cattle	None	Castillo et al., 2004

further harmonization to provide more reproducible and scientifically reliable data sets that are not confusing to the public (Snell *et al.*, 2012).

In summary, the exact biological importance of the naturally occurring DNA transfer into the animal remains questionable. Due to highly fragmented feed DNA and proteins present after digestion within the GIT, the significant effects of GM polymers on animals have never been published. Generally, it can be stated that recDNA fragments may survive the digestion process and might only be detected in the excretion of animals.

9.3 Special Issues Concerning Distribution of Transgenic Polymers

From the above-mentioned studies, it is well known that fragmented Cry-DNA and degraded Cry-proteins are detectable within the content of the animal's GIT and therefore may subsequently be excreted. This fact might indicate potential interactions of recDNA/recProteins with the environment, concerning, for example, horizontal gene exchange with the microbiota or the effects of Cry-peptides on the pattern of soil organisms. Currently, the hypothetical horizontal gene transfer of feed recDNA to intestinal bacteria or the mammalian genome must be considered very unlikely. This statement is supported by the observation that neither the presence of plant genes nor an expression of foreign genes has ever been observed in animals (Einspanier and Flachowsky, 2009; Flachowsky and Wenk, 2010). In addition, the potential horizontal gene flow among microorganisms was formerly discussed as a possible mechanism for spreading novel genetic material into the environment (Nielsen and Townsend, 2004). However, the horizontal transfer of novel genes from GM plants to microorganisms may be neglected under field conditions (Brigulla and Wackernagel, 2010). When searching for antibiotic-resistant genes, which were initially present in some first-generation

GM plants, it could be stated that similar antibiotic-resistant genes were frequently found in microorganisms of the conventional environment, as well as within the untreated animal. When investigating the specific effects of GM plant material on intestinal microbiota, no significant influence on the pattern of the microbial populations persisting in bovine rumen was found under Bt maize feeding (Einspanier *et al.*, 2004). Similar experiments investigating horizontal gene transfer in goats generated comparable results (Rizzi *et al.*, 2008); a study performed in pigs showed that the overall composition of the caecal microbe population was almost unchanged after GM feeding (Buzoianu *et al.*, 2012). Such results indicate that Bt maize feeding does not show obvious influences on the intestinal microbiota of different farm animals.

As shown by different authors (Clark *et al.*, 2005; Gruber *et al.*, 2011), excreted Cry-protein fragments persist during slurry storage but are degraded successively after spreading on the fields. Deduced from these reports, the delivery of excreted, partially digested Bt proteins into the normal agricultural environment will result in their total degradation within months.

Another interesting point deals with game animals, such as wild boar and deer, being considered as uncontrollable targets and distributors of GM plant material in the environment (Guertler *et al.*, 2008). Currently, there is no scientific indication for different behaviours of these animals concerning ingestion, digestion and excretion of fed recDNA or recProteins when compared with the formerly observed fates of recDNA and recProteins in farm animals.

An upcoming concern will be the introduction of newly developed GM crops of the next generation. In the case of transplastomic GM crops, the chance to detect thereof derived more enriched recDNA, for example in animal products like meat, milk or eggs, may arise.

One more item may concern potential synergistic and cross-effects between GM forage plants and animals under specific stress (stacked traits, diseased, weaned and juvenile animals, medical treatments, etc.) being difficult to investigate, as well as increasing the experimental complexity. Hence, derived complex data sets are currently not available, but it may be of further interest to focus on GM-fed animals with a premature or disturbed intestinal barrier.

Finally, the short-term 3-month trials routinely performed have frequently been discussed as not being sufficient with regards to the potential chronic effects of GM crops on food and health safety. Therefore, a few long-term as well as multi-generation studies have been initiated using commercialized GM plants (Table 9.2). From a recent review (Snell *et al.*, 2012, see also Chapter 8), one can summarize that available long-term as well as multi-generation studies performed with commercialized GM crops do not decipher new adverse health effects or significant biological mal-regulations within the fed animals. It has been stated finally that long-term experiments will not generate significant new information, as has been provided previously by extensive 90-day feeding rodent studies (see Chapter 8).

9.4 Summary and Outlook

A significant transfer of functional recDNA/recProteins through the animal digestive tract into meat, milk or eggs appears highly uncertain and could not be quantified until now (Fig. 9.1). As reviewed before, foreign DNA/proteins are routinely processed and degraded during digestion and will lose most of their biological input (see also Chapter 6).

To date, feeding animals with commercialized GM crops cannot be rated as a significant health risk, either for the animals themselves or for the consumers eating the secondary products produced thereof.

However, based on the majority of publications, a significant transfer of recDNA and recProtein into animal organs, and subsequently into secondary products like meat, milk and eggs, appears highly improbable. It may be stated that the passage of feed DNA fragments across the intestinal wall can be judged as a natural event, though the likelihood of detecting foreign DNA depends strongly on their abundance in the ingested feed. The appearance of recDNA may change if higher initial concentrations are present in feed; for example, when introducing transplastomic crops. Finally, the absolute concentration of GM material fed to animals, the animal species and the organ-specific distribution of feed components have to be acknowledged when considering the fate of ingested transgenic material. It is state of the art that a routine screening of secondary products derived from animals fed commercialized GM appears improbable. All recent publications have shown that recDNA/recProteins cannot be measured reliably within organs or secondary products like meat, milk or eggs derived from GM-fed animals. Such scientific observations are the rationale for legal regulations currently not

Table 9.2. Key long-term feeding studies (≥4 weeks) performed with commercialized Bt maize, ordered by duration of experiment.

Duration (weeks)	Species	Event (GM)	recDNA/Protein in animal tissues	Reference
188	Sheep	Cry1Ab (Bt 176)	None	Trabalza-Marinucci *et al.*, 2008
100	Cows	Cry1Ab (MON810)	None	Steinke *et al.*, 2010
100	Cows	Cry1Ab (MON810)	None	Guertler *et al.*, 2010
35	Bulls + cows	Cry1Ab (Bt 176)	None	Flachowsky *et al.*, 2007
31	Hens	Cry1Ab (Bt 176)	None	Flachowsky *et al.*, 2007
16	Pigs	Cry1Ab (MON810)	None	Walsh *et al.*, 2012

demanding the labelling of secondary products like meat, milk or eggs.

Since the first report of a transfer of feed-derived plant DNA through the GIT into animals in 2001, many feeding experiments have been conducted and have generated over the last decade essentially similar results (Beever and Kemp, 2000; Aumaitre *et al.*, 2002; Phipps *et al.*, 2006; Alexander *et al.*, 2007; EFSA, 2008), summarized as follows:

- Fragmented feed DNA and proteins are present in the intestinal tract. Feed DNA fragments may be transferred into the tissues of animals as a natural process.
- Evidence indicates that the presence of feed DNA fragments in animal tissues does not represent a safety risk to the animal or the consumer.
- When gene fragments from ingested DNA are found in organs, these foreign DNA fragments do not possess biological function and will not account for apparent effects in the animal, nor have they been found to be integrated in the animal genome.
- When finally reviewing all available data, there is no scientific evidence that milk, meat or eggs derived from animals fed recently commercialized GM forage are less safe for the consumer than those produced with conventional feed.

References

Aeschbacher, K., Messikommer, R., Meile, L. and Wenk, C. (2005) Bt176 corn in poultry nutrition: physiological characteristics and fate of recombinant plant DNA in chickens. *Poultry Science* 84, 385–394.

Alexander, T.W., Reuter, T., Aulrich, K., Sharma, R., Okine, E.K., Dixon, W.T., *et al.* (2007) A review of the detection and fate of novel plant molecules derived from biotechnology in livestock production. *Animal Feed Science and Technology* 133(1–2), 31–62.

Aulrich, K., Reuter, T. and Flachowsky, G. (2002) The fate of foreign DNA in farm animals fed with genetically modified plants. *Proceedings of the Society of Nutrition Physiology* 11, 187–188.

Aumaitre, A., Aulrich, K., Chesson, A., Flachowsky, G. and Piva, G. (2002) New feeds from genetically modified plants: substantial equivalence, nutritional equivalence, digestibility, and safety for animals and the food chain. *Livestock Production Science* 74, 223–238.

Beever, D.E. and Kemp, C.F. (2000) Safety issues associated with the DNA in animal feed derived from genetically modified crops. A review of scientific and regulatory procedures. *Livestock Feeds and Feeding* 70, 175–182.

Beever, D.E., Glenn, K. and Phipps, R.H. (2003) A safety evaluation of genetically modified feedstuffs for livestock production; the fate of transgenic DNA and proteins. *Asian-Australasian Journal of Animal Sciences* 16(5), 764–772.

Bendiek, J. and Grohmann, L. (2006) GVO-Kontrolle von Lebensmitteln, Futtermitteln und Saatgut: eine bundesweite Übersicht. *Journal of Consumer Protection and Food Safety* 1, 241–245.

Brigulla, M. and Wackernagel, W. (2010) Molecular aspects of gene transfer and foreign DNA acquisition in prokaryotes with regard to safety issues. *Applied Microbiology Biotechnology* 86(4), 1027–1041.

Broll, H., Zagon, J., Butschke, A., Leffke, A., Spiegelberg, A., Boehme, H., *et al.* (2005) The fate of DNA of transgenic inulin synthesizing potatoes in pigs. *Journal of Animal and Feed Sciences* 14, 337–340.

Buzoianu, S.G., Walsh, M.C., Rea, M.C., O'Sullivan, O., Crispie, F., Cotter, P.D., *et al.* (2012) The effect of feeding Bt MON810 maize to pigs for 110 days on intestinal microbiota. *PLoS One* 7(5), e33668.

Calsamiglia, S., Hernandez, B., Hartnell, G.F. and Phipps, R. (2007) Effects of corn silage derived from a genetically modified variety containing two transgenes on feed intake, milk production, and composition, and the absence of detectable transgenic deoxyribonucleic acid in milk in Holstein dairy cows. *Journal of Dairy Science* 90, 4718–4723.

Castillo, A.R., Gallardo, M.R., Maciel, M., Giordano, J.M., Conti, G.A., Gaggiotti, M.C., *et al.* (2004) Effects of feeding rations with genetically modified whole cottonseed to lactating dairy cows. *Journal of Dairy Science* 87, 1778–1785.

Chowdhury, E.H., Kuribara, H., Hino, A., Sultana, P., Mikami, O., Shimada, N., *et al.* (2003a) Detection of corn intrinsic and recombinant DNA fragments and Cry1Ab protein in the gastrointestinal contents of pigs fed genetically modified corn Bt11. *Journal of Animal Science* 81(10), 2546–2551.

Chowdhury, E.H., Shimada, N., Murata, H., Mikami, O., Sultana, P., Miyazaki, S., *et al.* (2003b) Detection of Cry1Ab protein in gastrointestinal contents but not visceral organs of genetically modified Bt11-fed calves. *Veterinary and Human Toxicology* 45, 72–75.

Clark, B.W., Phillips, T.A. and Coats, J.R. (2005) Environmental fate and effects of *Bacillus thuringiensis* (Bt) proteins from transgenic crops: a review. *Journal of Agricultural and Food Chemistry* 53(12), 4643–4653.

Duan, J.J., Marvier, M., Huesing, J., Dively, G. and Huang Z.Y. (2008) A meta-analysis of effects of Bt crops on honey bees (Hymenoptera: Apidae). *PLoS ONE* 3(1), e1415. DOI:10.1371/journal.pone.0001415.

EFSA (2008) Safety and nutrition assessment of GM plants and derived food and feed. *Food and Chemical Toxicology* 46, S2–S70.

EFSA (2011) Panel on Genetically Modified Organisms (GMO); Scientific opinion on guidance for risk assessment of food and feed from genetically modified plants. *EFSA Journal* 9(5), 2150, 1–37.

Einspanier, R. (2006) DNA-based methods for detection of genetic modifications. In: Heller, K. (ed.) *Genetically Engineered Foods: Methods and Detection*, 2nd edn. Wiley-VCH, Weinheim, Germany, pp. 163–185.

Einspanier, R. and Flachowsky, G. (2009) Biotechnology: potential absorption and biological functions of digested fragments of DNA or proteins in food animals fed genetically modified crops. In: Pond, W.G. and Bell, A.W. (eds) *Encyclopedia of Animal Science*. Taylor and Francis, London, DOI:10.1081/E-EAS-120045416.

Einspanier, R., Klotz, A., Kraft, J., Aulrich, K., Poser, R., Schwägele, F., *et al.* (2001) The fate of forage plant DNA in farm animals: a collaborative case-study investigating cattle and chicken fed recombinant plant material. *European Food Research and Technology* 212, 129–134.

Einspanier, R., Lutz, B., Rief, S., Berezina, O., Zverlov, V., Schwarz, W., *et al.* (2004) Tracing residual recombinant feed molecules during digestion and rumen bacterial diversity in cattle fed transgene maize. *European Food Research Technology* 218(3), 269–273.

El-Sanhoty, R., Abd El-Rahman, A.A. and Boegl, K.W. (2004) Quality and safety evaluation of genetically modified potatoes Spunta with Cry V gene: comparitional analysis, determination of some toxins, antinutrients compounds and feeding study in rats. *Nahrung* 48, 13–18.

Fagerstroem, T., Dixelius, C., Magnusson, U. and Sundstroem, J.F. (2012) Stop worrying; start growing – risk research on GM crops is a dead parrot: it is time to start reaping the benefits of GM. *EMBO Reports* DOI:10.1038/embor.2012.59.

Flachowsky, G. and Wenk, C. (2010) The role of animal feeding trials for the nutritional and safety assessment of feeds from genetically modified plants – present stage and future challenges. *Journal of Animal and Feed Sciences* 19, 149–170.

Flachowsky, G., Halle, I. and Aulrich, K. (2005) Long term feeding of Bt-corn: a ten generation study with quails. *Archives of Animal Nutrition* 59, 449–451.

Flachowsky, G., Aulrich, K., Boehme, H. and Halle, I. (2007) Studies on feeds from genetically modified plants (GMP) – contributions to nutritional and safety assessment. *Animal Feed Science and Technology* 133, 2–30.

Gruber, H., Paul, V., Guertler, P., Spiekers, H., Tichopad, A., Meyer, H.H.D., *et al.* (2011) Fate of Cry1Ab protein in agricultural systems under slurry management of cows fed genetically modified maize (*Zea mays* L.) MON810 a quantitative assessment. *Journal of Agricultural and Food Chemistry* 59(13), 7135–7144.

Guertler, P., Lutz, B., Kuehn, R., Meyer, H.H.D., Einspanier, R., Killermann, B., *et al.* (2008) Fate of recombinant DNA and Cry1Ab protein after ingestion and dispersal of genetically modified maize in comparison to rapeseed by fallow deer (*Dama dama*). *European Journal of Wildlife Research* 54, 36–43.

Guertler, P., Paul, V., Steinke, K., Wiedemann, S., Preißinger, W., Albrecht, C., *et al.* (2010) Long-term feeding of genetically modified corn (MON810) – fate of cry1Ab DNA and recombinant protein during the metabolism of the dairy cow. *Livestock Science* 131, 250–259.

Hendriksma, H.P., Härtel, S. and Steffan-Dewenter, I. (2011) Testing pollen of single and stacked insect-resistant Bt maize on in vitro reared honey bee larvae. *PLoS ONE* 6(12), e28174. DOI:10.1371/journal.pone.0028174.

ISAAA [International Service for the Acquisition of Agri-Biotech Applications], Clive James (2011) Global status of commercialized biotec-GM crops: 2011. *ISAAA Brief No 43*. ISAAA Ithaca, New York.

ISAAA, Clive James (2013) Global status of commercialized biotec/GM crops: 2012 *ISAAA Brief No 44*. ISAAA, Ithaca, New York.

Klaften, M., Whetsell, A., Webster, J., Grewal R., Fedyk, E., Einspanier, R., *et al.* (2004) Development of polymerase chain reaction methods to detect plant DNA in animal tissues. In: Bhalgat, M.M., Ridley, W.P., Felsot, A.S. and

Seiber, J.N. (eds) *Agricultural Biotechnology: Challenges and Prospects*, ACS Symposium Series 866. American Chemical Society, Washington, DC, pp. 83–99.

Klotz, A., Mayer, J. and Einspanier, R. (2002) Degradation and possible carry over of feed DNA monitored in pigs and poultry. *European Food Research and Technology* 214, 271–275.

Lutz, B., Wiedemann, S., Einspanier, R., Mayer, J. and Albrecht, C. (2005) Degradation of Cry1Ab protein from genetically modified maize in the bovine gastrointestinal tract. *Journal of Agricultural and Food Chemistry* 53,1453–1456.

Nielsen, K.M. and Townsend, J.P. (2004) Monitoring and modeling horizontal gene transfer. *Nature Biotechnology* 22, 1110–1114.

OECD [Organisation for Economic Co-operation and Development] (1998) Section 4 (Part 408), Health effects: repeated dose 90-day oral toxicity study. In: *Rodents, Guideline for the Testing of Chemicals*. OECD, Paris.

Paul, V., Guertler P., Wiedemann, S. and Meyer, H.D.D. (2010) Degradation of Cry1Ab protein from genetically modified maize (MON810) in relation to total dietary feed proteins in dairy cow digestion. *Transgenic Research* 19, 683–689.

Phipps, R.H., Einspanier, R. and Faust, M.A. (2006) Safety of meat, milk, and eggs from animals fed crops derived from modern biotechnology. Council for Agricultural Science and Technology (CAST), Issue Paper No 34.

Reuter, T. and Aulrich, K. (2003) Investigations on genetically modified maize (Bt-maize) in pig nutrition: fate of feed ingested foreign DNA in pig bodies. *European Food Research and Technology* 216, 185–192.

Rizzi, A., Brusetti, L., Arioli, S., Nielsen, K.M., Tamagnini, I., Tanburini, A., *et al.* (2008) Detection of feed-derived maize DNA in goat milk and evaluation of the potential of horizontal transfer to bacteria. *European Food Research and Technology* 227, 1699–1709.

Sanden, M., Bruce, I.J., Rahman, M.A. and Hemre, G.-I. (2004) The fate of transgenic sequences present in genetically modified plant products in fish feed, investigating the survival of GM soybean DNA fragments during feeding trials in Atlantic salmon, *Salmo salar* L. *Aquaculture* 237, 391–405.

Sanden, M., Johannessen, L.E., Bergal, K.G., Sissener, N. and Hemre, G.-I. (2011) Uptake and clearance of dietary DNA in the intestine of Atlantic salmon (*Salmo salar* L.) fed conventional or genetically modified soybeans. *Aquaculture Nutrition* 17, e750–e759.

Scheideler, S.E., Hileman, R.E., Weber, T., Robeson, T. and Hartnell, G.F. (2008) The *in vivo* digestive fate of the Cry3Bb1 protein in laying hens fed diets containing MON 863 corn. *Poultry Science* 87, 1089–1097.

Sharma, R., Damgoard, D., Alexander, T.W., Dugan, M.E.R., Aalhus, J.L., Stanford, K., *et al.* (2006) Detection of transgenic and endogenous plant DNA in digesta and tissues of sheep and pigs fed Roundup Ready canola meal. *Journal of Agricultural and Food Chemistry* 54, 1699–1709.

Snell, C., Bernheim, A., Berge, J.B., Kuntz, M., Pascal, G., Paris, A., *et al.* (2012) Assessment of the health impact of GM plant diets in long-term and multigenerational animal feeding trials: a literature review. *Food and Chemical Toxicology* 50(3–4), 1134–1148.

Steinke, K., Guertler, P., Paul, V., Wiedemann, S., Ettle, T., Albrecht, C., *et al.* (2010) Effects of long-term feeding of genetically modified corn (event MON810) on the performance of lactating dairy cows. *Journal of Animal Physiology and Animal Nutrition* 94(5), e185–193.

Tony, M.A., Butschke, A., Broll, H., Grohmann, L., Zagon, J., Halle, I., *et al.* (2003) Safety assessment of Bt 176 maize in broiler nutrition: degradation of maize-DNA and its metabolic fate. *Archives of Animal Nutrition* 57, 235–252.

Trabalza-Marinucci, M., Brandi, G., Rondini, C., Avellini, L., Giammarini, C., Costarelli, S., *et al.* (2008) A three-year longitudinal study on the effects of a diet containing genetically modified Bt176 maize on the health status and performance of sheep. *Livestock Science* 113, 178–190.

Tudisco, R., Lombardi, P., Bovera, F., d'Angelo, D., Cutrignelli, M.I., Mastelione, V., *et al.* (2006) Genetically modified soya bean in rabbit feeding: detection of DNA fragments and evaluation of metabolic effects by enzymatic analysis. *Animal Science* 82, 193–199.

Tudisco, R., Mastellone, V., Cutrignelli, M.I., Lombardi, P., Bovera, F., Mirabella, N., et al. (2010) Fate of transgenic DNA and evaluation of metabolic effects in goats fed genetically modified soybean and in their offsprings. *Animal* 4(10), 1662–1671.

Walsh, M.C., Buzoianu, S.G., Gardiner, G.E., Rea, M.C., Gelencser, E., Janosi, A., *et al.* (2011) Fate of transgenic DNA from orally administered Bt MON810 maize and effects on immune response and growth in pigs. *PLoS One* 6, e27177.

Walsh, M.C., Buzoianu, S.G., Rea, M.C., O'Donovan, O., Gelencsér, E., Ujhelyi, G., *et al.* (2012) Effects of feeding Bt MON810 maize to pigs for 110 days on peripheral immune response and digestive fate of the cry1Ab gene and truncated Bt toxin. *PLoS One* 7(5), e36141.

Wiedemann, S., Lutz, B., Kurtz, H., Schwarz, F.J. and Albrecht, C. (2006) *In situ* studies on the time-dependent degradation of recombinant corn DNA and protein in the bovine rumen. *Journal of Animal Science* 84, 135–144.

Yonemochi, C., Ikeda, T., Harada, C., Kusama, T. and Hanazumi, M. (2003) Influence of transgenic corn (CBH 351, named StarLink) on health condition of dairy cows and transfer of Cry9C protein and cry9C gene to milk, blood, liver and muscle. *Animal Science Journal* 74, 81–88.

10 Influence of Feeds from GM Plants on Composition/Quality of Food of Animal Origin

Gerhard Flachowsky*

Institute of Animal Nutrition, Friedrich-Loeffler-Institute (FLI), Federal Research Institute for Animal Health, Braunschweig, Germany

10.1 Introduction

The composition and quality of food of animal origin may be influenced substantially by animal feeding. Enhancing the content of various nutrients in food of animal origin provides an opportunity to increase nutrition levels in the human diet – especially those deemed to be marginal or insufficient.

It is well known that fatty acids (Flachowsky *et al.*, 2008; Glasser *et al.*, 2008; Mirghelenj *et al.*, 2009), as well some minerals (Franke *et al.*, 2009; Bennett and Cheng, 2010; Röttger *et al.*, 2012; Speight *et al.*, 2012) and vitamins (Naber and Squires, 1993; Sünder *et al.*, 1999; Leeson and Caston, 2003; Sahlin and House, 2006), are transferred from feed into food of animal origin (see Table 10.1). There are some differences between ruminants and non-ruminants because of the influences of the microbiota in the rumen (fermentation processes in the rumen, especially in the case of unsaturated fatty acids).

Some genetically modified (GM) traits could influence the composition and organoleptic quality of food of animal origin, such as the distribution of fat and lean in the carcass, fat grade and fatty acid composition, mineral and vitamin concentration in the body, in special organs or tissues, or in milk and eggs, yolk and meat colour, or other product quality measures such as tenderness, flavour, sensory properties or other characteristics of the final prepared food. Special studies/measurements are necessary to assess the influence of GM feed on the composition/quality of food of animal origin (ILSI, 2007; EFSA, 2012). Table 10.2 shows some end points for adequate measurements to investigate the influence of feed on the composition/quality of food of animal origin.

10.2 Studies with First-generation GM Crops

First-generation GM plants are characterized by the so-called substantial equivalence (plants with input traits). This means that there are no substantial differences in composition to that of their isogenic counterparts, apart from the introduced transgenic or recombinant DNA fragment(s) and the newly expressed protein(s). Therefore, the main components of feeds from first-generation transgenic plants do not influence the composition and quality of food of animal origin in comparison to its isogenic counterpart (see also Chapters 4 and 6).

*E-mail: gerhard.flachowsky@t-online.de

Table 10.1. Influence of animal nutrition on selected ingredients of food of animal origin (From Flachowsky, 2009.)

Nutrients	Milk	Meat[a]	Eggs
Protein/amino acids	(+)	−	−
Fat/fatty acids	+++	++	++
Major elements			
Calcium	−	−	−
Phosphorus	−	−	−
Magnesium	−	−	−
Trace elements			
Copper	(+)	(Liver: +++)	(+)
Iodine	+++	+	+++
Selenium	++	++	++
Zinc	+	+	+
Vitamins			
A	(+)	(Liver: +++)	+
D	+	+	+
E	(+)	(+)	+++
B vitamins	(+) (if rumen protected)	− until +	− until ++

Notes: +++ = very strong influence possible, transfer of supplementation into food >10%; ++ = strong influence possible, transfer 5–10%; + = moderate influence, transfer 1–5%; (+) = small influence, transfer <1%; − = no influence/no transfer from feed into food. [a]Differences between meat of ruminants and non-ruminants.

Table 10.2. Proposal of end points for comparative analysis of food of animal origin from animals fed with feed from second-generation GM plants and of GM animals. (From EFSA, 2012.)

Group of GM animals	Mammals	Birds	Aquaculture (e.g. fish, molluscs)	Insects (honey-bees)
Samples from the animal body	Tissues: • Meat, muscle (*Musculus longissimus dorsi*; *Musculus biceps femoris*) • Body fat • Blood • Some organs (liver, kidney, spleen, brain(?), etc.) • Residue body (meat and bonemeal as feed)	Tissues: • Meat, muscle (breast, thigh) • Abdominal fat • Blood • Some organs (liver, kidney, spleen, etc.) • Residue body (animal body meal as feed)	Edible fraction (e.g. fillet) • Residue body (e.g. fishmeal as feed)	−
Food/feed produced by animals	Milk	Eggs	Caviar	Honey

10.2.1 Recombinant DNA and newly expressed proteins

Recombinant DNA (see Chapter 9) is mainly degraded during ensiling (Hupfer *et al.*, 1999; Aulrich *et al.*, 2004), feed processing (Gawienowski *et al.*, 1999; Chiter *et al.*, 2000; Alexander *et al.*, 2002; Berger *et al.*, 2003) and in the digestive tract of animals (Alexander *et al.*, 2004, 2007; Einspanier *et al.*, 2004; Wiedemann *et al.*, 2006). It cannot be ruled out that gene fragments enter the intestinal epithelium and are absorbed by the host organism. Low copy DNA (e.g. recombinant DNA) has mainly not been detected, while multi-copy DNA (endogenous DNA; e.g. Rubisco DNA) has been detected in several body samples (Sanden *et al.*, 2011). Up to now, many authors have not found recombinant DNA fragments in animal body samples, milk or eggs (see Einspanier and Flachowsky, 2009, and Chapter 9), but some authors (e.g. Mazza *et al.*, 2005; Sharma *et al.*, 2006; Tudisco *et al.*, 2006, 2010; Chainark *et al.*, 2008; Ran *et al.*, 2009) have amplified traces of recombinant DNA fragments in animal tissues or milk.

Similar studies have been done with newly expressed proteins (see Chapter 9). Such proteins are degraded during feed processing/conservation and in the digestive tract, as demonstrated in *in vitro* studies and in digestion and feeding experiments (Lutz *et al.*, 2005; Wiedemann *et al.*, 2006; Calsamiglia *et al.*, 2007; Guertler *et al.*, 2008; Scheideler *et al.*, 2008; Paul *et al.*, 2010; Buzoianu, 2011; Gruber *et al.*, 2011; Walsh *et al.*, 2012). It cannot be excluded that small protein fragments (e.g. peptides) may pass the intestinal epithelium.

Some authors want a labelling of food of animal origin after consumption of GM feed (e.g. Seralini *et al.*, 2011; Antoniou *et al.*, 2012). Such wishes cannot be scientifically justified because of the inconsistency of detection of recombinant DNA and newly expressed protein in animal tissues, milk and eggs.

10.2.2 Product composition and quality

Many studies have been conducted to compare the quality of livestock products obtained from animals fed with diets containing feed from first-generation GM plants (with input traits) with those obtained from their isogenic counterparts. Some results are exemplarily demonstrated in Tables 10.3–10.11.

Milk composition (e.g. fat, protein, lactose) and milk quality were not influenced after the inclusion of first-generation GM feed in dairy cattle diets (Table 10.3).

Similar results have been reported from a long-term study recently completed (Table 10.4). Authors found no significant influence on milk composition during the 25-month experimental period.

More detailed analysis of milk from cows fed with high portions of isogenic or Bt maize (Table 10.5) did not show any significant differences.

Table 10.3. Influence of feeding maize silage and maize grain from isogenic and Bt maize on feed intake, milk production and milk composition of dairy cows (average of two experiments; six or eight cows per treatment; duration of experiment, 63 days; maize portion in ration: Experiment 1: 41.8% of DM maize silage; 34.1% of cracked maize; Experiment 2: 59.6% maize silage; 19.9% cracked maize). (From Donkin *et al.*, 2003.)

Parameter	Control maize	Bt maize
Dry matter intake (kg/day)	24.3	24.1
Milk yield (kg FCM/day)[a]	33.2	33.2
Milk fat (%)	3.60	3.67
Milk protein (%)	3.10	3.14
Milk lactose (%)	4.62	4.64

Note: [a]FCM = fat-corrected milk.

Table 10.4. Milk composition of the first and second lactation of a long-term feeding study with dairy cows (n = 18 per treatment; 25 months with Bt maize (MON810, 63% of roughage, 41% of concentrate from maize)).[a] (From Spiekers et al., 2009.)

Lactation of experiment	First		P level	Second		P level
Parameter	Isogenic	Transgenic		Isogenic	Transgenic	
Milk yield (kg/day)	23.9	23.7	0.566	29.2	28.8	0.413
Milk fat (%)	3.95	4.03	0.015	3.75	3.86	0.056
Milk protein (%)	3.62	3.71	<0.001	3.59	3.56	0.299
Milk lactose (%)	4.83	4.82	0.155	4.74	4.80	0.006
Milk urea (mg/l)	164	181	0.001	180	175	0.523
Somatic cell count (1000/ml)	157	205	0.073	241	220	0.075

Note: [a]No fragments of Cry1Ab DNA in blood, milk, faeces and urine of cows; traces of Cry1Ab protein were detected in faeces, but not in blood, milk and urine (Guertler et al., 2009).

Table 10.5. Selected parameters of milk composition, protein fractions and fatty acid proportion in milk of dairy cows fed conventional or Bt 176 maize (12 cows per treatment; 70% of DM maize silage). (From Barriere et al., 2001.)

Parameter	Isogenic control	Transgenic maize, Bt 176
Milk yield (kg/day)	31.5	31.8
Milk fat (%)	3.69	3.68
Milk protein (%)	3.20	3.18
Total caseins (g/kg)	24.9	24.9
α-Lactalbumin (g/kg)	1.12	1.16
ß-Lactoglobulin (g/kg)	3.86	3.90
Curd yield (%)	15.2	15.0
Fatty acids (% of total):		
$C_4 - C_{12}$	14.3	15.2
Total C_{14}	13.5	14.0
Total C_{16}	41.1	41.0
$C_{18:0}$	8.8	8.5
$C_{18:1}$	16.5	15.9
$C_{18:2}$	2.1	2.1

Some authors (e.g. Aulrich et al., 2001; Berger et al., 2003; Erickson et al., 2003) have also reported that the inclusion of first-generation GM feed in the diets of beef cattle did not affect body composition, marbling score, fat depth, area of Musculus longissimus dorsi or yield grade (Table 10.6; see also Chapter 6; Table 6.3).

Erickson et al. (2003) fed high portions of glyphosate-tolerant maize (GA21, NK603) to feedlot steers in three studies and measured some meat parameters. They did not find any significant difference between transgenic maize, its isogenic counterpart and two commercial lines (Table 10.7).

Feeding studies with feed from GM plants with input traits in small ruminants show similar results to those described with lactating cows and bulls/steers. Stanford et al. (2003) fed transgenic Roundup Ready (RR) canola meal to lambs (from 21.5 to 45 kg body weight) and compared it with the parenteral line and two commercial lines. The inclusion level of canola meals in the experimental diets was only 6.5%. Carcass composition, chemical composition of Musculus longissimus dorsi and meat quality parameters were not influenced significantly by the canola sources.

Similar ruminant feeding studies with pigs did not show any significant effects of feeds from first-generation GM plants on carcass composition and quality. Ryan et al. (2004) compared two studies with RF maize

Table 10.6. Influence of corn rootworm protected maize (MON863) on slaughtering data of feedlot steers (initial body weight: 457 kg; duration of experiment: 102 days). (From Berger *et al.*, 2003.)

Parameter	Commercial control RX740	Commercial control DK647	Isogenic control RX670	Transgenic maize MON863
Body weight gain (kg/day)	1.39	1.49	1.53	1.49
Carcass weight (kg)	367	374	377	374
Marbling score	484	470	489	493
Musculus longissimus dorsi area and 12th rib (cm^2)	97.3	99.5	95.6	97.2
Fat (cm)	0.85	0.89	0.99	0.92
Yield grade	1.9	1.9	2.3	2.1

Table 10.7. Influence of RR maize (GA21 or NK603) in rations of finishing steers on parameters of meat quality and muscle composition (Experiment 1: 75%; Experiment 2: 73%; Experiment 3: 79.5% maize in finishing diet; 92, 94 and 144 days feeding in Experiments 1, 2 and 3, respectively). (From Erickson *et al.*, 2003.)

Parameter	Commercial 1	Commercial 2	Isogenic control	Transgenic maize (Experiment 1: GA21; Experiments 2 and 3: NK603)
Final body weight (kg):				
Experiment 1	561	575	571	562
Experiment 2	565	570	564	566
Experiment 3	585	575	590	577
Carcass weight (kg):				
Experiment 1	347	356	354	348
Experiment 2	346	348	345	346
Experiment 3	349	364	371	364
Marbling score:				
Experiment 1	492	517	485	517
Experiment 2	545	555	559	549
Experiment 3	533	544	539	541
Muscle composition[a] (moisture %):				
Experiment 1	72.3	72.1	73.5	72.3
Experiment 2	71.8	71.7	71.4	71.7
Experiment 3	75.6	75.1	75.2	76.2
Protein (%):				
Experiment 1	22.3	22.0	22.1	22.0
Experiment 2	23.8	23.9	24.0	23.7
Experiment 3	20.5	20.0	20.2	19.6
Fat (%):				
Experiment 1	3.9	4.0	3.5	4.1
Experiment 2	4.0	4.1	4.6	4.3
Experiment 3	3.4	3.9	3.8	3.5

Note: [a]Experiments 1 and 2: *Musculus longissimus dorsi*; Experiment 3: Brachiocephalicus muscle.

Table 10.8. Influence of RR maize (NK603) in grower I and II (68.1 and 74.2%) and finisher (78.1 and 81.8% maize of diet) diets of two studies with pigs on parameters of meat quality and composition (final weight: 116 and 119 kg in Experiments 1 and 2, respectively). (From Hyan *et al.*, 2004.)

Parameter	Commercial 1	Commercial 2	Isogenic control	Transgenic maize (NK603)
Average weight gain (g/day):				
Experiment 1	910	909	912	912
Experiment 2	985	945	942	954
Musculus longissimus area (cm^2):	58.6	55.3	56.6	56.6
Experiment 1	43.4	41.9	42.9	42.2
Experiment 2				
Muscle composition (moisture %):				
Experiment 1	72.4	72.3	72.7	72.5
Experiment 2	73.4	73.2	73.3	73.2
Protein:				
Experiment 1	23.5	23.7	23.8	23.5
Experiment 2	23.1	22.9	23.4	23.4
Fat:				
Experiment 1	3.06	3.08	2.20	2.39
Experiment 2	2.46	2.83	2.78	2.77

(NK603) with similar non-transgenic maize and two commercial maize lines (Table 10.8). There was no significant effect of GM maize on carcass yield, composition and quality parameters.

Cromwell *et al.* (2002) compared the effect of glyphosate-tolerant soybean meal with its near-isogenic, non-GM counterpart on sensory qualities, loss in cooking and sheer forces in pork. The authors did not report any significant effect on the criteria measured (Table 10.9). No effects on meat composition and quality were reported by Reuter *et al.* (2002) if they fed Bt maize to pigs. Later, Cromwell *et al.* (2005) compared diets with high amounts of glufosinolate herbicide-tolerant (LibertyLink) rice with conventional rice in diets for growing/finishing pigs (72.8, 80.0 and 85.8% in growing, early-finishing and late-finishing phase) and did not observe any significant effects on carcass quality.

Taylor *et al.* (2004) compared the effects of RR (Event RT73) canola meal with a non-transgenic control and six commercial canola meals in broilers and observed no significant influence on slaughtering results

Table 10.9. Effect of glyphosate-tolerant (FF) soybean meal on parameters of pork quality. (From Cromwell *et al.*, 2002.)

Sensory parameter[a]	Control	GM
Juiciness	5.52	5.53
Tenderness	5.91	6.10
Off flavour	7.08	7.13
Flavour intensity	5.74	5.35
Connective tissue	3.95	3.33
Cook loss (%)	32.0	30.3

Note: [a]Panel scale 1–8 (8 is best).

and body composition, as shown for some parameters in Table 10.10. Most parameters of GM-fed animals were in the range of the commercial varieties.

Stadnik *et al.* (2011) investigated the influence of GM maize (MON810 and a GM soybean meal (Roundup Ready; see Swiatkiewicz *et al.*, 2010a,b, 2011 for experimental design) on the physical-chemical properties of breast and thigh muscles of broilers, as shown for the breast muscle in Table 10.11. Apart from a (L*) colour parameter for breast muscle no further parameter was affected significantly.

Table 10.10. Selected parameters of carcass yield and composition of broilers fed glyphopsate-tolerant canola meal (RT173), non-transgenic control and six commercial canola meals (starter: 25; grower/finisher: 20% of diet; 100 broilers per treatment; mean values of combined males and females). (From Taylor *et al.*, 2004.)

Parameter	Control	Range of six commercial varieties	Glyphosate-tolerant RT73
Chill weight (kg/bird)	1.55	1.54–1.60	1.60
Breast meat (% of chill weight)	25.17	25.04–25.63	24.93
Thigh meat (% of chill weight)	15.55	16.44–16.89	16.66
Breast meat analysis:			
Moisture (%)	75.10	74.96–75.37	75.25
Protein (%, as-is basis)	23.71	23.65–23.97	23.74
Fat (%, as-is basis)	0.86	0.76–0.92	0.82

Table 10.11. Effects of GM maize (MON810) and RR soybeans (MON-40-30-2) on physico-chemical properties of male broilers' breast muscle ($n = 20$). (From Stadnik *et al.*, 2011.)

Treatment/parameter	Control	+ GM soybean meal	+ GM maize	+ GM soybean meal + GM maize
pH	5.65	5.79	5.76	5.77
Water holding capacity (%)	49.9	55.2	49.9	52.5
Oxidation – reduction potential (mV)	363.1	350.3	333.9	355.7
Colour parameters:				
L*	56.2	53.6	56.3	56.4
a*	0.79	0.34	1.40	0.56
B*	9.6	9.7	9.7	9.9
TBARS (mg/kg)[a]	1.12	1.04	0.72	0.74

Note: [a]Thiobarbituric acid reactive substances (TBARS) as parameter for lipid oxidation.

by the incorporation of transgenic maize and soybean meal into broiler diets.

Further studies with broilers and various GM plants (e.g. soybeans, maize, wheat, canola, etc.) were carried out and did not show any significant differences in body composition and quality parameters of meat (Aulrich *et al.*, 2001; Kan and Hartnell, 2004a,b; FASS, 2013).

Some authors report no significant influence of feed from GM plants with input traits on the body composition and meat quality of fish (e.g. Brown *et al.*, 2003, in rainbow trout). Further studies can be found in FASS (2013).

10.3 Studies with Second-generation GM Crops

There is quite another situation with feeds from GM plants of the so-called second generation (GM plants with output traits). Such biofortified crops may be characterized with higher protein and/or fat content, changed amino acids or fatty acid patterns, changed content in vitamin precursors, vitamins and/or minerals and contribute towards alleviating malnutrition with important micronutrients (Bouis *et al.*, 2003; Welch and Graham, 2004; Nestel *et al.*, 2006; Pfeiffer and McClafferty, 2007; Mayer *et al.*, 2008; Maruyama *et al.*, 2011; see Chapter 7). Some of the nutrients may influence the composition and quality of food of animal origin (e.g. fatty acids); others may be deposed in various organs (e.g. vitamin A and Cu in the liver) or in animal products (e.g. iodine in milk and eggs; vitamin E in eggs).

Furthermore, the content of non-essential ingredients such as enzymes may be increased, or those of undesirable substances such as phytate, glucosinolates or mycotoxins may be reduced with certain influence on the composition and quality of food of animal origin. Special attention has been paid to modify the fatty acid pattern in GM plants (e.g. more oleic acid instead of linoleic and linolenic acids in oilseeds; Brown *et al.*, 1999) or to produce 'unusual' fatty acids in transgenic plants (Cahoon *et al.*, 2007). Genetic modification has facilitated the generation of oilseeds synthesizing non-native fatty acids. Napier (2007) differentiates between two particular classes of fatty acids:

1. Industrial fatty acids, which may contain functional groups such as hydroxyl, epoxy or acetylenic bonds.
2. Very long-chain polyunsaturated fatty acids normally found in fish oils and marine microorganisms.

Industrial fats are not used in animal nutrition, but fats with long-chain polyunsaturated fatty acids may influence the composition of body fat after feeding (see Section 10.3.2).

10.3.1 Increased content of amino acids

Some feeds are characteristically deficient in important essential amino acids such as lysine, methionine, threonine and/or tryptophane. The first limiting amino acid in maize is lysine, and this requires combination with feeds rich in lysine and/or lysine supplementation for optimal performance and carcass characteristics of birds and pigs (NRC, 1994, 1998; GfE, 1999, 2008; Corzo *et al.*, 2006). Therefore, one of the objectives

of traditional plant breeding was an increase of lysine in maize. Even in the 1960s, information about a mutant that changed protein fractions and increased the lysine content of maize endosperm was being reported (Mertz *et al.*, 1964; Nesar *et al.*, 1965; Wolf *et al.*, 1967). Later, high-lysine mutants were known as Opaque-2 and Floury-2 (Sodek and Wilson, 1971; Christianson *et al.*, 1974). Later, the cultivation stopped because of some disadvantages, such as more sensitivity to some diseases, problems during harvest and storage, as well as lower yields (Hoffer *et al.*, 1985).

Modern biotechnology also deals with this topic and provides alternatives to dietary amino acid supplementation in diets. The development of lysine maize (see Chapter 7, Table 7.4) was a step in the direction of improved broiler feeding. Lucas *et al.* (2007) used high portions of lysine maize (LY038 and LY038 × MON810) in broiler diets (59.2% in starter, 66.1% in grower/finisher diets). They compared these (see Table 7.4 for composition) with unsupplemented and lysine-supplemented control maize and five commercial varieties (unsupplemented and lysine supplemented). Carcass characteristics and body composition were not significantly different between the lysine maize and the lysine-supplemented control maize, but the unsupplemented control chicks showed lower body weights (Table 10.12). This means that the bioefficacy of the increased lysine in LY038 and LY038 × MON810 grain was no different from that of lysine in conventional maize diets supplemented with L-Lys HCl.

Table 10.12. Carcass characteristics of broilers fed with control maize, lysine-supplemented control maize and lysine maize. (From Lucas *et al.*, 2007.)

Parameter	Control	Control + lysine	LY038	LY038 × MON810
Chilled weight (g)	1156	1601	1591	1561
Breast weight (g)	222	347	349	333
Breast meat composition:				
Moisture (%)	77.6	77.6	77.3	77.3
Protein (%, as-is basis)	19.2	19.8	20.0	19.6
Fat (%, as-is basis)	2.17	1.58	2.35	1.85

10.3.2 Expression and use of stearidonic acid in soybeans (C18:4 n-3; stearidonic acid, SDA)

The changed composition of GM plants and derived feed may also influence the composition of food of animal origin, as has been demonstrated exemplarily for soybeans with a modified fatty acid pattern. The introduction of two new genes effecting the expression of Δ6 and higher expression of Δ15-desaturases and the biosynthesis of a highly unsaturated fatty acid with four double bonds (see Chapter 7, Fig. 7.1), which also occurs in echium oil (about 10% SDA; Surette *et al.*, 2004). This stearidonic acid (SDA; C18:4 n-3) may be transferred into the body fat of non-ruminants, or may be used as a precursor for longer fatty acid chains (e.g. C20:5 n-3, eicosapentaenoic acid, EPA; and C22:6 n-3, docosahexaenoic acid, DHA) not only in non-ruminants (see Tables 10.13 and 10.14), fish (Cleveland *et al.*, 2012) and ruminants but also in humans (James *et al.*, 2003; Harris *et al.*, 2007, 2008; Banz *et al.*, 2012). The efficacy of conversion of SDA to EPA has varied between 3:1 and 6:1 in studies (James *et al.*, 2003; Harris *et al.*, 2007; Whelan *et al.*, 2012).

Stearidonic soybean oil contains between 20 and 30% SDA (see Chapter 7, Table 7.6).

Rymer *et al.* (2011) added 45 (grower) and 50 g SDA oil containing 24% of SDA (see Table 7.6 for composition) per kg finisher broiler diet and compared this with conventional soybean oil and fish oil. The authors did not observe any significant influence of SDA oil on feed intake, weight gain and feed conversion rate in the animals, but they found higher concentrations of SDA as well as C_{20} and C_{22} polyunsaturated fatty acids in various body fats (Table 10.13).

Table 10.14 shows the influence of added oils on the body pool size of n-3 fatty acids in broilers.

Finally, Rymer *et al.* (2011) investigated the influence of three oil sources (soybean oil, SDA oil and fish oil) on the sensory attributes of breast and leg meat. Aroma, taste and aftertaste of freshly cooked breast meat were not influenced significantly by oil sources. Fishy aromas, tastes and aftertastes were associated with higher concentrations of n-3 fatty acids in the meat (see Table 10.14), being most noticeable in leg meat after fish oil supplementation, but also in reheated leg meat of the SDA group. More studies seem to be necessary to overcome this weakness.

Similar results of the transfer of SDA in milk are described by Bernal-Santos *et al.* (2010) in lactating cows after duodenal

Table 10.13. Concentrations of some n-3 fatty acids (mg/100 g fresh tissue) in body samples of broilers (unsupplemented control or supplemented with fish oil or SDA oil). (From Rymer *et al.*, 2011.)

Sample	Control	+45 (grower) and 50 g fish oil (finisher diet) per kg feed	+45 (grower) and 50 g SDA oil (finisher diet) per kg feed
Fat content (g/kg):			
Breast meat	39.9	29.9	39.4
Leg meat	68.0	67.4	70.7
Skin	481.0	422.0	465.0
Breast meat:			
C18:4 n-3	3	13	231
C20:5 n-3	12	49	28
C22:6 n-3	7	107	14
Leg meat:			
C18:4 n-3	10	36	442
C20:5 n-3	5	141	53
C22:6 n-3	8	185	21
Skin:			
C18:4 n-3	111	286	3673
C20:5 n-3	31	1058	317
C22:6 n-3	21	921	78

infusion of SDA soybean oil; by Kitessa and Young (2011) after feeding of rumen-protected SDA oil to dairy cows (see Table 10.15); by Mejia *et al.* (2010) in laying hens; and by Forster *et al.* (2011) in pacific white shrimp. Some authors (e.g. Gibbs *et al.*, 2010) consider the introduction of SDA oils in animal feed as a change in order to increase the intake of long-chain n-3 PUFA (polyunsaturated fatty acids) of humans.

However, for some polyunsaturated fatty acids upper limits also exist for human nutrition and one should be careful with supplementing such oils in animal nutrition (Miles *et al.*, 2004; Harris *et al.*, 2007; Schubert *et al.*, 2007; Hammond *et al.*, 2008). Therefore, animal body composition may also be an end point (see Table 10.2) of

animal feeding studies to measure the transfer of some ingredients of second-generation GM plants into animal tissues, milk or eggs. Animal body samples or products from animals such as milk, eggs, etc., should be considered and analysed adequately. Furthermore, efficiency studies with second-generation GM plants may also be used as or combined with studies to measure the digestibility/bioavailability of the newly expressed substances.

10.3.3 Conjugated linoleic acids

Apart from the transfer of fatty acids into animal products (see above), some fatty acids (e.g. conjugated fatty acids, CLA) may

Table 10.14. Calculation of the influence of added oils on the intake and pool size of n-3 fatty acids in edible tissues (breast and leg meat with skin) in broilers. (From Rymer *et al.*, 2011.)

Fatty acids	Control	Supplemented fish oil	Supplemented SDA oil
Intake of n-3 fatty acids (mg/broiler):			
C18:3	17,383	6,415	19,717
C18:4	0	3,133	51,219
C20:5	0	13,724	0
C22:5	3	4,919	7
C22:6	6	13,611	0
Pool size (mg/broiler) of fatty acids in edible tissues:			
C18:3	3,840	1,662	4,884
C18:4	227	690	9,645
C20:5	155	2,627	975
C22:5	234	1,777	1,126
C22:6	131	3,110	331

Table 10.15. Concentrations (mg/l) of some fatty acids in whole milk on the initial and the final days (10th day) of supplementation with ruminally protected SDA oil (offered about 40 g C18:4 n-3 per cow and day; SDA, intake about 30 g; $n = 5$). (From Kitessa and Young, 2011.)

Fatty acid	Initial value (Day 1)	Final values (Day 10)
C18:1 n-9, oleic acid	8,880	18,582
C18:2 n-6, linoleic acid	600	746
C18:3 n-3, linolenic acid	463	877
C18:4n-3, stearidonic acid	18	144
C20:5 n-5, EPA	13	76
C22:5 n-3, DPA	45	65
Total n-3	559	1,162
Total n-6	622	886

influence the fatty deposition in both the animal body and the human body (Akahoshi *et al.*, 2005; Ostrowska *et al.*, 2005) or should have specific health effects (Lunn and Theobald, 2006).

Such influences have been reported with a genetically modified rapeseed (Kohno-Murase *et al.*, 1994; Hornung *et al.*, 2002; Iwabuchi *et al.*, 2003) containing the conjugated linolenic acid isomer *cis*-9,*trans*-11,*cis*-13-octadecatrienoic acid (punicic acid) in the oil. The oil was added to the diets of mice and effected a significant decrease of white adipose tissue from 0.94 (control) to 0.61 g/100 g of body weight (+0.25% punicic acid; Koba *et al.*, 2007). Punicic acid was also found in the fatty acid profiles of liver triglicerides and liver phospholipids of mice fed with GM rapeseed oil. This result shows that it would be possible to produce CLA via plants.

10.3.4 Higher concentration in micronutrients (e.g. the nutrient precursor, ß-carotene)

Micronutrient deficiencies such as trace elements (e.g. iodine, iron, zinc) and vitamins (e.g. A, E, B_{12} and further B vitamins) in humans and animals are a serious problem in many countries (see Chapters 7 and 12). Green biotechnology may contribute towards overcoming this situation via biofortification (DellaPenna, 1999; Zimmermann *et al.*, 2004).

Interesting results have been reported with the so-called 'Golden Rice', rich in the vitamin A precursor, ß-carotene (Dawe *et al.*, 2002; see Chapters 7 and 12). Table 7.8 shows a study to assess the bioconversion of ß-carotene into vitamin A in a model animal (Mongolian gerbils). Different diets were fed after a depletion period. The results show that the retinol concentration in the liver of gerbils fed with carotene-rich maize was similar to those animals fed with maize poor in carotene and supplemented with adequate amounts of ß-carotene. This means, in this case, that ß-carotene from maize is almost identically converted into vitamin A as supplemented ß-carotene, but to a lower

extent than vitamin A (see Table 7.8). Apart from specific storage or indicator organs (e.g. liver), other organs or tissues are seldom influenced in composition by biofortified GM plants.

10.4 Conclusions

Feeds from first-generation GM plants did not influence significantly the composition and quality of food of animal origin. There exists no scientific advice that recombinant DNA and newly expressed proteins show chemical and physiological properties in animals other than endogenous products. Feeds from second-generation GM crops (with output traits) may influence the composition and quality of feeds of animal origin, especially in the case of fatty acids. Minerals and vitamins from biofortified plants can be stored in some organs or may be excreted via milk and/or eggs.

References

Akahoshi, A.A., Koba, K.K., Enmoto, R.R., Nishimura, K.K., Honda, Y.Y., Minami, M.M., *et al.* (2005) Combined effects of dietary protein type and fat level on the body fat-reducing activity of conjugated linoleic acid (CLA) in rats. *Bioscience, Biotechnology and Biochemistry* 69, 2409–2415.

Alexander, T.W., Sharma, R., Okine, E.K., Dixon, W.T., Forster, R.J., Stanford, K., *et al.* (2002) Impact of feed processing and mixed ruminal culture on the fate of recombinant EPSP synthase and endogenous canola plant DNA. *FEMS Microbiology Letters* 214, 263–269.

Alexander, T.W., Sharma, R., Deng, M.Y., Whetsell, A.J., Jennings, J.C., Wang, Y., *et al.* (2004) Use of quantitative real-time and conventional PCR to assess the stability of the *cp4 epsps* transgene from Roundup Ready canola in the intestinal, ruminal and fecal content of sheep. *Journal of Biotechnology* 11, 255–266.

Alexander, T.W., Reuter, T., Aulrich, K., Sharma, R., Okine, E.K., Dixon, W.T., *et al.* (2007) A review of the detection and fate of novel plant molecules derived from biotechnology in livestock production. *Animal Feed Science and Technology* 133, 31–62.

Antoniou, M., Robinson, C. and Fagan, J. (2012) *GMO Myths and Truths*. Earth Open Source, London, 123 pp.

Aulrich, K., Böhme, H., Daenicke, R., Halle, I. and Flachowsky, G. (2001) Genetically modified feeds in animal nutrition. 1st Communication: *Bacillus thuringiensis* (Bt) corn in poultry, pig and ruminant nutrition. *Archives of Animal Nutrition* 54, 183–195.

Aulrich, K., Pahlow, G. and Flachowsky, G. (2004) Influence of ensiling on the DNA-degradation in isogenic and transgenic corn. *Proceedings of the Society of Nutrition Physiology* 13, 112 (Abstract).

Banz, W.J., Davis, J.E., Clough, R.W. and Cheatwood, J.L. (2012) Stearidonic acids: Is there a role in the prevention and management of Type 2 Diabetes Mellitus? *Journal of Nutrition* 142, 635S–640S.

Barriere, Y., Verite, R., Brunschig, P., Surault, F. and Emile, J.C. (2001) Feeding value of corn silage estimated with sheep and dairy cows is not altered by genetic incorporation of Bt176 resistance to Ostrinia nubilalis. *Journal of Dairy Science* 84, 1863–1871.

Bennett, D.E. and Cheng, K.M. (2010) Selenium enrichment in table eggs. *Poultry Science* 89, 2166–2172.

Berger, B., Aulrich, K., Fleck, G. and Flachowsky, G. (2003) Influence of processing of isogenic and transgenic rapeseed on DNA-degradation. *Proceedings of the Society of Nutrition Physiology* 12, 108 (Abstract).

Berger, L.L., Robbins, N.D., Sewell, J.R., Stanisiewski, E.P. and Hartnell, G.F. (2003) Effects of feeding diets containing corn grain with corn rootworm protection (event MON863), control or conventional varieties on steer feedlot performance and carcass characteristics. *Journal of Animal Science* 81 (Suppl. 1), 214, Abstract M150.

Bernal-Santos, B., O'Donell, S.M., Vicine, J.L., Hartnell, G.F. and Bauman, D.E. (2010) Hot topic: Enhancing omega-3 fatty acids in milk fat of dairy cows by using stearidonic acid-enriched soybean oil from genetically modified soybeans. *Journal of Dairy Science* 93, 32–37.

Bouis, H.E., Chassy, B.M. and Ochanda, J.O. (2003) Genetically modified food crops and their contribution to human nutrition and food quality. *Trends in Food Science and Technology* 14, 191–209.

Brown, P., Gettner, S. and Somerville, C. (1999) Genetic engineering of plant lipids. *Annual Review of Nutrition* 19, 197–216.

Brown, P.B., Wilson, K.A., Jonker, Y. and Nickson, T.E. (2003) Glyphosate tolerant canola meal is equivalent to the parental line in diets fed to rainbow trout. *Journal of Agricultural and Food Chemistry* 51, 4268–4272.

Buzoianu, S.G. (2011) Fate of transgenic DNA and protein in pigs fed genetically modified Bt maize and effects on growth and health. PhD thesis, Waterford Institute of Technology, Waterford, Ireland, 307 pp.

Cahoon, E.B., Shockey, J.M., Dietrich, C.R., Gidda, S.K., Mullen, T.T. and Dyer, J.M. (2007) Engineering oilseeds for sustainable production of industrial and nutritional feedstocks: solving bottlenecks in fatty acid flux. *Current Opinion in Plant Biology* 10, 236–244.

Calsamiglia, S., Hernandez, B., Hartnell, G.F. and Phipps, R. (2007) Effect of corn silage derived from genetically modified variety containing two transgenes on feed intake, milk production, and composition, and the absence of detectable transgenic deoxyribonucleic acid in milk of Holstein dairy cows. *Journal of Dairy Science* 90, 4718–4723.

Chainark, P., Satoh, S., Hirono, J., Aoki, T. and Endo, M. (2008) Availability of genetically modified feed ingredient: investigations of ingested foreign DNA in rainbow trout *Oncorhynchus mykiss*. *Fisheries Science* 74, 380–390.

Chiter, A., Forbes, M. and Blair, E. (2000) DNA stability in plant tissues: implications for the possible transfer of genes from genetically modified foods. *FEBS Letters* 481, 164–168.

Christianson, D.D., Khoo, U., Nielsen, H.C. and Wall, J.S. (1974) Influence of *Opaque-2* and *Floury-2* genes on formation of proteins in particulates of corn endosperm. *Plant Physiology* 53, 851–855.

Cleveland, B.J., Francis, D.S. and Turchini, G.M. (2012) Echium oil provides no benefit over linseed oil for (n-3) long-chain PUFA biosynthesis in rainbow trout. *Journal of Nutrition* 142, 1449–1455.

Corzo, A., Dozier, W.A. III and Kidd, M.T. (2006) Dietary lysine needs of late-developing heavy broilers. *Poultry Science* 85, 457–461.

Cromwell, G.L., Lindemann, M.D., Randolph, J.H. Parker, G.R., Coffey, R.D., Laurent, K.M. et al. (2002) Soybean meal from Roundup Ready or conventional soybeans in diets for growing finishing swine. *Journal of Animal Science* 80, 708–715.

Cromwell, G.L., Henry, B.J., Scott, A.L., Gerngross, M.F., Dusek, D.L. and Fletcher, D.W. (2005) Glufosinate herbicide-tolerant (Liberty Link) rice vs. conventional rice in diets for growing finishing swine. *Journal of Animal Science* 83, 1068–1074.

Dawe, D., Robertson, R. and Unnevehr, L. (2002) Golden rice: what role could it play in alleviation of vitamin A deficiency? *Food Policy* 27, 541–560.

DellaPenna, D. (1999) Nutritional genomics: manipulating plant micronutrients to improve human health. *Science* 285, 375–379.

Donkin, S.S., Velez, J.C., Totten, A.K., Stanisiewski, E.P. and Hartnell, G.F. (2003) Effects of feeding silage and grain from glyphosate-tolerant or insect-protected corn hybrids on feed intake, ruminal digestion, and milk composition in dairy cattle. *Journal of Dairy Science* 86, 1780–1788.

EFSA (2012) Guidance on the risk assessment of food and feed from genetically modified animals and on animal health and welfare aspects. *EFSA Journal* 10(1), 2501, 43 pp.

Einspanier, R. and Flachowsky, G. (2009) Biotechnology: potential absorption and biological functions of digested fragments of DNA or proteins in food animals fed genetically modified crops. In: Pond, W.G. and Bell, A.W. (eds) *Encyclopedia of Animal Science*. Taylor and Francis, London, DOI:10.1081/E-EAS-120045416.

Einspanier, R., Lutz, B., Rief, S., Berezina, O., Zverlov, V., Schwarz, W., *et al.* (2004) Tracing residual recombinant feed molecules during digestion and rumen bacterial diversity in cattle fed transgene maize. *European Food Research Technology* 218(3), 269–273.

Erickson, G.E., Robbins, N.D., Simon, J.J., Berger, L.L., Klopfenstein, T.J., Stanisiewski, E.P., *et al.* (2003) Effect of feeding glyphosate-tolerant (Roundup-Ready events GA21 or nk603) corn compared with reference hybrids on feedlot steer performance and carcass characteristics. *Journal of Animal Science* 81, 2600–2608.

FASS [Federation of Animal Science Societies] (2013) References – Feeding transgenic crops to livestock; dated 31 May 2013.

Flachowsky, G. (2009) Concentration and transfer of desired feed ingredients in food of animal origin. Conference Nutztierernährung morgen: Gesunde Tiere – effiziente und nachhaltige Erzeugung – wertvolle Produkte. Tagungsbericht, 2–4 September 2009. Schriftenreihe, ETH Zürich, Switzerland, pp. 204–218. [In German.]

Flachowsky, G., Schulz, E., Kratz, R. and Glodek, P. (2008) Effects of different dietary fat sources on the fatty acid profile of backfat and intramuscular fat in pigs of various sire breeds. *Journal of Animal and Feed Science* 17, 363–371.

Forster, I.P., Dominy, W.G., Obaldo, L.C., Hartnell, G.F., Sanders, E.F., Hickman, T.C., *et al.* (2011) The effect of soybean oil containing stearidonic acid on growth performance, n-3 fatty acid deposition and sensory characteristics of pacific white shrimp (*Litopenaeus vannamei*). *Aquaculture Nutrition* 17, 200–213.

Franke, K., Meyer, U., Wagner, H. and Flachowsky, G. (2009) Influence of various iodine supplementations and two different iodine species on the iodine content of milk of cows fed with rapeseed meal or distillers grain with solubles as protein source. *Journal of Dairy Science* 92, 4514–4523.

Gawienowski, M.C., Eckhoff, S.R., Yang, P., Raypati, P.J., Binder, T. and Briskin, D.P. (1999) Fate of maize DNA during steeping, wet milling, and processing. *Cereal Chemistry* 76, 371–374.

GfE [Gesellschaft für Ernährungsphysiologie; Society of Nutrition Physiology] (1999) *Recommendations for Energy and Nutrient Requirements for Laying Hens and Broilers*. DLG-Verlag, Frankfurt/Main, Germany, 185 pp. [In German.]

GfE (2008) *Recommendations for the Supply of Energy and Nutrients to Pigs*. DLG-Verlag, Frankfurt/Main, Germany, 245 pp.

Gibbs, R.A., Rymer, C. and Givens, D.I. (2010) Long-chain n-3 PUFA: intakes in the UK and the potential of a chicken meat prototype to increase them. *Proceedings of the Nutrition Society* 69, 144–155.

Glasser, F., Ferlay, A. and Chilliard, Y. (2008) Oilseed lipid supplements and fatty acid composition of cow milk: a meta-analysis. *Journal of Dairy Science* 91, 4687–4703.

Gruber, H., Paul, V., Guertler, P., Spiekers, H., Tichopad, A., Meyer, H.H.D., *et al.* (2011) Fate of Cry1Ab protein in agricultural systems under slurry management of cows fed genetically modified maize (*Zea mays* L.) MON810: a quantitative assessment. *Journal of Agricultural and Food Chemistry* 59(13), 7135–7144.

Guertler, P., Lutz, B., Kuehn, R., Meyer, H.H.D., Einspanier, R., Killermann, B., *et al.* (2008) Fate of recombinant DNA and Cry1Ab protein after ingestion and dispersal of genetically modified maize in comparison to rapeseed by fallow deer (*Dama dama*). *European Journal of Wildlife Research* 54, 36–43.

Guertler, P., Paul, V., Albrecht, C. and Meyer, H.H.D. (2009) Sensitive and highly specific quantitative real-time PCR and ELISA for recording a potential transfer of novel DNA and Cry1Ab protein from feed into bovine milk. *Analytical and Bioanalytical Chemistry* 393, 1629–1638.

Hammond, B.G., Lemen, J.K., Ahmed, G., Miller, K.D., Kirkpatrick, J. and Fleeman, T. (2008) Safety assessment of SDA soybean oil: results of a 28-day study and a 90-day/one generation reproduction feeding study in rats. *Regulatory Toxicology and Pharmacology* 52, 311–323.

Harris, W., DiRienzo, M., Sands, S., Geupge, C., Jones, P. and Eapen, A. (2007) Stearidonic acid increases the red blood cells and heart eicosapentaenoic acid content in dogs. *Lipids* 42, 325–333.

Harris, W., Kris-Etherton, P. and Harris, K. (2008) Intakes of long chain omega-3 fatty acid associated with reduced risk for death from coronary heart disease in healthy adults. *Current Atherosclerosis Report* 10, 503–509.

Hoffmann, W., Madra, A. and Plarre, W. (1985) *Textbook of Breeding of Agricultural Plants*, Vol 2, Special Part, 2nd edn. Paul Parey, Singhofen, Germany, 424 pp.

Hornung, E., Pernstich, C. and Feussner, I. (2002) Formation of conjugated Δ11Δ13-double bonds by Δ12-linoleic acid (1,4)-acyl-lipid-desaturase in pomegranate seeds. *European Journal of Biochemistry* 269, 4852–4859.

Hupfer, C., Mayer, J., Hotzel, H., Sachs, K. and Engel, K.-H. (1999) The effect of ensiling on PCR-based detection of GM-Bt maize. *European Food Research Technology* 209, 301–304.

Hyan, Y., Bressner, G.E., Ellis, M., Lewis, A.J., Fischer, R., Stanisiewski, E.P., *et al.* (2004) Performance of growing-finishing pigs fed diets containing Roundup Ready corn (event nk603), a non-transgenic genetically similar corn, or conventional corn lines. *Journal of Animal Science* 82, 571–580.

ILSI (2007) *Best Practices for the Conduct of Animal Studies to Evaluate Crops Genetically Modified for Output Traits*. International Life Science Institute, Washington, DC, 194 pp.

Iwabuchi, M., Kohno-Murase, J. and Imamura, J. (2003) Δ12-Oleate desaturase-related enzymes associated with formation of conjugated trans Δ11, cis-Δ13 double bonds. *Journal of Biology and Chemistry* 278, 4603–4610.

James, M.J., Ursin, V.M. and Cleland, L.G. (2003) Metabolism of stearidonic acid in human subjects: comparison with the metabolism of other n-3 fatty acids. *American Journal of Clinical Nutrition* 77, 1140–1145.

Kan, C.A. and Hartnell, G. (2004a) Evaluation of broiler performance when fed Roundup-Ready wheat (Event MON 71800), control, and commercial wheat varieties. *Poultry Science* 83, 1325–1334.

Kan, C.A. and Hartnell, G.F. (2004b) Evaluation of broiler performance when fed insect-protected, control and commercial varieties of dehulled soybean meal. *Poultry Science* 83, 2029–2038.

Kitessa, S.M. and Young, P. (2011) Enriched milk fat with n-3 polyunsaturated fatty acids by supplementing grazing dairy cows with ruminally protected echium oil. *Animal Feed Science and Technology* 170, 35–44.

Koba, K., Imamura, J., Akoshoshi, A., Kohno-Murase, J., Nishizono, S., Iwabuchi, M., *et al.* (2007) Genetically modified rapeseed oil containing cis-9,trans-11,cis-13-octadecatrienoic acid affects body fat mass and lipid metabolism in mice. *Journal of Agricultural and Food Chemistry* 55, 3741–3748.

Kohno-Murase, J., Murase, M., Ichikawa, H. and Imamura, J. (1994) Effects of an antisens napin gene on seed storage compounds in transgenic *Brassica napus* seeds. *Plant Molecular Biology* 26, 1115–1124.

Leeson, S. and Caston, L.J. (2003) Vitamin enrichment in eggs. *Journal of Applied Poultry Research* 12, 24–26.

Lucas, D.M., Taylor, M.L., Hartnell, G.F., Nemeth, M.A., Glenn, K.C. and Davis, S.W. (2007) Broiler performance and carcass characteristics when fed diets containing lysine maize (Ly038 or Ly038 × Mon 810), control, or conventional reference maize. *Poultry Science* 86, 2152–2161.

Lunn, J. and Theobald, H.E. (2006) The health effects of dietary unsaturated fatty acids. *Nutrition Bulletin* 31, 178–224.

Lutz, B., Wiedemann, S., Einspanier, R., Mayer, J. and Albrecht, C. (2005) Degradation of Cry1Ab protein from genetically modified maize in the bovine gastrointestinal tract. *Journal of Agricultural and Food Chemistry* 53, 1453–1456.

Maruyama, N., Mikami, B. and Utsumi, S. (2011) The development of transgenic crops to improve human health by advanced utilization of seed storage proteins. *Bioscience, Biotechnology and Biochemistry* 75, 823–828.

Mayer, J.E., Pfeiffer, W.H. and Beyer, P. (2008) Biofortified crops to alleviate micronutrient malnutrition. *Current Opinion in Plant Biology* 11, 166–170.

Mazza R., Soave M., Morlacchini, M., Piva, G. and Marocco, A. (2005) Assessing the transfer of genetically modified DNA from feed to animal tissues. *Transgenic Research* 14, 775–785.

Mejia, L., Jacobs, C.M., Utterback, P.L., Parsons, C.M., Rice, D., Sanders, C., *et al.* (2010) Evaluation of the nutritional equivalency of soybean meal with the genetically modified trait DP-3O5423-1 when fed to laying hens. *Poultry Science* 89, 2634–2639.

Mertz, E.T., Bates, L.S. and Nelson, O.E. (1964) Mutant gene that changes protein composition and increases lysine content of maize endosperm. *Science* 145, 279–280.

Miles, E.A., Banerjee, T., Dooper, M., M'Rabet, L., Graus, Y.M.F. and Calder, P.C. (2004) The influence of different combinations of gamma-linolenic acid, stearidonic acid and EPA on immune function in healthy young male subjects. *British Journal of Nutrition* 91, 893–903.

Mirghelenj, S.A., Golian, A. and Taghizadeh, V. (2009) Enrichment of chicken meat with long chain omega-3 fatty acids through dietary fish oil. *Research Journal of Biological Science* 4, 604–608.

Naber, E.C. and Squires, M.W. (1993) Vitamin profiles of eggs as indicators of nutritional status in the laying hens: diet to egg transfer and commercial flock survey. *Poultry Science* 72, 1046–1053.

Napier, J.A. (2007) The production of unusual fatty acids in transgenic plants. *Annual Review of Plant Biology* 58, 295–319.

Nelson, E.E., Mertz, E.T. and Bates, L.S. (1965) Second mutant gene affecting the amino acid pattern of maize endosperm proteins. *Science* 150, 1469–1470.

Nestel, P., Bouis, H.E., Meenakshi, J.V. and Pfeiffer, W. (2006) Biofortification of staple food crops. *Journal of Nutrition* 136, 1064–1067.

NRC [National Research Council] (1994) *Nutrient Requirements of Poultry*, 9th edn. National Academy Press, Washington, DC.

NRC (1998) *Nutrient Requirements of Swine*, 10th edn. National Academy Press, Washington, DC, 212 pp.

Ostrowska, E., Cross, R.F., Warner, R.D., Muralitharan, M., Bauman, D.E. and Dunshea, F.F. (2005) Dietary conjugated linoleic acid improves carcass leanness without altering meat quality in the growing pig. *Australian Journal of Experimental Agriculture* 45, 691–697.

Paul, V., Guertler P., Wiedemann, S. and Meyer, H.D.D. (2010) Degradation of Cry1Ab protein from genetically modified maize (MON810) in relation to total dietary feed proteins in dairy cow digestion. *Transgenic Research* 19, 683–689.

Pfeiffer, W.H. and McClafferty, B. (2007) HarvestPlus: breeding crops for better nutrition. *Crop Science* 47, S88–S105.

Ran, T., Mei, L., Lei, W., Aihua, L., Ru, H. and Jie, S. (2009) Detection of transgenic DNA in tilapias (*Oreochromis niloticus*, GIFT strain) fed genetically modified soybeans (Roundup Ready). *Aquaculture Research* 40, 1350–1357.

Reuter, T., Aulrich, K. and Berk, A. (2002) Investigations on genetically modified maize (BT-maize) in pig nutrition: fattening performance and slaughtering results. *Archives of Animal Nutrition* 56, 319–326.

Röttger, A.S., Halle, I., Wagner, H., Breves, G., Dänicke, S. and Flachowsky, G. (2012) The effects of iodine level and source on iodine carry-over in eggs and body tissues of laying hens. *Archives of Animal Nutrition* 66, 385–401.

Rymer, C., Hartnell, G.F. and Givens, D.I. (2011) The effect of feeding modified soyabean oil enriched with C18:4n-3 to broilers on the deposition of n-3 fatty acids in chicken meat. *British Journal of Nutrition* 105, 866–878.

Sahlin, A. and House, J.D. (2006) Enhancing the vitamin content of meat and eggs: implications for the human diet. *Canadian Journal of Animal Science* 86, 181–195.

Sanden, M., Johannessen, L.E., Berdal, K.G., Sissener, N. and Hemre, G.I. (2011) Uptake and clearance of dietary DNA in the intestine of Atlantic salmon (*Salmo salar* L.) fed conventional or genetically modified soybeans. *Aquaculture Nutrition* 17, E750–E759.

Scheideler, S.E., Hileman, R.E., Weber, T., Robeson, T. and Hartnell, G.F. (2008) The *in vivo* digestive fate of the Cry3Bb1 protein in laying hens fed diets containing MON 863 corn. *Poultry Science* 87, 1089–1097.

Schubert, R., Kitz, R., Beermann, C., Rose, M.A., Baer, P.C., Zielen, S. and Boehles, H. (2007) Influence of low-dose polyunsaturated fatty acids supplementation on the inflammatory response of healthy adults. *Nutrition* 23, 724–730.

Seralini, G.E., Mesnage, R., Clair, E., Gress, S., Vendomois, J.S. de and Cellier, D. (2011) Genetically modified crops safety assessments: present limits and possible improvements. *Environmental Science Europe* 23, 10, DOI:10.1186/2190-4715-23-10.

Sharma R., Damgaard, D., Alexander, T.W., Dugan, M.E.R., Aalhus, J.L., Stanford, K., *et al.* (2006) Detection of transgenic and endogenous plant DNA in digesta and tissues of sheep and pigs fed Roundup Ready canola meal. *Journal of Agricultural and Food Chemistry* 54, 1699–1709.

Sodek, L. and Wilson, C.M. (1971) Amino acid composition of proteins isolated from normal, *opaque-2*, and *floury-2* corn endosperm by a modified Osborne procedure. *Journal of Agricultural and Food Chemistry* 19, 1144–1150.

Speight, S.M., Estienne, M.J., Harper, A.F., Barb, C.R. and Pringle, T.D. (2012) Effects of organic selenium supplementation on growth performance, carcass measurements, tissue selenium concentrations, characteristics of reproductive organs, and testis gene expression profile in boars. *Journal of Animal Sciences* 90, 533–542.

Spiekers, H., Meyer, H.H.D. and Schwarz, F.J. (2009) Effects of long term utilization of genetically modified maize (MON 810) in dairy cattle feeding on performance and metabolism parameters. *Schriftenreihe, Bayerische Landesanstalt für Landwirtschaft* (LfL), Vol 18, 105 pp. [In German.]

Stadnik, J., Karwowska, M., Dolatowski, Z.J., Swiatkiewicz, S. and Kwiatek, K. (2011) Effect of genetically modified insect resistant corn (MON 810) and glyphosate tolerant soybean meal (Roundup Ready) on physico-chemical properties of broilers' breast and thigh muscles. *Bulletin of the Veterinary Institute in Pulawy* 55, 541–546.

Stanford, K., Aalhus, J.L., Dugan, M.E.R., Wallins, G.L., Sharma, R. and McAllister, T.A. (2003) Effects of feeding transgenic canola on apparent digestibility, growth performance and characteristics of lambs. *Journal of Animal Science* 83, 299–305.

Sünder, A., Halle, I. and Flachowsky, G. (1999) Vitamin-E hypervitaminosis in laying hens. *Archives of Animal Nutrition* 52, 185–194.

Surette, M., Edens, M. and Chilton, F. (2004) Dietary echium oil increases plasma and neutrophil long-chain (n-3) fatty acids and lowers serum triacylglycerols in hyper-triglyceridemic humans. *Journal of Nutrition* 134, 1406–1411.

Swiatkiewicz, S., Swiatkiewicz, M., Koreleski, J. and Kwiatek, K. (2010a) Nutritional efficiency of genetically-modified insect resistant corn (MON810) and glyphosate-tolerant soybean meal (Roundup Ready) for broilers. *Bulletin of the Veterinary Institute in Pulawy* 54, 43–48.

Swiatkiewicz, S., Twardowska, M., Markowsky, J., Mazur, M., Sieradzki, Z. and Kwiatek, K. (2010b) Fate of transgenic DNA from Bt corn and Roundup Ready soybean meal in broilers fed GMO feed. *Bulletin of the Veterinary Institute in Pulawy* 54, 237–242.

Swiatkiewicz, M., Hanczakowska, E., Twardowska, M., Mazur, M., Kwiatek, K., Kozaczynski, W., *et al.* (2011) Effect of genetically modified feeds on fattening results and transfer of transgenic DNA to swine tissues. *Bulletin of the Veterinary Institute in Pulawy* 55, 121–125.

Taylor, M.L., Stanisiewski, E.P., Riordan, S.G., Nemeth, M.A., George, B. and Hartnell, G. (2004) Comparison of broiler performance when fed diets containing Roundup Ready (Event RT73), non-transgenic control, or commercial canola meal. *Poultry Science* 83, 456–461.

Tudisco, R., Lombardi, P., Bovera, F., d'Angelo, D., Cutrignelli, M.I., Mastelione, V., *et al.* (2006) Genetically modified soya bean in rabbit feeding: detection of DNA fragments and evaluation of metabolic effects by enzymatic analysis. *Animal Science* 82, 193–199.

Tudisco, R., Mastellone, V., Cutrignelli, M., Lombardi, P., Bovera, F., Mirabella, N., *et al.* (2010) Fate of transgenic DNA and metabolic effects in goats fed genetically modified soybean and in their offsprings. *Animal* 4, 1662–1671.

Walsh, M.C., Buzoianu, S.G., Gardiner, G.E., Rea, M.C., Ross, R.P., Cassidy, J.P., *et al.* (2012) Effects of short term feeding of Bt MON810 maize on growth performance, organ morphology and function in pigs. *British Journal of Nutrition* 107, 364–371.

Welch, R.M. and Graham, R.D. (2004) Breeding for micronutrients in staple food crops from human nutrition perspectives. *Journal of Experimental Botany* 55, 353–364.

Whelan, J., Gouffon, J. and Zhao, Y. (2012) Effects of dietary stearidonic acid on biomarkers of lipid metabolism. *Journal of Nutrition* 142, 630S–634S.

Wiedemann, S., Lutz, B., Kurtz, H., Schwarz, F.J. and Albrecht, C. (2006) In situ studies on the time-dependent degradation of recombinant corn DNA and protein in the bovine rumen. *Journal of Animal Science* 84, 135–144.

Wolf, M.J., Khoo, U. and Seckinger, H.S. (1967) Subcellular structure of endosperm protein in high-lysine and normal corn. *Science* 157, 556–557.

Zimmermann, R., Stein, A. and Qaim, M. (2004) Agricultural technology to fight micronutrient malnutrition? A health economics approach. *Agrarwirtschaft* 53(2), 67–76. [In German.]

11 Feed Additives Produced by GM Microorganisms (GMMs)

Atte von Wright*

University of Eastern Finland, Institute of Public Health and Clinical Nutrition, Finland

11.1 Introduction

Genetically modified microorganisms (GMMs) are powerful biotechnological tools in producing enzymes, medically important proteins and chemicals. The possibilities of metabolic engineering and targeted mutagenesis of desired genes have further increased the usefulness of GMMs in their various industrial applications.

Regarding the potential fields of application of GMMs in the production of feeds, they could be used as such as animal probiotics, digestibility enhancers, silage additives or preservatives. Although this remains a possibility, the more immediate applications include the use of GMM-produced enzymes and amino acids for technological and nutritional purposes. In addition, GMM biomasses obtained as side streams (co-product of feed additives) of biotechnological industries could be used after nutritional and safety assessment as feed ingredients.

In the present chapter, these potential and actual applications are reviewed, paying particular attention to the safety, efficacy and regulatory aspects of GMMs in feed uses. The focus is mainly on the practices in the European Union (EU).

11.2 What is a GMM?

The history of genetic modification, or recombinant DNA technology, actually started with the introduction of GMMs in the early 1970s, when the basic cloning techniques, combining the use of restriction enzymes, DNA ligases and suitable vector plasmids to create recombinant plasmids that could then be introduced into recipient cells by different transformation techniques, were developed (Cohen *et al.*, 1973). Subsequently, the techniques have been considerably refined with the design of expression and integration vectors, allowing optimal function and stability of the cloned genes. Because of the advantages of microorganisms in biotechnology (easy cultivation and containment, cheap substrates, efficient downstream processing), GMMs soon found wide applications, both in traditional microbiological biotechnology and in complete novel applications, such as the production of hormones and vaccines.

In the legal context, the use of terminology is sometimes different from the strictly scientific conventions. Thus, according to Article 2 of EU directive 2009/41/EC (OJEU, 2009) "micro-organism" means any

*E-mail: atte.vonwright@uef.fi

microbiological entity, cellular or non-cellular, capable of replication or of transferring genetic material, including viruses, viroids, and animal and plant cells in culture' and '"genetically modified micro-organism" (GMM) means a micro-organism in which the genetic material has been altered in a way that does not occur naturally by mating and/or natural recombination'. It should be noted that, in addition to actual microorganisms (bacteria, fungi, protozoa, microalgae, viruses and viroids), cultured animal and plant cells, when they are genetically modified, are also legally considered as GMMs.

11.3 The Traditional Use of Microorganisms and Derived Products in Feeds

11.3.1 Microorganisms added directly into feed

Live microorganisms have been used in feeds for technological purposes (silage additives) and as zootechnological additives or animal probiotics intended to affect the performance of production animals or pets favourably. In the current EU Register of Feed Additives (European Commission, 2012), there are nearly 100 microorganisms (mainly lactic acid bacteria) registered as technological additives (silage starters). In addition, there are around 70 authorized microbial products (Bacilli, Lactobacilli, Enterococci, yeasts) that are used as animal probiotics. The target animal categories include poultry, piglets and pigs for fattening, ruminants, salmonids and even shrimps.

In the case of silage additives, the mechanism of action is simply acid production, which inhibits the microbial deterioration of the ensiled material. With animal probiotics, although there is evidence of improved animal performance, the mechanisms of action are still unknown. In the case of yeasts, which are particularly efficient in ruminants, the effects may be based on the consumption of residual oxygen in the rumen and on the resulting

enhancement of the anaerobic microbiota. The current status of animal probiotics has been reviewed extensively by Bomba et al. (2011).

11.3.2 Microbial enzymes

Several enzymes are used in animal nutrition to improve the digestibility and nutritive value of feeds, as recently summarized by Bedford and Partridge (2010). The most important of those are enzymes that degrade complex polysaccharides or anti-nutritive compounds, although other types of enzymes, such as proteases or oligosaccharide-degrading enzymes, are also sometimes used.

The endosperms of most cereals used in animal nutrition are rich in arabinoxylans and different types of beta-glucans. The tendency of these polysaccharides to form viscous colloids decreases their digestibility, particularly in monogastric animals like pig and poultry. The application of enzymes like xylanases and glucanases to cereal-based feeds is an old and well-established methodology to improve the nutritional value of cereal-based feeds (Annison, 1992; Bedford and Classen, 1993).

Many of the enzymes involved in the breakdown of complex plant polysaccharides are of fungal origin (van den Brink and de Vries, 2011), and strains of *Aspergillus* and *Trichoderma* have been used extensively for the biotechnological production of both xylanases and glucanases.

Phytases, or enzymes that remove orthophosphates from phytic acid or myo-inosito hexakisphosphates, are another example of enzymes widely used in animal nutrition (Pallauf and Rimbach, 1997). Much of the dietary phosphorus can be incorporated in phytic acid and thus unavailable to the animal. Moreover, phytic acids are efficient chelators of zinc, iron, calcium and magnesium, thus affecting the bioavailability of these minerals. Phytases occur widely in nature among microorganisms and plants but not in animals (see Chapters 7 and 12). Biotechnologically produced phytases are predominantly of fungal origin, although several bacterial species also produce their own phytases (Haefner *et al.*, 2005; Yao *et al.*, 2011). Although some cereals, such as rye, wheat and barley, have their own phytases, some other cereals, such as maize, lack phytase activity (Eeckhout and de Pape, 1994). Moreover, plant phytases may be inactivated during feed processing, and thus supplementation of external phytase often increases the nutritional value of feeds significantly.

The EU Register of Feed Additives (European Commission, 2012) currently includes tens of different enzyme preparations aimed at monogastric animals, most of them endo-1,3(4)-beta-glucanases either alone or in combination with endo-1,4-beta-xylanases or endo-1,3(4)-beta-glucanase, and different phytases. However, other types of enzymes (such as amylases and proteinases) are also being used, and some enzyme preparations are also intended for ruminants.

11.3.3 Microbially produced amino acids

The microbial production of amino acids is a well-established technology, and currently feed uses cover approximately more than half of the global market value of amino acids (Leuchtenberger *et al.*, 2005). Quantitatively, the most important of these is L-lysine as a limiting amino acid in pigs and poultry. In the EU Register of Feed Additives (European Commission, 2012), there are several authorizations for L-lysine, L-methionine, L-histidine, L-valine and L-threonine.

11.3.4 Microbial biomasses

Microbial biomasses (usually obtained as side streams of various biotechnological processes) are used to a certain extent as protein-rich feed ingredients. Although biomasses are thus not categorized as feed additives, they can anyway form a means to introduce GMM-derived materials in animal feed. While conventional microbial bio-

masses do not require specific authorization in the EU, GMM-based biomasses do (see Section 11.4.2).

11.4 The Safety Assessment of GMM-derived Feed Additives and Feed Ingredients

Almost from the beginning, recombinant DNA technology has created safety concerns. Particularly, the potential unintended effects and unforeseen environmental consequences have frequently been pointed out as the potential risks of genetically modified organisms (GMOs). In 1975, for the first time in history, the scientific community in the USA imposed voluntary restrictions on the freedom of research by imposing certain safety measures (Asilomar Conference, 1975; Berg *et al.*, 1975), which were followed in 1976 by specific guidelines from the National Institute of Health (NIH, 1976). Since then, both national and international legislation has been widely introduced to address the safety aspects of GMOs in research and in their various applications.

Regarding the specific applications to food and feed, different countries have adopted different regulatory approaches. One specific case is the USA, where GM foods and feeds are not subjected to any specific legislation, because of the basic assumption that the technology as such does not introduce any specific safety concerns that could not be addressed by the general food and feed legislation. Thus, the regulatory responsibilities in the USA are divided between the US Department of Agriculture (USDA), the US Environmental Protection Agency (EPA) and the US Food and Drug Administration (FDA). While USDA and the FDA oversee the agricultural and environmental aspects, the task of the FDA is to ensure the safety of food and feed according to the Federal Food, Drug and Cosmetic Act.

In contrast to the USA, the EU has introduced a specific and detailed regulatory framework to ensure the safety of GM foods and feed both for the consumer and for the environment (and in the case of feed additives, for the target animal also). In the EU, the relevant legal documents are Directives 2009/41/EC (OJEU, 2009) and 2001/18/EC (OJEC, 2001), on the contained use and deliberate release of GMOs, and particularly regarding the authorization of genetically modified (GM) food and feed, Regulation 1829/2003 EC (OJEU, 2003a).

In the authorization process, the European Food Safety Authority (EFSA) is in the central role as an independent body for safety assessment (see also Chapter 3). The actual assessment is performed by scientific panels, whose members are independent experts. After receiving the EFSA's opinion, the European Commission makes the final decision on authorization, with the help of the Standing Committee on the Food Chain and Animal Health (SCFCAH). The SCFCAH is formed by representatives of the member states, and a qualified majority (representing the majority of the member states and more than half of the EU population) has to be reached for a decision. If the SCFCAH cannot reach a decision, the matter is shifted to the Council of the European Union, consisting of the relevant cabinet ministers of the member states. If no conclusion is achieved even there, the European Commission has the final word.

Regarding GM feeds, the central EFSA scientific panels are the Panel on Genetically Modified Organisms (GMO) and the Panel on Additives and Products or Substances used in Animal Feed (FEEDAP). According to EFSA policy, the GMMs and derived products that are used for feeds are evaluated by FEEDAP but take into account the specific guidelines of the GMO Panel. The specific requirements for the safety assessment of GMMs are defined in a recent guidance document (EFSA, 2011a).

11.4.1 The safety requirements for GMMs and derived products as defined by the EFSA

The EFSA guidance document divides GMM products into four categories:

Category 1. Chemically defined purified compounds and their mixtures in which both GMMs and newly introduced genes have been removed (e.g. amino acids, vitamins).

Category 2. Complex products in which both GMMs and newly introduced genes are no longer present (e.g. cell extracts, most enzyme preparations).

Category 3. Products derived from GMMs in which GMMs capable of multiplication or of transferring genes are not present, but in which newly introduced genes are still present (e.g. heat-inactivated starter cultures).

Category 4. Products consisting of or containing GMMs capable of multiplication or of transferring genes (e.g. live starter cultures for fermented foods and feed).

From the point of safety assessment Categories 1 and 2 represent the simplest cases. The basic requirement is the demonstration of the absence of either GMMs or the recombinant DNA in the product. Categories 3 and 4 are more complicated, Category 4 being the most challenging. Generally, toxicological studies may be needed and the potential of horizontal gene transfer has to be assessed. With Category 4 products, the ability of a GMM itself to survive and multiply in different receiving environments and the resulting consequences also have to be assessed. Regarding feed use, it is also stated in the guideline that in each case the product has, in addition to the safety aspects related to the GMM, also to fulfil the general safety requirements for feed additives as defined in Regulation 1831/2003EC (OJEU, 2003b).

11.4.2 The general safety requirements for microbial feed additives

The basic outline for the general assessment of microorganisms and enzymes (and which is basically adapted to other microbially derived feed additives) is given in the guidance document formulated by the former Scientific Committee on Animal

Nutrition[1] (SCAN; European Commission, 2001). As all feed additives, microbial products should also be assessed for target animal safety, user safety, consumer safety and environmental safety, as well as for efficacy in target animals. Detailed guidance documents on how to perform the safety and efficacy assessments have been published by the FEEDAP Panel.

The tolerance test (with, if possible, at least tenfold overdoses) to establish the safety for target animals should be performed according to the instructions in the technical guidance on tolerance and efficacy studies in target animals (EFSA, 2008a). The aim is to provide a limited evaluation of the short-term toxicity of the additive and to establish a margin of safety if the additive is consumed at higher doses than are recommended.

The safety for the user (the person handling the feed) should be established according to the technical guidance on studies concerning the safety of the additive for users/workers (EFSA, 2008b). The studies include tests for respiratory toxicity (in case the additive contains more than 1% on a weight basis of particles with a diameter of 50 μm), skin and eye irritation and skin sensitization, all performed using the commercial formulation. Microbial additives, as proteinaceous substances, are considered automatically as respiratory sensitizers and the general recommendation is to treat them accordingly.

Regarding the consumer, the safety concern is the potential contamination of animal products by unknown microbial metabolites produced during the manufacturing process. Accordingly, both genotoxicity tests (assays for point mutations and clastogenicity) and 90-day repeated dose feeding studies on laboratory animals are formally required, unless the product is intended for companion animals only (EFSA, 2008c).

The environmental safety of microorganisms is usually assessed case by case.

With conventional microorganisms, the primary question is whether the intended use is going to increase significantly the levels of the microorganism in the receiving environment. If a viable GMM is to be used as a feed additive, then the environmental safety assessment should be done according to the principles laid out in the EFSA technical guidance of 2011a mentioned above.

To facilitate the safety assessment of microorganisms, the concept of a qualified presumption of safety (QPS) has been introduced (EFSA, 2007a). According to the QPS approach, a microorganism that has an established safety record can be notified to the EFSA without the studies for the target animal, consumer or environmental safety formally required in the guidance documents cited above. This applies also to microbial products, such as enzymes, derived from a QPS microorganism. The EFSA Panel on Biological Hazards (BIOHAZ) updates annually the list of QPS organisms. So far, no GMM has a QPS status and therefore feed additives produced by GMMs are, in principle, subject to the full safety assessment, even in the cases when the parental organism of the GMM is on the QPS list.

Microbial biomasses derived from GMMs are a special case. Formally, microbial biomasses used as feed ingredients do not need any specific notification. If, however, the biomasses consist of GMMs, their safety has to be assessed by the EFSA. According to the relevant guideline (EFSA, 2011b), the general principles of the GMM guidance (EFSA, 2011a) regarding the presence of viable GMMs and recombinant DNA in the product apply. Additionally, the document gives detailed instructions on the compositional analysis and for the experimental design to define the safe use level for target animals. Assessment of the user and consumer safety is done according to the principles outlined for microbial feed additives.

11.5 Examples of GMM Feed Additives Assessed and Authorized in the EU

No live GMM to be used as a feed additive has so far been notified in the EU for authorization. Given the controversial nature of the public acceptance of GMOs in the EU and the complicated environmental safety assessment associated with a deliberate release of GMMs, the likelihood of such a notification in the near future is remote. Moreover, as noted in Section 11.3.1, the mechanisms of probiotic action are still rather unknown and consequently the intentional enhancement of probiotic properties by genetic means waits for future breakthroughs in the understanding of the interactions between probiotic microorganisms and the host. Theoretically, GMMs with enhanced antimicrobial properties or enzymatic activities could be used in silage fermentations to control spoilage microorganisms or enhance acid production. Even then, the difficulties with public acceptance of such products undoubtedly discourage development of these applications.

In contrast to live GMMs, the prospects of GMM biomasses entering into the market as feed ingredients are much more likely, due to the prevalence of GMMs in different biotechnological processes and the consequent formation of such biomasses as industrial side streams. GMM biomasses have already been notified to the EFSA, although none has been authorized in the EU, yet.

The most relevant GMM products in the EU, so far, are enzymes and amino acids. Since 2006, the EFSA has assessed at least nine enzymes and two amino acids (L-valine and L-isoleucine) produced by GMMs, while several others are in the process of assess-ment. The complete assessments are summarized in Table 11.1.

11.6 Conclusions

Although live GMMs currently are not used directly in feeds as additives or ingredients, GMM products, like enzymes and amino acids, already have an established position as feed additives. It is to be expected that in the near future more and more of these types of product will be introduced. Genetic engineering will, undoubtedly, lead to even more versatile biotechnological products with optimized properties for each particular application.

Safety of the biotechnological products is, of course, of primary importance to consumers, regulators, industry and users. While there have been many public concerns related to GMOs, no adverse effects associated with these relatively highly purified GMM products containing no production organisms or recombinant DNA (Category 1 and 2 products) have been observed, either during the risk assessment or the subsequent use of the products. The adverse effects, such as skin or eye irritation observed in some cases, are specific for the product itself and its formulation, not a result of it being produced by a GMM. While this situation should not trigger complacency or undue relaxation of safety standards, it should be taken as an indication that the current safety assessment practices are effective.

Note

1 SCAN was an expert committee under the Directorate General for Health and Consumers that performed the tasks of the present FEEDAP Panel before the establishment of the EFSA.

Table 11.1. Summaries of the EFSA assessment of feed products produced using GMMs.

Product	Production organism	Target animals	Cloned gene(s)	Safety studies	Efficacy studies	Remarks
Natuphos ® (EFSA, 2006)	*Aspergillus niger*	Piglets, pigs for fattening, sows, poultry	3-Phytase from another *A. niger* strain	The standard set of studies performed.[a] No cause for concern identified.	Enhancement of phosphorus absorption in turkeys demonstrated.	The enzyme identical to another previously authorized enzyme produced by a conventional *A. niger* strain.
Danisco Xylanase G/L (EFSA, 2007b)	*Trichoderma reesei*	Poultry	Endo-1,4-beta-xylanase modified to enhance thermotolerance	The standard set of studies performed. No cause for concern identified.	Efficacy demonstrated in all target animal categories.	
Natugrain ® TS (EFSA, 2008d)	*Aspergillus niger*	Piglets and poultry	Endo-1,4-beta-xylanase and endo-1,4-beta-glucanase	The standard set of studies performed. No cause for concern identified.	Efficacy demonstrated in all target animal categories.	
Quantum™ Phytase (EFSA, 2008e)	*Pichia pastoris*	Poultry and piglets	6-Phytase	The standard set of studies performed. No cause for concern identified.	Efficacy demonstrated in all target animal categories.	
Econase XT P/L (EFSA, 2008d)	*Trichoderma reesei*	Poultry and piglets	Endo-1,4-beta-xylanase	The standard set of studies performed. No cause for concern identified.	Efficacy demonstrated in all target animal categories.	
Avizyme 1505 (EFSA, 2009)	*Trichoderma reesei* *Bacillus amyloliquefaciens* *Bacillus subtilis*	Poultry	Endo-1,4-beta-xylanase (in *T. reesei*), alpha-amylase (in *B. amyloliquefaciens*), subtilisin (in *B. subtilis*)	The standard set of studies performed (tests for consumer safety with each enzyme separately). No cause for concern identified.	Efficacy demonstrated in all target animal categories.	Both *Bacillus* strains contain antibiotic-resistance genes. However, since the production organisms and the recombinant DNA were absent from the final product, this was not considered as a safety concern.
Optiphos (EFSA, 2011c)	*Pichia pastoris*	Poultry, piglets and pigs	6-Phytase	The standard set of studies performed. No cause for concern identified.	Efficacy demonstrated in all target animal categories.	

Biogalactosidase BL (EFSA, 2011d)	*Saccharomyces cerevisiae* *Aspergillus niger*	Chickens for fattening	Alpha-galactosidase (in *S. cerevisiae*), beta-glucanase (*A. niger*)	The standard set of studies performed for each enzyme separately (except for sensitization, which was tested with glucanase only). Both enzymes were mildly irritant to skin and eyes, and glucanase was also a mild sensitizer.	Efficacy demonstrated in all target animal categories.	Although the product is considered as an irritant to skin and eyes and also a dermal irritant, it has been authorized.
Ronozyme RumiStar (EFSA, 2012)	*Bacillus licheniformis*	Dairy cows	Alpha-amylase	The standard set of studies performed, except for sensitization tests.	No conclusive evidence of efficacy demonstrated.	In the absence of test data, the product is considered as a potential sensitizer.
L-Valine feed grade (EFSA, 2008f)	*Escherichia coli*	All animal species	Two gene cassettes from *E. coli* strains increasing the capacity of L-valine production and enabling the use of variable carbon sources for production.	Absence of virulence factors in the production strain demonstrated, as well as absence of the production strain and recombinant DNA in the product. Genotoxicity and 90-day oral toxicity studies provided and the skin and eye irritation, sensitization and inhalation toxicity provided.	No actual efficacy studies provided.	The safety studies were all performed with >98% pure product, although the purity in the product specification was given as >95%. The product was authorized with a minimum purity specification of 98%.
L-Isoleucine (EFSA, 2010)	*Escherichia coli*	All animal species		Absence of virulence factors in the production strain demonstrated, as well as absence of the production strain and recombinant DNA in the product. A short-term study on piglets with a tenfold overdose performed. Genotoxicity and 90-day oral toxicity and irritation, sensitization and inhalation toxicity studies provided.	Demonstration of bioequivalence with pharmaceutical grade L-leucine in piglets.	

Note: [a]Absence of the production organism(s) and recombinant DNA demonstrated. Tolerance studies in target animals done, tests for skin and eye irritation and skin sensitization performed for user safety, as well as genotoxicity studies and 90-day rodent feeding studies to ensure consumer safety.

References

Annison, G. (1992) Commercial enzyme supplementation of wheat based diets raises ileal glycanase activities and improves apparent metabolizable energy, starch and pentosan digestibilities in broiler chickens. *Animal Feed Science and Technology* 38, 105–121.

Bedford, M.R. and Classen, H.L. (1993) An in vitro assay for prediction of broiler intestinal viscosity and growth when fed rye-based diets in the presence of exogenous enzymes. *Poultry Science* 72, 137–143.

Bedford, M.R. and Partridge, G.G. (ed.) (2010) *Enzymes in Farm Animal Nutrition*, 2nd edn. CAB International, Wallingford, UK, 319 pp.

Berg, P., Baltimore, D., Brenner, S., Roblin, R.O. and Singer, M.F. (1975) Summary Statement of the Asilomar Conference on Recombinant DNA. *Proceedings of the National Academy of Sciences USA* 72, 1981–1984.

Bomba, A., Nemcová, R., Strojný, L. and Mudronová, L. (2011) Probiotics for farm animals. In: Lahtinen, S., Ouwehand, A., Salminen, S. and von Wright, A. (eds) *Lactic Acid Bacteria. Microbiological and Functional Aspects.* CRC Press, Boca Raton, Florida, pp. 633–670.

Cohen, S.N., Chang A.C.Y., Boyer, H.W. and Helling, R.B. (1973) Construction of biologically functional bacterial plasmids *in vitro. Proceedings of the National Academy of Sciences USA* 70, 3240–3244.

Eeckhout, W. and De Pape, M. (1994) Total phosphorus, phytate phosphorus and phytase activity in plant feedstuffs. *Animal Feed Science and Technology* 47, 19–29.

EFSA (2006) Opinion of the Scientific Panel on Additives and Products or Substances used in Animal Feed and of the Scientific Panel on Genetically Modified Organisms on the safety and efficacy of the enzyme preparation Natuphos ® (3-phytase) produced by *Aspergillus niger. EFSA Journal* 369, 1–19.

EFSA (2007a) Opinion of the scientific committee on introduction of a qualified perception of safety (QPS) approach for assessment of selected microorganisms referred to EFSA. *EFSA Journal* 587, 1–16.

EFSA (2007b) Safety and efficacy of Danisco Xylanase G/l (endo1,4-beta-xylanase) as a feed additive for chickens for fattening, laying hens and ducks for fattening. Scientific Opinion of the Panel on Additives and Products or Substances used in Animal Feed and of the Panel on Genetically Modified Organisms. *EFSA Journal* 548, 1–18.

EFSA (2008a) Technical guidance. Tolerance and efficacy studies in target animals prepared by the Panel on Additives and Products or Substances used in Animal Feed. *EFSA Journal* 778, 1–13.

EFSA (2008b) Technical guidance. Studies concerning the safety of use of the additive for users/workers prepared by the Panel on Additives and Products or Substances used in Animal Feed. *EFSA Journal* 802, 1–2.

EFSA (2008c) Technical guidance for establishing the safety of additives for the consumer. Prepared by the Panel on Additives and Products or Substances used in Animal Feed. *EFSA Journal* 801, 1–12.

EFSA (2008d) Safety and efficacy of Natugrain ® TS (endo-1,4-β-xylanase and endo-1,4-β-glucanase) as a feed additive for piglets (weaned), chickens for fattening, laying hens, turkeys for fattening and ducks. Scientific Opinion of the Panel on Additives and Products or Substances used in Animal Feed and of the Panel on Genetically Modified Organisms. *EFSA Journal* 914, 1–19.

EFSA (2008e) Safety and efficacy of the product Quantum™ Phytase 5000 L and Quantum™ Phytase 2500 D (6-phytase) as a feed additive for chickens for fattening, laying hens, turkeys for fattening, ducks for fattening and piglets (weaned). Scientific Opinion of the Panel on Additives and Products or Substances used in Animal Feed and of the Panel on Genetically Modified Organisms. *EFSA Journal* 627, 1–27.

EFSA (2008f) Efficacy and safety of L-valine from a modified *E. coli* K12 for all animal species. Scientific Opinion of the Panel on Additives and Products or Substances used in Animal Feed and of the Panel on Genetically Modified Organisms. *EFSA Journal* 695, 1–21.

EFSA (2009) Safety and Efficacy of Avizyme 1505 (endo-1,4-β-xylanase, α-amylase, subtilisin) as a feed additive for chickens and ducks for fattening. Scientific Opinion of the Panel on Additives and Products or Substances used in Animal Feed and of the Panel on Genetically Modified Organisms. *EFSA Journal* 1156, 1–25.

EFSA (2010) Scientific opinion on the safety and efficacy of L-isoleucine for all animal species. EFSA Panel on Additives and Products or Substances used in Animal Feed and Panel on Genetically Modified Organisms. *EFSA Journal* 1425, 1–19.

EFSA (2011a) Guidance on the risk assessment of genetically modified microorganisms and their products intended for food and feed use: EFSA Panel on Genetically Modified Organisms (GMO). *EFSA Journal* 2193, 1–54.

EFSA (2011b) Guidance on the assessment of microbial biomasses for use in animal nutrition: EFSA Panel on Additives and Products or Substances used in Animal Feed (FEEDAP). *EFSA Journal* 2117, 1–8.

EFSA (2011c) Scientific opinion on the safety and efficacy of Optiphos® (6-phytase) as a feed additive for chickens and turkeys for fattening, chickens reared for laying, turkeys reared for breeding, laying hens, other birds for fattening and laying, weaned piglets, pigs for fattening and sows. EFSA Panel on Additives and Products or Substances used in Animal Feed and Panel on Genetically Modified Organisms. *EFSA Journal* 2414, 1–2.

EFSA (2011d) Scientific opinion on the safety and efficacy of Biogalactosidase BL (alpha-galactosidase and beta-glucanase) as feed additive for chickens for fattening. EFSA Panel on Additives and Products or Substances used in Animal Feed and Panel on Genetically Modified Organisms. *EFSA Journal* 2451, 1–19.

EFSA (2012) Scientific opinion of the safety and efficacy of Ronozyme RumiStar (alpha-amylase) as a feed additive for dairy cows. EFSA Panel on Additives and Products or Substances used in Animal Feed (FEEDAP). *EFSA Journal,* 2777, 1–13.

European Commission (2001) Guidelines for the Assessment of Additives in Feeding Stuffs, Part II. Enzymes and Micro-organisms (http://ec. europa.eu/food/fs/sc/scan/out68_en.pdf, accessed 27 April 2012).

European Commission (2012) European Union Register of Feed Additives Pursuant of Regulation (EC) No 1831/2003. Appendices 3c and 4. Annex: List of Additives, edn 145. Health and Consumers Directorate General, Brussels.

Haefner, S., Knietsch, A., Scholten, E., Braun, J., Lochscheidt, M. and Zelder, O. (2005) Bio-technological production applications of phytases. *Applied Microbiology and Biotechnology* 68, 588–597.

Leuchtenberger, W., Huthmacher, K. and Drauz, K. (2005) Biotechnological production of amino acids and derivatives: current status and prospects. *Applied Microbiology and Biotechnology* 69, 1–8.

National Institutes of Health [NIH] (1976) Guidelines for research involving recombinant DNA molecules. *Federal Register* 41, 27911–27943.

OJEU (2009) Directive 2009/41/EC of the European Parliament and of the Council of 6 May 2009 or the contained use of genetically modified microorganisms. *Official Journal of the European Union* L125/75–97.

OJEU (2001) Directive 2001/18/EC of the European Parliament and of the Council of 12 March 2001 on the deliberate release into the environment of genetically modified organisms. *Official Journal of the European Union* L106/1–37.

OJEU (2003a) Regulation (EC) No 1829/2003 of The European Parliament and of the Council of 22 September 2003 on genetically modified food and feed. *Official Journal of the European Union* 46, L 268/1–23.

OJEU (2003b) Regulation (EC) No 1831/2003 of the European Parliament and of the Council of 22 September 2003 on additives for use in animal nutrition. *Official Journal of the European Union* 46, L 268/29–43.

Pallauf, J. and Rimbach, G. (1997) Nutritional significance of phytic acid and phytase. *Archiv für Tierernahrung* 50, 301–319.

van den Brink, J. and de Vries, R.P. (2011) Fungal enzyme sets for plant polysaccharide degradation. *Applied Microbiology and Biotechnology* 91, 1477–1492.

Yao, M.-Z., Zhang, Y.-H., Lu, W.L., Hu, M.-Q., Wang, W. and Liang, A.-H. (2011) Phytases: crystal structures, protein engineering and potential biotechnological applications. *Journal of Applied Microbiology* 112, 1–14.

12 The Pipeline of GM Crops for Improved Animal Feed: Challenges for Commercial Use

Pascal Tillie,* Koen Dillen and
Emilio Rodríguez-Cerezo

*European Commission – Joint Research Centre – Institute for
Prospective Technological Studies, Seville, Spain*

12.1 Introduction

One of the consequences of economic development is that people consume more animal products per capita – the so-called 'Westernization' of diets. As this goes hand in hand with a growing world population, the demand for animal products has increased by 51% in the past 20 years (FAOSTAT, 2013) and will keep increasing at a significant rate in the near future. This translates into an even faster increasing demand for agricultural crops, since the conversion rate of vegetable calories to animal calories is, on average, higher than 3 to 1. Estimates show that about one-quarter of plant calories is used to feed animals (Chaumet *et al.*, 2011). Increasing feed efficiency is thus an essential objective to relax the pressure on arable land due to animal product consumption. At farm level, nutritional efficiency is an important determinant of farmers' profits. But it also goes beyond farm economics, since increasing efficiency in resource use would also release the pressure on fossil energy and biological resources, as well as reducing some of the negative environmental externalities associated with livestock production.

Plant breeding can increase the efficiency of livestock production at two levels: by raising the number of calories produced by area of land and by improving the rate of conversion of vegetable calories into animal calories. Biotechnology offers new possibilities for the improvement of plants, such as organ-specific expression of proteins or expression of characters derived from other species, etc. The potential benefits of genetically modified (GM) plants in the field of animal nutrition are: improving the nutritional value of feed; reducing manure excretion through a higher net energy value; and lowering nitrogen and phosphorus pollution (Cunningham, 2005). Here, we review the pipeline for those GM events that involve potential benefits for the animal sector and that are likely to be commercialized in the future. This chapter focuses exclusively on quality traits, as they directly address the nutritional efficiency of feed. We also describe the prospects they offer and the challenges that remain to be addressed. This chapter is organized according to the main traits currently under research for animal nutrition: low phytate content, amino acid rich, improved digestibility and enhanced oil content (see also Chapter 7).

12.2 Research Methodology for the Pipeline Survey

The information that is provided in this chapter, in order to describe the pipeline of

*E-mail: pascal.tillie@ec.europa.eu

GM events that are relevant for feed use, has been collected in various ways: Internet searches; screening of the official regulatory agency pipeline for the USA,[1] the European Union (EU),[2] Brazil and Argentina; from information publicly available from the major biotech companies; from queries on the ISI Web of Science database for publications in peer-reviewed journals using the appropriate keywords; from reviewing the main field trials registry;[3] and from searches for relevant patents within the worldwide collection of published applications[4] using relevant keywords and publication authors' names following the methodology of Parisi *et al.* (2013). Only the events for which a proof of concept exists – i.e. an article, a patent or a field trial – were then included in a specified database where the relevant information for this chapter was gathered. Also, when one type of proof of concept was found (for instance, a publication regarding a specific event), the others were searched systematically (in this case, corresponding field trials and patents). The purpose was to establish a typology of events according to their advancement in the research pipeline. All events were classified depending on their proximity to market, using five categories adapted from Stein and Rodríguez Cerezo (2009):

- Commercialized: when the event is already marketed in at least one country.
- Commercial pipeline: events that have been authorized for cultivation in at least one country but are not yet marketed.
- Regulatory pipeline: events in the regulatory process for being marketed in at least one country.
- Advanced development: events for which there are multiple-location field trials and more than one proof of concept.
- Early development: events for which there is only one proof of concept.

Information on the existence of feed trials, feeding trials or patents is displayed in Tables 12.1–12.4 (see also Chapters 7 and 10). Altogether, about 110 events relevant to animal nutrition have been identified; of those, only two are already commercialized,

seven are in the commercial pipeline, a couple are in the regulatory pipeline and about one-quarter are in advanced development. The pipeline also confirms the dominant position of US research: about half of the events are developed in this country, while the second developer is the EU-27, with 14 events, followed by China (12) and Japan (10).

12.3 New Events in the Pipeline of GM Crops for Animal Nutrition

12.3.1 Low-phytate crops to improve phosphorus nutrition

Phytate – a salt form of phytic acid – is the main storage form of phosphorus (P) in plant seeds and represents the major source of flux of P into the environment: plants take up P from the soil and transfer a significant amount to seeds, where 75% of the total P is stored in the phytate form (Raboy, 2001). Altogether, the amount of P that crops incorporate each year into phytate is equivalent to nearly 65% of the quantity of P in mineral fertilizer used worldwide (Lott *et al.*, 2000). However, phytate is considered an anti-nutrient and an undesirable component of feedstuffs for a number of reasons:

1. It is a strong chelator of mineral cations such as iron, calcium, zinc or magnesium and therefore prevents the use of these essential minerals by humans and animals eating the plants. It may also preclude the availability of proteins by reacting with them.

2. The digestibility of phytate is also very poor and it constitutes a bad source of inorganic and available P for non-ruminant animals, including humans, since they lack dephosphorylation enzymes.

3. Hence, livestock such as poultry, swine and fish excrete large quantities of undigested phytate that eventually contribute to water pollution and generate eutrophication.

4. While the P content of their manure is high, the feed of monogastric (i.e. non-ruminants) animals needs to be supplemented with inorganic P to meet their nutritional requirements. Although this is a rather cheap solution, this practice increases the amount of P leaching into the environment (Hamada *et al.*, 2005; Gontia *et al.*, 2012).

Besides inorganic P supplementation, another solution to improve the P content of feedstuffs is to add the microbial enzyme phytase to the feed. The action of this enzyme on the phytate–protein complex, one of the primary storage forms of phytate in seed, releases up to 50% of the P content of seeds, as well as bound metal cations (Raboy, 2001). Phytase supplementation therefore improves the bioavailability of essential elements for animals and reduces P pollution through animal manure. However, the introduction of phytase in feedstuffs increases feeding costs significantly and imposes some restrictions to diet formulation. Other approaches for improving P and mineral availability in animal feed include:

1. The activation of endogenous phytase in grains prior to feed processing.
2. The transformation of plants with a mutant *lpa* gene responsible for a low phytic acid phenotype.
3. The introduction of a transgene for the production of exogenous phytase in crops.
4. The genetic modification of livestock for the production of phytase (Brinch-Pedersen *et al.*, 2002; see also Chapter 7).

We focus here on the pipeline for GM plants with a low-phytate phenotype aimed at improving the quality of animal feed.

The pipeline for low-phytate crops

There are about 20 GM events with low phytate content currently in the pipeline, at different development stages. They follow two main approaches. The large majority of events come from the transformation of a target plant with an exogenous phytase *phy* transgene, whose origin can be either bacterial or fungal, or from yeast, while some other events are based on the silencing of the phytate biosynthetical pathway. This distinction is important, since if both approaches lead to plants with a low-phytate phenotype, only those with a high phytase level can be used as a feed additive to hydrolyse the phytate contained in other crops. Two crops, maize and soybean, account for half of the events under research, an indication that these transformations are mostly animal feed-oriented. The rest of the events concern wheat, barley, lucerne, rapeseed and rice.

The first low-phytate GM event to reach the market is likely to be the GM maize developed by a Chinese biotech company (see Table 12.1). This transgenic maize transformed by a fungal gene, *phyA2*, displays a 50-fold increase in phytase expression, is stable over several generations, has no impaired germination (Chen *et al.*, 2008) and, most important, is the first GM maize to go through the five stages of the regulatory process in China. This event has been introduced in two maize hybrids, and their commercialization in China is pending approval from the government. Another maize expressing an *E. coli* phytase gene went through various field trials in the USA, and feeding trials with weanling pigs have shown that it is as efficient as the supplementation of feed with phytase to improve the growth performance of a P-deficient diet (Nyannor *et al.*, 2007). Other events are under development in the USA and in Germany, relying on the introduction of a *phyA* gene (Drakakaki *et al.*, 2005), on the silencing expression of a transporter involved in the production of phytate (Shi *et al.*, 2007) or on the use of zinc-finger nuclease to disrupt the *IPK1* gene, which encodes an enzyme that catalyses the biosynthesis of phytate in maize seeds (Shukla *et al.*, 2009).

Soybean is the second crop that has received a lot of attention from researchers willing to reduce its phytate content. One of the first studies has already shown that the transformation of soybean with a fungal phytase *phyA* gene is an effective approach to improve P availability of feed, while reducing P excretion by 50% (Denbow *et al.*, 1998). Recently, the same approach also led

Table 12.1. Pipeline of events with a low-phytate phenotype and/or high phytase expression.

Crop	Developer	Country	Event name	Site modification	Phenotype modification	Development stage	Reference	Field trials	Feed trials	Patent year (if any)
Barley	Palacký University, Olomouc	Czech Rep		*phyA*	High phytase	4	Ohnoutkova, 2010	Yes	–	–
Barley	Aarhus University	Denmark		*HvPAPhy_a*	High phytase	4	Holme *et al.*, 2012	Yes	–	–
Lucerne	University of Wisconsin	USA		*phyA*	High phytase	5	Ullah *et al.*, 2002	–	–	1999
Lucerne	Samuel Roberts Noble Foundation	USA		*MtPHY1* and *MtPAP1*	P uptake	5	Ma *et al.*, 2012	Yes	–	2005
Maize	CAAS and Origin Agritech	China	B23-3-1	*phyA2*	High phytase	2	Chen *et al.*, 2008	Yes	–	2006
Maize	Syngenta	USA		*appA*	High phytase	4	Nyannor *et al.*, 2007	Yes	Yes	–
Maize	Aachen University	Germany		*phyA*	High phytase	5	Drakakaki *et al.*, 2005	–	–	–
Maize	Pioneer	USA		*lpa1*	Low phytate	5	Shi *et al.*, 2007	Yes	–	2005
Maize	Dow and Sangamo BioSciences	USA		*IPK1*	Low phytate	5	Shukla *et al.*, 2009	–	–	2006
Rapeseed	Syngenta, BASF and PlantZymes	Netherlands	MPS961-5	*phyA*	High phytase	3 – discontinued	Ponstein *et al.*, 2002	Yes	–	1990
Rapeseed	Shanghai Academy of Agricultural Sciences	China		*phyA*	High phytase	5	Peng *et al.*, 2006	–	Yes	2002
Rice	Yangzhou University	China		*PRSPhyl*	High phytase	5	Liu *et al.*, 2006	–	–	–
Rice	Zhejiang University	China		*appA*	High phytase	5	–	–	–	2008
Rice	Mitsui Chemicals	Japan		*phy*	High phytase	5	Hamada *et al.*, 2005	–	–	–
Rice	University of Tokyo	Japan		*RINO1*	Low phytate	5	Kuwano *et al.*, 2009	–	–	–
Rice	Taiwan Institute of Molecular Biology	Taiwan	Nat-AN	*appA* and *SrPf6*	High phytase	5	Hong *et al.*, 2004		Yes	–
Soybean	Dalian University of Technology	China		*phyA*	High phytase	5	Gao *et al.*, 2007	–	–	2002

Continued

Table 12.1. Continued

Crop	Developer	Country	Event name	Site modification	Phenotype modification	Development stage	Reference	Field trials	Feed trials	Patent year (if any)
Soybean	Tianjin University and Hebei University of Science and Technology	China		*AfPhyA*	P uptake	5	Li *et al.*, 2009	–	–	–
Soybean	Tianjin University and Hebei University of Science and Technology	China		*AfPhyA*	High phytase	5	Yang *et al.*, 2011	–	–	–
Soybean	BASF and Virginia Polytechnic Institute	USA		*phyA*	High phytase	5	Denbow *et al.*, 1998	–	–	1999
Soybean	Pioneer	USA		*lpa1*	Low phytate	5	Shi *et al.*, 2007	Yes	–	2005
Soybean	USDA and University of Missouri	USA		*appA*	High phytase	5	Bilyeu *et al.*, 2008	Yes	–	–
Wheat	Danish Institute of Agricultural Sciences and Novozymes	Denmark		*phyA*	High phytase	5	Brinch-Pedersen *et al.*, 2006	–	–	1999

Notes: Development stage: 1 = commercialized; 2 = commercial pipeline; 3 = regulatory pipeline; 4 = advanced development; 5 = early development. In this and the following tables, events are ordered by crop, development stage and country.

to a twofold increase of the phytase activity in transgenic soybean seeds (Gao *et al.*, 2007). However, the thermostability of the phytase expressed was not sufficient to make it eligible for commercial use, since the pelleting process of soybean requires a high temperature in order to inactivate some anti-nutrient compounds of the seeds. Other research conducted in China follows the same approach (Yang *et al.*, 2011), and the same team has also developed a transgenic soybean expressing phytase in its roots and excreting it into the surrounding soil in order to make phytate available for plant uptake (Li *et al.*, 2009).

Besides these, two promising events are under development in the USA, both of them having been tested in the field. The first relies on a technique that has also been used in maize: the silencing of transporter gene expression in an organ-specific manner led to a 15- to 30-fold increase of inorganic P concentration and to a dramatic reduction of the phytic acid content of soybean seeds (Shi *et al.*, 2007). The second event, based on the expression of an *E. coli* phytase, exhibits a nearly total conversion of phytate into inorganic P together with a very high level of phytase expression, in addition to a rather high thermostability of the enzyme (Bilyeu *et al.*, 2008). Seeds of this event used as an additive to feed were as effective as commercial phytase in reducing the phytate content of soybean meal and maize meal, paving the way to promising commercial applications. Up to now, though, none of these soybean events have entered the regulatory pipeline, suggesting that the conditions for a successful commercial release might not yet be satisfied.

In addition to maize and soybean, two GM phytase-rich barley events are currently in an advanced development stage with field trials in the EU (see Table 12.1). The first event was generated thanks to a cisgenic transformation – i.e. the phytase gene inserted derives from the same plant species – and showed a stable 2.8-fold increase in the phytase activity of the grain (Holme *et al.*, 2012). This level of activity is higher than that of the microbial phytase used as an additive in feed to make P available from phytate, which might be of great interest for those farmers who process their own feed from home-grown cereals. The second GM barley under development relies on the insertion of a fungal phytase gene in a spring barley line, and it also displays an increased phytase activity (Ohnoutkova *et al.*, 2010).

This last approach has also been applied to canola in order to raise its phytase concentration. In contrast to soybean, canola grains do not have to be toasted prior to their processing into feed, avoiding the requirement regarding the thermostability of the phytase protein. One GM canola event went through many field and feeding trials in the USA and displayed a high level of phytase in seeds (Ponstein *et al.*, 2002). However, this research seems to have been discontinued. Another event under development in China has proved to be as effective as microbial phytase in releasing P when mixed with feed (Peng *et al.*, 2006).

In recent years, Danish researchers have been working on improving a wheat line with a rationally designed thermostable phytase. Their results show that it is possible to accumulate heat-stable phytase in wheat that is still efficient in hydrolysing phytate and improving zinc and iron availability, even after a prolonged boiling process (Brinch-Pedersen *et al.*, 2006). This important outcome might pave the way for an effective improvement of P and mineral uptake in cereal food and feed nutrition. Efforts to improve the nutritive value of rice are also facing the thermostability of phytase issue (Liu *et al.*, 2006). It has been suggested that expressing a bacterial phytase in rice seeds would allow the use of rice as a feed additive (Hong *et al.*, 2004), but whether this approach is compatible with high-temperature processing requires further investigation. By contrast, a rice plant transformed with a yeast gene has proved to produce an elevated level of phytase with a relatively good stability at temperatures as high as 70°C (Hamada *et al.*, 2005). The authors of this study advocate the use of the whole rice plant as a silage crop or as a feed additive for monogastric animals.

Finally, lucerne (*Medicago sativa*) has been modified to produce phytase and has been tested in animal rations as leaf meal or as a juice dried on maize. A poultry feeding trial showed that this approach could halve the P content of manure compared to feeding with inorganic P supplements (Mueller *et al.*, 2008). Phytase expression can also be used to improve P uptake: recently, a US-based team developed a transgenic lucerne overexpressing phytase in its roots, which produced twice as much biomass as non-transgenic lines when grown on natural soil without P fertilization (Ma *et al.*, 2012).

Prospects and challenges for commercial use of low-phytate GM crops

As demonstrated in the previous section, the pipeline for low-phytate GM crops is very active, with many research teams involved and aiming to improve the most important crops in the world. However, as most of the studies point out, there are still a number of challenges to overcome, which may explain why, so far, few events have reached a status beyond advanced development. First, for plants with a low phytic acid (*lpa*) trait developed by knocking out genes involved in phytate biosynthesis, some negative effects on seed and plant growth have been described, leading to adverse yield effects (Raboy, 2001). Second, the industrial feed pelleting process requires heating of the seeds in order to eliminate some anti-nutrient compounds and to avoid possible *Salmonella* infections. This implies that the phytase enzymes present in the seeds must be able to retain activity after this process in order to be still useful when consumed by animals. The same limitations apply for those cereals destined for human consumption that require being boiled or baked. Finally, the economic incentive for farmers to pay a technology fee for a GM low-phytate crop is another important issue. For farmers raising monogastric livestock, it might be a way to reduce feeding costs, especially if they produce their own feed from self-produced crops. Otherwise, the adoption of these crops will depend on the premium the feed industry will be willing to pay for the modified crop, which in turn will depend on

the price of the inorganic P currently used in feed supplementation. Yet this additive is rather inexpensive, despite its negative environmental externalities. Thus, the use of feed with low phytate content could be economically rational in countries where environmental regulations controlling phosphate leaching are in place, limiting the use of conventional, high-phytate feed.

Once these limitations are overcome, transgenic plants with high phytase content would definitely enhance the uptake of minerals and phosphorus by animals, while at the same time reducing P pollution in soils. When accounting only for savings in feed costs due to the replacement of inorganic P, Johnson *et al.* (2001) calculated that the introduction of high-phytase maize in diets could translate into an added value of US\$4.6/t and that approximately 55 million tonnes (Mt) of such a maize could be consumed. This would generate an additional gross value of US\$260 million for US maize on world markets. This estimation, however, assumes that the production cost for low-phytate maize will remain the same as for conventional maize. It does not consider any possible technology fee, which will certainly determine the adoption of such an innovation by maize growers. On the other hand, the positive environmental effects of a low-phytate animal feed should also be considered as benefits of this technology, since it could lead to a reduction of up to 85% of P waste when provided to poultry, swine or fish (Raboy, 2001). Internalizing this external effect definitely would contribute to the adoption and success of such GM plants for animal nutrition.

12.3.2 Crops enriched in essential amino acids

Humans, as well as many farm animals, are unable to synthesize certain amino acids, called essential amino acids (EAAs) (see Chapters 1, 7 and 10). Since plants are the primary source of EAAs for animals, this is why intensive breeding efforts have focused on elevating the amino acid content in plants. In fact, some EAAs are found in very limiting amounts in plants, such as lysine (Lys) and tryptophan (Trp) in cereals and methionine (Met) or cysteine (Cys) in legume crops. The interest for such an enrichment varies by country and diet. In developing countries where plants provide people with most of their protein, this would help prevent malnutrition, while in developed countries the challenge is to improve the economic efficiency of the conversion of plant proteins into animal proteins, the latter representing the first source of EAAs in these countries (Ufaz and Galili, 2008).

Despite the potential high economic benefits, conventional breeding for high protein and EAA content is very difficult to achieve, because of the well-studied negative correlation between yield and protein content (Barneix, 2007). Over time, the efforts to breed high-yielding maize varieties, for instance, have generated a shift in grain composition from protein to starch (Scott *et al.*, 2006). The only success so far is the quality protein maize (QPM), richer in Lys and Trp, that was bred at the Maize and Wheat Improvement Center (CIMMYT) in the late 1990s and which has produced very positive results in many regions of the world (Vasal, 2000). The difficulties in conventional breeding arise from the fact that amino acids such as Lys, Trp or Met play an important role in plant development, so their high content is generally associated with abnormal plant growth and inferior agronomic traits.

GM techniques, in contrast, allow a seed-specific expression of traits that has proven to be a promising approach to overcome these limitations. GM techniques also allow the insertion of new quality traits in high agronomic performing lines of multiple species, making the breeding process shorter. None the less, such plant enhancements are especially relevant for feedstuffs destined for monogastric mammals, since ruminants need EAAs that in addition are resistant to rumen proteolysis. We review here the GM approaches that have proved to be suitable in elevating the content of Lys, Trp, Met and Cys in plants (see Table 12.2).

The pipeline for lysine-enriched plants

Lysine is considered to be the most limiting amino acid in cereals, especially for swine and poultry nutrition. Although maize is one of the most productive crops on a per hectare basis in terms of energy and yield, its nutritional quality is rather poor when it comes to amino acid content. Therefore, maize meal-based rations have to be supplemented with Lys in poultry or swine diets, generally from soybean meal or with synthetic Lys produced by fermentation (Johnson *et al.*, 2001; Huang *et al.*, 2008). Other crops such as rice or rapeseed suffer the same limitations. Hence, numerous GM researches have been focusing on the enhancement of Lys content in seeds used for livestock nutrition.

In maize, various research teams have developed Lys-rich GM events, with promising results. A rather simple approach is based on the expression in maize seeds of a bacterial enzyme (CordapA), which is involved in Lys biosynthesis but which has been made insensitive to Lys feedback inhibition (i.e. the mechanism that regulates Lys production caused by its own accumulation) by a mutation (Huang *et al.*, 2005). This modification led to the development of maize event LY038, authorized for cultivation in the USA in 2005. This event, as well as its stack with event MON810 (that makes the maize insect resistant), was tested in many field and feeding trials and turned out to be superior in poultry nutrition to its non-GM counterpart (Lucas *et al.*, 2007). However, these events were never commercialized and in 2009 the applications for approval submitted to the European Food Safety Authority (EFSA) were withdrawn by the applicant. No official reason was provided, but it is likely that the cost–benefit ratio of this innovation was not sufficient to justify its commercial release.

Another GM event obtained by silencing through RNAi (RNA interference), the expression of a gene involved in Lys catabolism (i.e. degradation), has resulted in a 30-fold increase of free Lys content in

Table 12.2. Pipeline of events with an enhanced content in amino acid.

Crop	Developer	Country	Event name	Site modification	Targeted AA	Development stage	Reference	Field trials	Feed trials	Patent year (if any)
Cassava	International Laboratory for Tropical Agricultural Biotechnology	USA		Zeolin	Met, Cys	4	Abhary *et al.*, 2011	Yes	–	2009
Lucerne	New Mexico State University	USA		β-zein and *AtCγS*	Met	5	Bagga *et al.*, 2005	–	–	1998
Lucerne	Migal–Galilee Technology Center	Israel		AtCGS	Met, Cys	5	Avraham *et al.*, 2005	–	–	2001
Maize	Monsanto	USA	LY038	*cordapA*	Lys	2	Huang *et al.*, 2005	Yes	Yes	2003
Maize	Monsanto	USA	LY038 × MON810	*cordapA + cry1Ab*	Lys	2	Huang *et al.*, 2005	Yes	Yes	2003
Maize	Monsanto	USA	MON93066	*cordapA + LKR/ SDH*	Lys	4	Frizzi *et al.*, 2008	Multi-location	Yes	2005
Maize	Monsanto	USA		Anthranilate synthase	Trp	4	–	Multi-location	–	2006
Maize	Iowa State University	USA		α-lactalbumin	Lys	4	Bicar *et al.*, 2008	Multi-location	–	–
Maize	China Agricultural University, Beijing	China		*sb401*	Lys	5	Yu *et al.*, 2004	Yes	–	
Maize	Southern Illinois University	USA		*gdhA*	Various	5	Guthrie *et al.*, 2004	–	Yes	1999
Maize	Monsanto	USA		α-zein reduction	Lys, Trp	5	Huang *et al.*, 2006	Yes	–	2005
Maize	Monsanto	USA		*LKR/SDH* silencing	Lys	5	Houmard *et al.*, 2007	–	–	–
Maize	BASF[a]	USA	Nutritionally enhanced	*glgC*	Lys and others	5 – discontinued	–	–	–	2004
Maize	University of Missouri-Columbia	USA		Lysyl-tRNA synthetase	Lys	5	Wu *et al.*, 2007	–	–	1998
Rapeseed	Yangzhou University	China		*LRP*	Lys	5	Wang *et al.*, 2011	–	–	–
Rice	National Agriculture and Bio-oriented Research Organization	Japan	HW1	*OASA1D*	Trp	4	Wakasa *et al.*, 2006	Yes	Yes	2005

Rice	National Agriculture and Bio-oriented Research Organization	Japan	HW5	*OASA1D*	Trp	4	Wakasa *et al.*, 2006	Yes	Yes	2005
Rice	National Agriculture and Bio-oriented Research Organization	Japan	KPD722-4	*OASA1D*	Trp	4	–	Multi-location	–	2005
Rice	National Agriculture and Bio-oriented Research Organization	Japan	KPD627-8	*OASA1D*	Trp	4	–	Multi-location	–	2005
Rice	National Agriculture and Bio-oriented Research Organization	Japan	KA317	*OASA1D*	Trp	4	–	Multi-location	Yes	2005
Rice	University of Missouri-Columbia	USA		Altered tRNA (Lys)	Lys	5	Wu *et al.*, 2003	–	–	1998
Rice	Sichuan Agricultural University	China		*sb401*	Lys	5	Li *et al.*, 2008	–	–	–
Sorghum	Pioneer	USA	ABS#1	Kafirin silencing	Lys	4	Zhao *et al.*, 2003	Multi-location	–	2008
Sorghum	Pioneer	USA	ABS#2	Stack	Lys	5	Zhao, 2007	Yes	–	2008
Sorghum	USDA and University of Nebraska	USA		HMW-GS	Various	5	Kumar *et al.*, 2012	–	–	–
Sorghum	USDA and University of Nebraska	USA		Kafirin silencing	Lys, Arg, Asp	5	Kumar *et al.*, 2012	–	–	–
Soybean	Monsanto	USA		–	–	5	Peng *et al.*, 2004	Yes	–	2001
Soybean	Kansas State University	USA		γ-zein	Met, Cys	5	Li *et al.*, 2005	–	–	–
Soybean	University of Illinois	USA		*ASA2*	Trp	5	Inaba *et al.*, 2007	–	–	–
Soybean	Japan Science and Technology Agency	Japan		*OASA1D*	Trp	5	Ishimoto *et al.*, 2010	–	Yes	2004
Soybean	National Agriculture and Bio-oriented Research Organization	Japan		*OASA1D*	Trp	5	Kita *et al.*, 2010	–	–	–
Wheat	Leibnitz Institute of Plant Genetics	Germany		HvSUT1	Phe, Tyr, Trp, Leu	4	Weichert *et al.*, 2010	Yes	–	–
Wheat	Agricultural Research Institute	Hungary		*Ama1*	Lys, Tyr	5	Tamás *et al.*, 2009	–	–	–
Wheat	Jiangsu Academy of Agricultural Science	China		*cflr*	Lys	5	Sun *et al.*, 2010	–	–	2009

Note: [a]This event is commonly referred to as 'nutritionally enhanced maize' and features both high amino-acid and oil content; however, this research project was discontinued in 2013.

maize (Houmard *et al.*, 2007), while the combination of both previous approaches in a single event (MON93066, see Table 12.2) has produced more than 4000 ppm free Lys in the endosperm of maize kernels, a 100-fold increase compared to wild-type maize (Frizzi *et al.*, 2008). This last event has been patented and several field trials are currently being conducted (see also Table 12.2).

Other approaches to enhance the Lys content of maize are being explored. One of these, first developed for rice (Wu *et al.*, 2003), consists of the introduction of a gene encoding for an enzyme (Lysyl-tRNA synthetase – AtKRS) involved in the formation of tRNA (transfer RNA) lysyl (Wu *et al.*, 2007). This resulted in the incorporation of Lys into maize zeins – maize proteins that normally have little or no Lys in the wild type – and elevated the Lys content of modified maize seed by up to 26%. Another Lys-rich maize event is under research with field trials in China: the introduction of a potato gene expressing a protein with high Lys content (*sb401*) led to a significant elevation (16–55%) of total Lys in the plant (Yu *et al.*, 2004). More recently, a US-based team developed transgenic maize lines expressing a porcine milk protein in the endosperm (Bicar *et al.*, 2008). Field cultivations confirmed that the resulting seeds had an improved amino acid balance – but no increase in total protein content – leading to an increase in Lys content of about 29–47%.

Regarding GM rice, two new events with increased Lys content are currently in the early development stage. The first is based on an original approach: the introduction in a rice plant of a gene encoding for an altered form of the tRNA specific to Lys, which is introduced instead of other amino acids during the protein biosynthesis process. The transformation generates a meaningful enrichment of Lys in rice prolamin proteins by 43–75%, resulting in an increase comprising between 1% and 6.6% of the overall Lys content of rice seeds (Wu *et al.*, 2003). The expression of the *sb401* gene in rice was also demonstrated to be a relevant approach (Li *et al.*, 2008).

Noteworthy are the efforts to improve the protein content of sorghum, generally following a common approach of downregulating the expression of sorghum prolamins, or kafirins, using RNAi silencing techniques (Zhao *et al.*, 2003; Zhao, 2007; Kumar *et al.*, 2012). These kafirins tend to be poorly digestible, whereas decreasing their expression enhances EAA content in grains. Among these new events are the high-Lys sorghum events developed for Africa in the framework of the African Biofortified Sorghum initiatives (Wambugu, 2007). Wild-type sorghum features between 35 and 90% lower Lys than other cereals, but event P898012, a transgenic biofortified sorghum, already contains up to 112% more Lys than the wild type, among other nutritional improvements (Taylor and Taylor, 2011; Kruger *et al.*, 2012). These new sorghum GM events have considerable potential to alleviate malnutrition in Africa, as well as to enhance the efficiency of sorghum as feedstuffs.

The nutritional quality of wheat as a feed for livestock is also limited by low levels in certain EAAs – especially Lys and threonine (Thr) – composing its storage proteins. Following previous attempts to elevate the content of wheat-specific gluten proteins by expressing additional copies of the corresponding genes (Shewry *et al.*, 2006), an amaranth (*Amaranthus hypochondriacus*) albumin gene has recently been introduced in a wheat line. The results indicate an increased content in EAAs, about 30% in the case of Lys (Tamás *et al.*, 2009). Another approach involving the transformation of wheat seeds with a chilli pepper (*Capsicum frutescens*) gene led to a 7.4% increase in Lys content (Sun *et al.*, 2010), whereas the overexpression of a barley sucrose transporter in wheat also showed increased EAA levels in Lys, leucine (Leu) and Trp, among others (Weichert *et al.*, 2010). The feasibility of this last approach to overcome the negative correlation between yield and grain protein content observed in conventional breeding of wheat has been confirmed by numerous greenhouse and field trials.

Recently, a new crop species has been enriched in Lys: the introduction of a winged bean (*Psophocarpus tetragonolobus*) high Lys protein gene resulted in a 16.7% increase of the Lys content of rapeseed seeds compared to non-transgenic lines (Wang *et al.*, 2011).

The pipeline for tryptophan-enriched plants

Tryptophan is generally considered the second most important EAA, since its deficiency in certain crops leads to nutritional and clinical disorders when used to feed animals (Henry *et al.*, 1992). Thus, crop enrichment in Trp would be beneficial for animal nutrition. The first high-Trp GM events were obtained through a rice enzyme (*OASA1D*) involved in Trp catabolism, which was modified in order to become Trp feedback-insensitive, leading to free Trp accumulation in rice (Wakasa *et al.*, 2006). Although rice is cultivated primarily for human consumption, its use as a feed product is growing. Five different Trp-rich GM rice events are currently in advanced development in Japan (see Table 12.2), and many field trials have been conducted that have confirmed a 193- to 311-fold increase of Trp accumulation in modified rice seeds (Wakasa *et al.*, 2006). Nutritional experiments with chickens have shown that the nutritive value of high-Trp GM event HW-1 is similar to that of non-GM rice supplemented with synthetic Trp (Takada and Otsuka, 2007).

The same gene has recently been used for the development of high-Trp soybean GM events, resulting in an elevation by at least a factor of 20 of the free Trp content of soy seeds. Transgenic seeds obtained, tested in a trout feeding trial, were at least as efficient as Trp-supplemented feed in increasing the body weight of fish (Ishimoto *et al.*, 2010). Modified high-Trp lines are also richer in other EAAs such as histidine (Kita *et al.*, 2010). Another GM soybean carrying a tobacco gene has been proved to display an increased Trp content (Inaba *et al.*, 2007). Non-GM soybean seeds already feature a high protein content (about 40%), making this crop an important staple source of vegetable protein for both humans and animals. Soybean meal, the coproduct of soybean remaining after oil extraction, is not limited in Trp or Lys content. However, since soybean meal is used extensively to supplement cereal grains like maize in animal rations, its enrichment in EAAs would still be beneficial and explains the research efforts in this direction.

Research is also being conducted to enhance the Trp content of maize, with at least two events under development involving numerous field trials in the USA. One of these has proved to have superior nutritional quality than the conventional QPM that has brought considerable improvements for human consumption in developing countries in the last decades (Huang *et al.*, 2006).

The pipeline for methionine- and cysteine-enriched plants

Legume crops, such as clover, lucerne or soybean, are generally poor in sulfur amino acids, Met and Cys. Besides being an important component of proteins, Met is involved in a wide range of biological processes in plants, including biosynthesis of the ethylene hormone, replication of DNA, development of the cell wall and production of secondary metabolites. Thus, attempts to elevate Met content in plants by manipulation of the genes involved in its synthesis are generally constrained by its extensive catabolism or because they result in abnormal phenotypes (Amir and Tabe, 2006). Nevertheless, some promising results have been obtained in *Arabidopsis* by eliminating the activity of an enzyme (HMT2) involved in Met metabolism and transport (Lee *et al.*, 2008).

Another approach to increase the nutritional value of legume plants is to express heterologous sulfur proteins that are naturally Met-rich in seeds. A sunflower protein gene was introduced in lupin (*Lupinus angustifolius*) and the nutritive value of the modified seeds was tested in various animal feed experiments (Molvig *et al.*, 1997). They confirmed the increased availability of Met and showed that the transgenic seeds were better than the control seeds for rats, poultry and sheep feeding, requiring less Met supplementation. Similar results were obtained with the insertion of a maize γ-zein protein gene into soybean, leading to an increase in Cys and Met content of at least 27% and 15%, respectively (Li *et al.*, 2005).

Cassava, which has the lowest content in proteins among major crops, was recently modified to express a storage protein (Abhary *et al.*, 2011). The resulting plant, tested in field trials, showed a fourfold increase in its protein content, and its roots were especially richer in Met and Cys (level enhanced 4.5-fold and ninefold, respectively), paving the way to its more efficient use as food and feed in sub-Saharan Africa, where it is extensively cultivated. However, other studies have shown that the accumulation of sulfur-rich protein is limited by the availability of free Met in seeds and that the expression of heterologous proteins generally comes at the expense of endogenous Met-rich proteins (Ufaz and Galili, 2008). Combining this approach with a transgene that increases free Met accumulation has the potential for considerable improvements, as shown by successful tests with lucerne (Avraham *et al.*, 2005; Bagga *et al.*, 2005).

Prospects and challenges for commercial use of EAA-enriched plants

The enrichment of plants in EAAs is a highly desirable goal, for humanitarian as well as economic reasons. On the one hand, conventional breeding seems to have reached its upper limit, while on the other hand GM techniques allow the improvement of varieties that already have high agronomic performance. Moreover, the growing world population is also showing an increasing preference for diets richer in animal proteins, making the improvement of the protein content in plants a highly desirable goal, since they are the primary source of EAAs for animals.

The GM pipeline for amino acid enrichment of plants is currently one of the most active, with more than 30 events identified. Apart from two events that have been approved for cultivation but have not been released, 11 other events are in the advanced

R&D pipeline, with numerous field trials performed and some feeding trials. Among the most promising for commercial use are the Trp-rich rice events under research in Japan and the high-Lys maize event, MON93066, developed in the USA. Also interesting are the events focusing on the needs of developing countries, namely sorghum events ABS#1 and ABS#2 and a cassava event.

Increasing the content of protein in maize was estimated by Johnson *et al.* (2001) as being the most valuable modification of this crop for feed use: he calculated that an 8% increase in protein content would bring to the entire maize sector in the USA an additional annual gross value of US$3.45 billion. The reasoning behind this is that using EAA-enriched crops in feedstuff processing would generate savings on feed cost for monogastric animals, since it would reduce the need for synthetic EAAs. However, this would occur only if both farmers and feed processors had an incentive to adopt this technology, i.e. the feed industry would be willing to pay a premium to growers that overcame the innovator fee and the induced segregation costs, because EAA-enriched plants would prove to be cheaper options for protein source than soybean meal or synthetic EAAs. Proteins are among the most valuable nutrients in feedstuffs for livestock; however, it is rather unclear which is the minimum EAA enrichment level that would make the use of the corresponding crop profitable. Together with the remaining technological challenges, these might be the most important constraints to the development of a commercial EAA-enriched GM crop. One specific use could be for those farmers that produce their own feed for their monogastric animals. They would benefit directly from the higher protein content of crops without facing any segregation or transaction costs.

12.3.3 Crops with a low lignin content and improved digestibility

For ruminant animals, forage plants represent the basis of the diet. The digestibility of the forage crop stem is thus a key aspect of its nutritive value, and by consequence also influences the profitability of the livestock farm. Forage digestibility is correlated negatively with concentration in lignin – a complex polymer that is a constituent of plant cell walls (Jung and Vogel, 1986). Lignification of plant tissues, which increases with plant maturation, slows down the hydrolysis of polysaccharides to simple sugars and prevents full cell wall digestion in the rumen of ruminant animals (Fu *et al.*, 2011; Jung *et al.*, 2012). Similar barriers due to lignin exist for the industrial conversion of sugars contained in cell wall poly-saccharides to ethanol, which is an important step for biofuel production.

Different strategies have been explored to improve the digestibility of forage, such as chemical pretreatment of fodders or conventional breeding of forage crops. While the first option is not economically rational, the latter has resulted in the development of varieties with improved digestibility. For maize, the cereal crop that is the most widely used as forage, conventional breeding has allowed cell wall digestibility to almost double (Barrière *et al.*, 2009). Spontaneous mutants in maize and sorghum, known as brown midrib (*bm*), which feature reduced lignin concentration and improved digesti-bility, have also been used to obtain a few commercial varieties (Oliver *et al.*, 2004). However, this mutation is recessive and thus the breeding of *bm* hybrids is laborious, since the mutation should be obtained in all copies of the gene carried in the plant genome.

Therefore, transgenic approaches offer new perspectives for increasing the digesti-bility of forages such as maize, sorghum, lucerne, tall fescue or switchgrass (see Chapter 7). GM techniques allow the improvement of cell wall digestibility, rather than total digestibility, by changing the composition of lignin or, even better, of the cell wall itself (Jung *et al.*, 2012).

The pipeline for low-lignin GM forage crops

The pipeline for improved digestibility crops is a rather active one, with about 15 different events under research (see Table 12.3): nine are purely forage crops (lucerne, tall fescue and ryegrass, among others) and six are cereal crops (maize, sorghum and rice). The interest in such plants has grown steadily in the recent years, as shown for instance by the high number of field trial requests in the USA. Interestingly, most of the research efforts to improve the digestibility of crops are concentrated in developed countries, mainly because of the prospects for biofuel production based on low-lignin GM crops in these countries. Only one event has reached the regulatory phase of the pipeline so far (lucerne event KK 179-5), but at least four more are in advanced development.

Lucerne is the plant that has received the most attention and that concentrates the largest number of events in the advanced development stage. Indeed, two teams based in the USA started research to improve its digestibility more than 10 years ago. Numerous field trials have been conducted since 2000 to test lucerne lines with a downregulation of various enzymatic path-ways that play a key role in lignin biosynthesis (Reddy *et al.*, 2005; Chen *et al.*, 2006). This has resulted in the production of transgenic lucerne lines with reduced lignin content and acceptable agronomic perform-ance, which display conclusive digestibility improvement in feeding trials (Reisen *et al.*, 2009). The first GM plant with a reduced lignin content to reach the commercial pipeline will likely be a product of these research efforts, since one event is currently in the regulatory pipeline. Stacks with herbicide-tolerant trait events can also be expected.

The downregulation of a lignin biosynthetic enzyme (CAD) of tall fescue (*Festuca arundinacea*), an important perennial forage crop for the cool season, is also reaching advanced development. While various field trials did not show a significant difference regarding the agronomic perform-ance of this GM tall fescue compared to the control plants, its lignin content was reduced significantly and this improved its digestibility by up to 9.5% (Chen *et al.*, 2003). More recently, another team using a different approach also reported promising

Table 12.3. Pipeline of events with a low-lignin phenotype.

Crop	Developer	Country	Event name	Site modification	Development stage	Reference	Field trials	Feed trials	Patent year (if any)
Lucerne	Forage Genetics and Monsanto	USA	KK 179-5	CCOMT	3	–	Yes	–	2011
Lucerne	Forage Genetics	USA		COMT	4	–	Yes	–	–
Lucerne	Forage Genetics	USA		CCOMT + HT	4	–	Yes	–	–
Lucerne	Samuel Roberts Noble Foundation	USA		CCoAOMT	5	Chen *et al.*, 2006	Yes	Yes	2000
Lucerne	Samuel Roberts Noble Foundation	USA		Cytochrome P450	5	Reddy *et al.*, 2005	–	Yes	2006
Maize	CRAG and Iden Biotechnology	Spain		CAD silencing	4	Fornalé *et al.*, 2012	Yes	–	–
Maize	Simon Fraser University and University of Florida	Canada and USA		COMT	5	He *et al.*, 2003	–	–	–
Maize	INRA-CNRS	France		COMT	5	Piquemal *et al.*, 2002	–	–	2001
Maize	Biogemma	France		Cinnamate 4-hydoxylase	5	–	–	–	2006
Rice	CAAS	China		*gh2*	5	Zhang *et al.*, 2006	–	–	–
Ryegrass	Institute of Grassland and Environmental Research	UK		Ferulic acid esterase	5	Buanafina *et al.*, 2006	–	–	–
Sorghum	USDA-ARS and University of Nebraska	USA	Atlas bmr-12	*bmr*	5	Funnell and Pedersen, 2006	Yes	–	–
Switchgrass	Samuel Roberts Noble Foundation	USA		COMT	5	Fu *et al.*, 2011	Yes	–	2006
Tall fescue	Samuel Roberts Noble Foundation	USA		CAD silencing and COMT	4	Chen *et al.*, 2003	Yes	–	2000
Tall fescue	Institute of Grassland and Environmental Research	UK		Ferulic acid esterase	5	Buanafina *et al.*, 2010	–	–	–

Note: Development stage: 1 = commercialized; 2 = commercial pipeline; 3 = regulatory pipeline; 4 = advanced development; 5 = early development.

results improving cell wall digestion in tall fescue (Buanafina *et al.*, 2010) or Italian ryegrass (*Lolium multiflorum*) (Buanafina *et al.*, 2006).

After lucerne, maize is the second plant that concentrates research efforts aiming at improving its digestibility, notably because of the growing demand of the biofuel industry; even so, it also has interest for forage use. Advanced research is currently being conducted in Spain to improve the nutritional value of maize using the CAD downregulation of lignin biosynthesis. Field trials have proven that this approach is suitable to improve the digestibility of maize, as well as its energetic value for biomass production (Fornalé *et al.*, 2012). Other teams in France and in North America are also undertaking research to improve the nutritive value of maize (Piquemal *et al.*, 2002; He *et al.*, 2003) or sorghum (Funnell and Pedersen, 2006). Interestingly, early research has also been conducted in order to improve the digestibility of rice, with promising results (Zhang *et al.*, 2006). Rice is an important staple crop and its by-products are used extensively for animal feed, although the lignin content of the stems is a limiting factor of this application.

Prospects and challenges for commercial use of low-lignin plants

The breeding of GM plants featuring a low lignin content suitable for commercial-ization is facing a number of limitations, which are linked principally to the role that lignin plays in other plant physiological pathways (Zhao and Dixon, 2011; Jung *et al.*, 2012). First, lignin influences plant fitness, through its effect on vigour, on susceptibility to lodging and to disease, on drought tolerance and on productivity. Plants with severe reduction in the lignin content also tend to dwarf. Second, knowledge of the regulation of the genes involved in the biosynthesis of lignin, and more generally in cell wall biosynthesis, still has to be improved. Finally, even the nature of the relationship between cell wall composition and forage digestibility is still incompletely known.

Despite these limitations, one low lignin event has reached the regulatory pipeline and others will probably follow. The lucerne event, KK 179-5, should be ready for commercialization before 2017. The potential for improved forage is very high. In the EU-27, maize used as fodder was covering 4.8 million hectares (Mha) in 2007, representing about 37% of the total maize area (Eurostat, 2013). Other forage plants – temporary grasses, sorghum and legume crops – occupy an additional 13.9 Mha, which represents, together with the silage maize area, about 11% of the EU-27 utilized agricultural area. Moreover, the forage plant area is likely to increase due to the steady growth of the world's demand for meat.

Low-lignin plants would bring different kinds of benefits to world agriculture. They would contribute to the elevation of the productivity of meat and dairy farms. It has been estimated that a 1% increase in forage digestibility would result in a 3.2% increase in the daily weight gain of beef cattle (Casler and Vogel, 1999). For the US dairy industry as a whole, a 10% increase in cell wall digestibility would generate additional meat and milk sales of about US$380 million yearly, decrease manure production by 2.3 Mt and reduce the needs for grain complementation of rations of about 3 Mt (Hatfield *et al.*, 1999).

For forage crops such as lucerne, a low-lignin variety would also allow farmers to get higher yield with the same forage quality, and at a cheaper production cost. Indeed, field trials have shown that low-lignin varieties of lucerne grown under a three-cut system (i.e. involving three cuttings of lucerne in one crop season) produce a higher amount of forage than conventional varieties grown under a four-cut system, since the forage can be harvested later when the plant is mature and has the highest growth without suffering loss of quality (Undersander, 2010). Last, but not least, low-lignin maize, switchgrass or tall fescue have huge potential benefits for bioethanol production. It has been reported that a low-lignin maize line would increase bioethanol production by up to 51% per unit of land compared to a conventional maize (Fornalé *et al.*, 2012). This potential use of low-lignin plants is one of the main drivers of the research efforts in this area.

New ways of improving the cell wall digestibility of forage crops are still under research. The identification of all genes – over 750 – involved in the formation of the cell wall, and the understanding of their role, is essential. This would facilitate manipulation of the polysaccharide content of cell walls, which would make them more susceptible to hydrolysis. Additionally, controlling the expression of the modified genes in specific organs would allow the tissues that are more recalcitrant to rumen digestion, such as secondary xylem, to be targeted (Jung *et al.*, 2012). These modifications should not be at the expense of the agronomic performance of the plants.

12.3.4 Crops with a modified fatty acid profile or content

Oils and fats are important constituents of the human diet and come mostly from plants. According to their fatty acid (FA) composition, oils can have beneficial or negative effects. The consumption of oils/fats rich in satured FAs – such as palm oil – and of trans-FAs – resulting from the hydrogenation of soybean or canola oil, for instance – has been linked to cardiovascular disease. Conversely, deficiency in omega-3 long chain polyunsaturated FAs (PUFAs) is associated with many diseases of the Western diet, including cardiovascular or cognitive disorders (see Damude and Kinney, 2008, for a complete review of the issue).

As humans cannot synthesize two essential PUFAs (linoleic acid and linolenic acid) endogenously, they have to be obtained from food intake (Wallis *et al.*, 2002). Conventional breeders always have had interest in enhancing the oil content of plants and a limited number of high-oil crops are available, but breeding efforts have generally faced adverse effects on yield or protein content, or other deleterious effects. Biotechnological tools open up new prospects, since they allow modification of the oil content or FA composition by

targeting specific organs or development stages of the oil crops. The potential high benefits from enhanced oil varieties explain the dynamism of the research pipeline, with some events already in the market and several major companies involved in the development of new ones.

Most of the events in the pipeline are addressing human nutrition needs directly (Swiatkiewicz and Arczewska-Wlosek, 2011); however, some of them also have relevant applications for animal nutrition. A high-oil crop-based ration reduces the amount of feedstuffs needed to raise animals, since oil contained in grains provides 2.25 times more metabolizable energy than starch. For a given amount of feed, the use of high-oil crops would thus translate into an increased daily weight gain of animals and a reduced production of manure (Van Deynze *et al.*, 2004). Moreover, GM oil crops with high oleic acid (a monounsatured FA) content can reduce low-density lipoprotein cholesterol and might also increase high-density lipoprotein cholesterol: in sum, they are protective against coronary heart disease. The characteristics of high-oleic crops are thus directed mainly at consumer health; however, raising the level of oleic acid in feed also improves the quality of animal products, since the fatty profile of animals reflects the kind of fat that they ingest.

Finally, GM oil crops could also address the deficiency in omega-3 PUFAs of Western diets. As most vegetable oils are a poor source of omega-3, the primary sources for human diets are marine fish oils. However, as the demand for this non-renewable resource has increased dramatically and resulted in overfishing and the depletion of fish stocks, it has in turn generated an important development of aquaculture (Tocher, 2009). Yet, farmed fish are not able to produce omega-3 PUFAs and they need to find them in their diet, which generates the paradoxical and unsustainable situation that most of the world fish captures end up being converted into feed for farmed fishes (Naylor *et al.*, 2000, 2009). Therefore, an oil crop rich in omega-3 PUFAs would contribute to the urgent need for alternative sources of omega-3 PUFAs for aquaculture and the human diet.

The pipeline for oil crops with enhanced fatty acid profile

There are two main types of GM events with improved oil profile relevant for animal nutrition in the pipeline: the high-oleic events and the omega-3 events (Table 12.4). Both pipelines are very active, with some events already on the market or close to reaching it. Not surprisingly, almost all events in the pipeline are oilseed crops, namely soybean and rapeseed, and a GM high-oleic soybean has already been commercialized (event DP-3Ø5423-1). With a higher level of monounsaturated but no trans-FAs, it has been designed to provide the food industry with an oil that has a better oxidative stability without requiring hydrogenation. Integrated in the diet of swine, such high-oleic FAs would make the pork fat firmer, and this would ease the processing and the storage of the pork products. However, broilers fed with this event did not show any difference in daily weight gain with respect to those fed with a non-GM counterpart (McNaughton et al., 2008). At least three other high-oleic soybean events are currently in the commercial pipeline, after receiving all the necessary approvals from the US authorities: one is a stack of the previous one with a glyphosate-tolerant trait; one is an older version of the same and will not be commercialized (DD-Ø26ØØ5-3); and the last one, which relies on another biosynthetic pathway to obtain the desired high-oleic trait, will soon be commercialized (MON-877Ø5-6). Some research has also been conducted to improve the oleic content of rapeseed (Böhme et al., 2007) and cotton (Liu et al., 2002). However, in general, the feeding trials conducted on farm animals to assess the performance of high-oleic GM crops have shown no or very limited impacts, at least in terms of productivity gains (Böhme et al., 2007; McNaughton et al., 2008).

An oil crop with an enhanced content in omega-3 PUFAs is already available on the market (see Table 12.4): this is a GM safflower that accumulates up to 70% of γ-linolenic acid (GLA), one of the highest levels observed of newly produced PUFAs in a transgenic plant (Nykiforuk et al. 2012). But this crop is intended essentially for human food use. However, a GM soybean with an increased level of stearidonic acid (SDA – a precursor of omega-3 PUFA) has been deregulated by the USDA and its commercialization is pending its safety approval (MON87769). This soybean delivers oil with an SDA content of about 15–30% of total FAs, a promising improvement (Hammond et al., 2008; see Chapters 7 and 10). Another soybean event is currently in the advanced research stage and has been through numerous field trials in the USA, with published results showing GLA levels up to 28% of total FAs, while it is absent from control lines (Sato et al., 2004). Other active programmes include a PUFA-enriched rapeseed with an SDA content up to 26% of total FAs, compared to 0% in the control lines (Ursin, 2003), and a soybean that expresses a fungal bifunctional desaturase that results in more than 70% of α-linolenic acid (ALA – a omega-3 PUFA) compared to 53% in linseed oil, the best vegetable source of ALA (Damude et al., 2006).

Prospects and challenges for commercial use of crops with enhanced fatty acid profile

As already mentioned, the prospects for high-oleic crops for use as feed appear to be rather limited to improving the quality of the animal fat. Therefore, the use of such enhanced crops for feed could be restricted to some niche markets for high-quality meat products for which consumers are willing to pay a premium. The benefits from omega-3-enriched GM crops, on the other hand, are likely to be more important for the livestock sector. It is now acknowledged that transgenic techniques make it possible to assemble omega-3 PUFA pathways in oil crops (Damude and Kinney, 2008). If the public health authority recommendations keep being increasingly heeded, the demand for aquaculture products is likely to keep growing, and so will the demand for omega-3-rich oil, raising its price. GM crops

Table 12.4. Pipeline of events with an enhanced oil profile.

Crop	Developer	Country	Event name	Site modification	Phenotype modification	Development stage	Reference	Field trials	Feed trials	Patent year (if any)
Cotton	CSIRO Plant Industry	Australia		*ghFAD2-1*	High oleic	5	Liu *et al.*, 2002	–	–	–
Cotton	CSIRO Plant Industry	Australia		*ghSAD-1*	Ω-3 FAs	5	Liu *et al.*, 2002	–	–	–
Rapeseed		Germany	TM-5	Acyl-thioesterase	High oleic	5	Böhme *et al.*, 2007	Yes	Yes	–
Rapeseed	Monsanto	USA	SDA-canola	Δ6 and Δ12 FA desaturases	Ω-3 FAs	5	Ursin, 2003	Yes	–	2002
Safflower	SemBioSys Genetics and Arcadia BioSciences	Canada and USA		Δ6-desaturase	Ω-3 FAs	1	Nykiforuk *et al.*, 2012	–	–	2005
Soybean	Pioneer	USA	305423	*gm-fad2* and *gm-hra*	High oleic	1	McNaughton *et al.*, 2008	Multi-location	–	2006
Soybean	Pioneer	USA		*gm-fad2-1* + *cp4epsps*	High oleic + HT	2	McNaughton *et al.*, 2008	Multi-location	–	–
Soybean	Monsanto	USA	MON87705	*fatb1-A* + *fad2-1A* + *cp4epsps*	High oleic + HT	2	–	Multi-location	–	2008
Soybean	Pioneer	USA	260-05	*gm-fad2-1*	High oleic	2 – discontinued	–	Multi-location	–	–
Soybean	Monsanto	USA	MON87769		Ω-3 FAs	2	Hammond *et al.*, 2008	Multi-location	–	2008
Soybean	Monsanto	USA	MON87754	*dgat-2A*	High oleic	3 – discontinued	–	Multi-location	–	–
Soybean	DuPont and University of Nebraska	USA		Δ12 FA desaturase	High oleic	4	Buhr *et al.*, 2002	Multi-location	–	2000
Soybean	DuPont and University of Nebraska	USA		Δ6-desaturase	Ω-3 FAs	4	Sato *et al.*, 2004	Multi-location	–	2008
Soybean	DuPont	USA		Δ12/ω3 desaturases	Ω-3 FAs	5	Damude *et al.*, 2006	–	–	2006
Soybean	University of Nebraska	USA		Δ6-desaturase and Δ15-desaturase	Ω-3 FAs	5	Eckert *et al.*, 2006	Yes	–	–

Note: Development stage: 1 = commercialized; 2 = commercial pipeline; 3 = regulatory pipeline; 4 = advanced development; 5 = early development.

enriched in omega-3 could provide a credible alternative to the dwindling world fish stocks that are currently used to meet the demand of aquaculture for fish oil. Some anti-nutritional compound of plant oil may hinder its use as a fish feed, but recent experiments with farmed fish have shown that it is possible to replace up to 70% of fish meal with soybean meal, with some adjustments in the ration without affecting the overall fish growth (NOAA-USDA, 2011). Omega-3-enhanced crops are a much more renewable resource than forage fish. With the current price of fish oil, and provided that the nutritional limitations to the incorporation of plant oil in fish and animal diets are overcome, there is little doubt this feed application will be profitable.

However, some technical challenges remain for the development and commercialization of a major oil crop with an enhanced content of omega-3 PUFAs. The acyl–lipid metabolism is extremely complex; in model plants it requires more than 120 enzymatic reactions, and at least 600 genes are involved; moreover, the PUFA biosynthesis pathways vary from one oil crop to another and a better genomic characterization of these is still needed. In order to develop a variety that would become a commercially viable alternative to fish oils, breeders are now focusing on optimizing the level of PUFAs in oilseeds to bring it closer to that found in fish oils, while avoiding the presence of undesirable intermediate compounds (Ruiz-Lopez *et al.*, 2012). Once this objective is completed, this will pave the way to a possible sustainable alternative to marine resource depletion, while at the same time providing an improvement to human nutritional health.

12.4 Discussion and Conclusions

The pipeline for GM crops enhanced for animal nutrition is a rather active one, with almost 100 events under research in many countries of the world. This reflects both the importance of feed markets for GM crops and the potential important improvements that can be brought to the quality of feedstuffs. However, as this chapter has shown, very few events are, as yet, available for farmers, and those that may reach the market within the next 5 years (i.e. before 2018) are very few (Table 12.5). In spite of the legitimate expectations of the actors in the chain, there is still a long way to go before GM crops with animal nutrition-designed traits dominate the market.

Some events seem to be quite advanced in the research stage, such as low-lignin lucerne, high-Trp rice, omega-3 soybean or low-phytase barley, while others need further research. However, the biotechnological issues will not be the only constraints to the commercial success of GM varieties developed for animal nutrition. There will be regulatory issues regarding the fact that these are events with a modified composition, for which the substantial equivalence rule might not apply, and on top of this there will be economic and market issues. Indeed, for those feed quality traits and contrary to the agronomic traits of the first generation of GM crops, the adopter of the innovative technology – the farmer, grower of the GM plant – is not necessarily the one who benefits directly from the innovation (unless he or she is also a livestock farmer, but this is a specific case). This means that he or she needs to be convinced to adopt the modified crops, by the way of an economic incentive, in this case a premium that would be paid in addition to the price of the alternative conventional crops. This also implies that a segregation system should be in place to ensure that the identity of the crop with a specific trait is preserved in the supply chain until it reaches the final user. This requires dedicated infrastructures, especially if various identity preservation schemes have to be operated simultaneously. The costs incurred by the segregation of these crops, added to the technology fee to be paid to the innovator, are likely to represent a limit to the adoption of these new events by farmers and a question of their economic profitability. The market will only retain those that clearly bring

Table 12.5. Summary of events in the latest stages of the pipeline.

Crop	OECD unique identifier	Development stage	Event name	Commercial name	Trait	Developer	Developer country	Status in USA	Status in EU	Status in Japan
Soybean	DP-3Ø5423-1	1	305423	Treus-Plenish	High oleic	Pioneer	USA	Commercialized – 2011	Food and feed application; additional data request – 2012	All uses – 2010
Safflower		1	–	Sonova 400	Omega-3	Arcadia BioSciences	USA	Commercialized	–	–
Maize	BVLA430101	2			Phytase expression	CAAS and Origin Agritech	China	No application	No application	No application
Maize[a]	REN-ØØØ38-3	2	LY038	Mavera	High lysine	Monsanto	USA	All uses – 2006	Application withdrawn – 2009	All uses – 2007
Maize[a]	REN-ØØØ38-3 × MON-ØØ81Ø-6	2	LY038 × MON810	Mavera YieldGard	High-lysine + herbicide tolerance	Monsanto	USA	All uses – 2006	Application withdrawn – 2009	All uses – 2007
Soybean	DP-3Ø5423-1 × MON-Ø4Ø32-6	2	DP305423 × GTS 40-3-2		High oleic + herbicide tolerance	Pioneer	USA	All uses – 2009	Food and feed application; additional data request – 2012	All uses – 2012
Soybean	MON-877Ø5-6	2	MON87705	Vistive Gold	High oleic	Monsanto	USA	All uses – 2011	Imports and domestic use – 2012	Ongoing application
Soybean[a]	DD-Ø26ØØ5-3	2	260-05		High oleic	Pioneer	USA	All uses – 1997	No application	All uses – 2007
Lucerne	MON-00179-5	3	KK179-5		Low lignin	Forage Genetics and Monsanto	USA	No application	No application	Imports and domestic uses – 2012
Rapeseed[a]		3	MPS961-5	PhytaSeed	Phytase expression	BASF	USA	Food and feed – 1999	No application	N/A
Soybean	MON-87769-7	3	MON87769		Omega-3	Monsanto	USA	Cultivation 2012, food and feed under assessment	Food and feed application; additional data request – 2012	Ongoing application

Note: Development stage: 1 = commercialized; 2 = commercial pipeline; 3 = regulatory pipeline; 4 = advanced development; 5 = early development. [a]Events whose development is currently discontinued. The information regarding the regulatory status of the events reported in this table was updated in February 2013.

important gains to the nutritional efficiency of feedstuffs.

Under some circumstances, however, the burden of the costs associated with the segregation of crops can be overstepped. This is the case when farmers produce their own feedstuffs on-farm, such as silage crops or cereals mixed for swine or poultry. Then, the benefits of the GM modification of feed crops are delivered directly to the final user, avoiding many costs. This is also the purpose of the implementation of integrated supply chains: when crop products are too specialized, farmers lose flexibility and cannot control market outputs. Then, it makes sense for them to enter an integrated scheme, usually organized locally, in order to reduce transaction costs. Some niche markets might also constitute a possible output, when premiums exceed segregation costs. The case of the feedstuffs used by aquaculture might be illustrative of this: as the price of fish oil has increased dramatically in the last decade, due to the booming demand for farmed fish, any land-based and renewable resource substitute – such as GM omega-3 seed oil – might be profitable. In general, more research on the economic profitability of GM crops for animal nutrition is thus desirable to better assess their potential market and economic impacts.

Acknowledgements

The authors are very grateful to Claudia Parisi and Mauro Vigani for their helpful comments in the preparation of this chapter.

Notes

[1] The three US regulatory agencies – the Food and Drug Administration (FDA), the Animal and Plant Health Inspection Service (APHIS) of the US Department of Agriculture (USDA) and the Environmental Protection Agency (EPA) – have a unified website that features a complete database of GM crop reviews: http://usbiotechreg.epa.gov/usbiotechreg/.

[2] This refers to the European Food Safety Authority (EFSA) Register of Questions: http://register ofquestions.efsa.europa.eu/roqFrontend/.

[3] The USDA Field Tests of GM Crops for the USA (http://www.isb.vt.edu/search-release-data.aspx), the GMO Register for experimental releases of the EU (http://gmoinfo.jrc.ec.europa.eu/gmp_browse.aspx) and the Japanese Biosafety Clearing House (http://www.bch.biodic.go.jp/english/lmo.html).

[4] This was performed through the European Patent Office webpage: http://www.epo.org/.

References

Abhary, M., Siritunga, D., Stevens, G., Taylor, N.J. and Fauquet, C.M. (2011) Transgenic biofortification of the starchy staple cassava (*Manihot esculenta*) generates a novel sink for protein. *PLoS ONE* 6, e16256.

Amir, R. and Tabe, L. (2006) Molecular approaches to improving plant methionine content. In: Pawan, K.J., Jaiwal, K. and Rana, P.S. (eds) *Plant Genetic Engineering: Metabolic Engineering and Molecular Farming II*. Studium Press LLC, Houston, Texas, pp. 1–26.

Avraham, T., Badani, H., Galili, S. and Amir, R. (2005) Enhanced levels of methionine and cysteine in transgenic alfalfa (*Medicago sativa* L.) plants over-expressing the Arabidopsis cystathionine γ-synthase gene. *Plant Biotechnology Journal* 3, 71–79.

Bagga, S., Potenza, C., Ross, J., Martin, M., Leustek, T. and Sengupta-Gopalan, C. (2005) A transgene for high methionine protein is posttranscriptionally regulated by methionine. *In Vitro Cellular and Developmental Biology – Plant* 41, 731–741.

Barneix, A.J. (2007) Physiology and biochemistry of source-regulated protein accumulation in the wheat grain. *Journal of Plant Physiology* 164, 581–590.

Barrière, Y., Guillaumie, S., Pichon, M. and Emile, J.C. (2009) Breeding for silage quality traits in cereals. In: Carena, M.J. (ed.) *Cereals – Handbook of Plant Breeding*. Springer, New York, pp. 367–394.

Bicar, E., Woodman-Clikeman, W., Sangtong, V., Peterson, J., Yang, S., Lee, M., *et al.* (2008) Transgenic maize endosperm containing a milk protein has improved amino acid balance. *Transgenic Research* 17, 59–71.

Bilyeu, K.D., Zeng, P., Coello, P., Zhang, Z.J., Krishnan, H.B., Bailey, A., *et al.* (2008) Quantitative conversion of phytate to inorganic phosphorus in soybean seeds expressing a bacterial phytase. *Plant Physiology* 146, 468–477.

Böhme, H., Rudloff, E., Schöne, F., Schumann, W., Hüther, L. and Flachowsky, G. (2007) Nutritional assessment of genetically modified rapeseed synthesizing high amounts of mid-chain fatty acids including production responses of growing-finishing pigs. *Archives of Animal Nutrition* 61, 308–316.

Brinch-Pedersen, H., Sørensen, L.D. and Holm, P.B. (2002) Engineering crop plants: getting a handle on phosphate. *Trends in Plant Science* 7, 118–125.

Brinch-Pedersen, H., Hatzack, F., Stöger, E., Arcalis, E., Pontopidan, K. and Holm, P.B. (2006) Heat-stable phytases in transgenic wheat (*Triticum aestivum* L.): deposition pattern, thermostability, and phytate hydrolysis. *Journal of Agricultural and Food Chemistry* 54, 4624–4632.

Buanafina, M.M. de O., Langdon, T., Hauck, B., Dalton, S. and Morris, P. (2006) Manipulating the phenolic acid content and digestibility of Italian ryegrass (*Lolium multiflorum*) by vacuolar-targeted expression of a fungal ferulic acid esterase. In: McMillan, J., Adney, W., Mielenz, J. and Klasson, K.T. (eds) *Proceedings of the Twenty-Seventh Symposium on Biotechnology for Fuels and Chemicals.* Humana Press, pp. 416–426.

Buanafina, M.M. de O., Langdon, T., Hauck, B., Dalton, S., Timms-Taravella, E. and Morris, P. (2010) Targeting expression of a fungal ferulic acid esterase to the apoplast, endoplasmic reticulum or golgi can disrupt feruloylation of the growing cell wall and increase the biodegradability of tall fescue (*Festuca arundinacea*). *Plant Biotechnology Journal* 8, 316–331.

Buhr, T., Sato, S., Ebrahim, F., Xing, A., Zhou, Y., Mathiesen, M., *et al.* (2002) Ribozyme termination of RNA transcripts down-regulate seed fatty acid genes in transgenic soybean. *The Plant Journal* 30, 155–163.

Casler, M.D. and Vogel, K.P. (1999) Accomplishments and impact from breeding for increased forage nutritional value. *Crop Science* 39, 12–20.

Chaumet, J.-M., Ghersi, G. and Rastoin, J.-L. (2011) Food consumption in 2050. In: Paillard, S., Treyer, S. and Dorin, B. (eds) *Agrimonde : Scenarios and Challenges for Feeding the World in 2050.* Quae Editions, Versailles, France.

Chen, F., Srinivasa Reddy, M.S., Temple, S., Jackson, L., Shadle, G. and Dixon, R.A. (2006) Multi-site genetic modulation of monolignol biosynthesis suggests new routes for formation of syringyl lignin and wall-bound ferulic acid in alfalfa (*Medicago sativa* L.). *The Plant Journal* 48, 113–124.

Chen, L., Auh, C.K., Dowling, P., Bell, J., Chen, F., Hopkins, A., *et al.* (2003) Improved forage digestibility of tall fescue (*Festuca arundinacea*) by transgenic down-regulation of cinnamyl alcohol dehydrogenase. *Plant Biotechnology Journal* 1, 437–449.

Chen, R., Xue, G., Chen, P., Yao, B., Yang, W., Ma, Q., *et al.* (2008) Transgenic maize plants expressing a fungal phytase gene. *Transgenic Research* 17, 633–643.

Cunningham, E.P. (2005) Gene-based technologies for livestock industries in the 3(RD) millennium. In: Makkar, H.P.S. and Viljoen, G. (eds) *Applications of Gene-Based Technologies for Improving Animal Production and Health in Developing Countries.* Springer, Dordrecht, Netherlands, pp. 41–51.

Damude, H.G. and Kinney, A.J. (2008) Enhancing plant seed oils for human nutrition. *Plant Physiology* 147, 962–968.

Damude, H.G., Zhang, H., Farrall, L., Ripp, K.G., Tomb, J.F., Hollerbach, D., *et al.* (2006) Identification of bifunctional delta12/omega3 fatty acid desaturases for improving the ratio of omega3 to omega6 fatty acids in microbes and plants. *Proceedings of the National Academy of Sciences* 103, 9446–9451.

Denbow, D.M., Grabau, E.A., Lacy, G.H., Kornegay, E.T., Russell, D.R. and Umbeck, P.F. (1998) Soybeans transformed with a fungal phytase gene improve phosphorus availability for broilers. *Poultry Science* 77, 878–881.

Drakakaki, G., Marcel, S., Glahn, R.P., Lund, E.K., Pariagh, S., Fischer, R., *et al.* (2005) Endosperm-specific co-expression of recombinant soybean ferritin and Aspergillus phytase in maize results in significant increases in the levels of bioavailable iron. *Plant Molecular Biology* 59, 869–880.

Eckert, H., LaVallee, B., Schweiger, B., Kinney, A., Cahoon, E. and Clemente, T. (2006) Co-expression of the borage Δ6 desaturase and the Arabidopsis Δ15 desaturase results in high accumulation of stearidonic acid in the seeds of transgenic soybean. *Planta* 224, 1050–1057.

Eurostat (2013) Farm Structure Survey 2007. European Commission (http://epp.eurostat.ec.europa.eu/portal/page/portal/agriculture/farm_structure/database, accessed 16 January 2013).

FAOSTAT (2013) FAOSTAT Production database. Statistics Division of the Food and Agriculture Organization of the United Nations (http://faostat3.fao.org/home/index.html, accessed 16 January 2013).

Fornalé, S., Capellades, M., Encina, A., Wang, K., Irar, S., Lapierre, C., *et al.* (2012) Altered lignin biosynthesis improves cellulosic bioethanol production in transgenic maize plants down-regulated for cinnamyl alcohol dehydrogenase. *Molecular Plant* 5, 817–830.

Frizzi, A., Huang, S., Gilbertson, L.A., Armstrong, T.A., Luethy, M.H. and Malvar, T.M. (2008) Modifying lysine biosynthesis and catabolism in corn with a single bifunctional expression/silencing transgene cassette. *Plant Biotechnology Journal* 6, 13–21.

Fu, C., Mielenz, J.R., Xiao, X., Ge, Y., Hamilton, C.Y., Rodriguez, M., *et al.* (2011) Genetic manipulation of lignin reduces recalcitrance and improves ethanol production from switchgrass. *Proceedings of the National Academy of Sciences* 108, 3803–3808.

Funnell, D.L. and Pedersen, J.F. (2006) Reaction of sorghum lines genetically modified for reduced lignin content to infection by *Fusarium* and *Alternaria* spp. *Plant Disease* 90, 331–338.

Gao, X., Wang, G., Su, Q., Wang, Y. and An, L. (2007) Phytase expression in transgenic soybeans: stable transformation with a vector-less construct. *Biotechnology Letters* 29, 1781–1787.

Gontia, I., Tantwai, K., Prasad, L., Rajput, S. and Tiwari, S. (2012) Transgenic plants expressing phytase gene of microbial origin and their prospective application as feed. *Food Technology and Biotechnology* 50, 3–10.

Guthrie, T.A., Apgar, G.A., Griswold, K.E., Lindemann, M.D., Radcliffe, J.S. and Jacobson, B.N. (2004) Nutritional value of a corn containing a glutamate dehydrogenase gene for growing pigs. *Journal of Animal Science* 82, 1693–1698.

Hamada, A., Yamaguchi, K.-I., Ohnishi, N., Harada, M., Nikumaru, S. and Honda, H. (2005) High-level production of yeast (*Schwanniomyces occidentalis*) phytase in transgenic rice plants by a combination of signal sequence and codon modification of the phytase gene. *Plant Biotechnology Journal* 3, 43–55.

Hammond, B.G., Lemen, J.K., Ahmed, G., Miller, K.D., Kirkpatrick, J. and Fleeman, T. (2008) Safety assessment of SDA soybean oil: results of a 28-day gavage study and a 90-day/one generation reproduction feeding study in rats. *Regulatory Toxicology and Pharmacology* 52, 311–323.

Hatfield, R.D., Ralph, J. and Grabber, J.H. (1999) Cell wall structural foundations: molecular basis for improving forage digestibilities. *Crop Science* 39, 27–37.

He, X., Hall, M.B., Gallo-Meagher, M. and Smith, R.L. (2003) Improvement of forage quality by downregulation of maize O-methyltransferase. *Crop Science* 43, 2240–2251.

Henry, Y., Sève, B., Colléaux, Y., Ganier, P., Saligaut, C. and Jégo, P. (1992) Interactive effects of dietary levels of tryptophan and protein on voluntary feed intake and growth performance in pigs, in relation to plasma free amino acids and hypothalamic serotonin. *Journal of Animal Science* 70, 1873–1887.

Holme, I.B., Dionisio, G., Brinch-Pedersen, H., Wendt, T., Madsen, C.K., Vincze, E., *et al.* (2012) Cisgenic barley with improved phytase activity. *Plant Biotechnology Journal* 10, 237–247.

Hong, C.-Y., Cheng, K.-J., Tseng, T.-H., Wang, C.-S., Liu, L.-F. and Yu, S.-M. (2004) Production of two highly active bacterial phytases with broad pH optima in germinated transgenic rice seeds. *Transgenic Research* 13, 29–39.

Houmard, N.M., Mainville, J.L., Bonin, C.P., Huang, S., Luethy, M.H. and Malvar, T.M. (2007) High-lysine corn generated by endosperm-specific suppression of lysine catabolism using RNAi. *Plant Biotechnology Journal* 5, 605–614.

Huang, S., Kruger, D.E., Frizzi, A., D'Ordine, R.L., Florida, C.A., Adams, W.R., *et al.* (2005) High-lysine corn produced by the combination of enhanced lysine biosynthesis and reduced zein accumulation. *Plant Biotechnology Journal* 3, 555–569.

Huang, S., Frizzi, A., Florida, C., Kruger, D. and Luethy, M. (2006) High lysine and high tryptophan transgenic maize resulting from the reduction of both 19- and 22-kD α-zeins. *Plant Molecular Biology* 61, 525–535.

Huang, S., Frizzi, A. and Malvar, T.M. (2008) *Genetically Engineered High Lysine Corn*. ISB News report – January 2008.

Inaba, Y., Brotherton, J., Ulanov, A. and Widholm, J. (2007) Expression of a feedback insensitive anthranilate synthase gene from tobacco increases free tryptophan in soybean plants. *Plant Cell Reports* 26, 1763–1771.

Ishimoto, M., Rahman, S., Hanafy, M., Khalafalla, M., El-Shemy, H., Nakamoto, Y., *et al.* (2010) Evaluation of amino acid content and nutritional quality of transgenic soybean seeds with high-level tryptophan accumulation. *Molecular Breeding* 25, 313–326.

Johnson, L.A., Hardy, C.L., Baumel, C.P., Yu, T.H. and Sell, J.L. (2001) Identifying valuable corn quality traits for livestock feed. *Cereal Foods World* 46, 472–481.

Jung, H.G. and Vogel, K.P. (1986) Influence of lignin on digestibility of forage cell wall material. *Journal of Animal Science* 62, 1703–1712.

Jung, H.-J.G., Samac, D.A. and Sarath, G. (2012) Modifying crops to increase cell wall digestibility. *Plant Science* 185–186, 65–77.

Kita, Y., Nakamoto, Y., Takahashi, M., Kitamura, K., Wakasa, K. and Ishimoto, M. (2010) Manipulation of amino acid composition in soybean seeds by the combination of deregulated tryptophan biosynthesis and storage protein deficiency. *Plant Cell Reports* 29, 87–95.

Kruger, J., Taylor, J.R.N. and Oelofse, A. (2012) Effects of reducing phytate content in sorghum through genetic modification and fermentation on in vitro iron availability in whole grain porridges. *Food Chemistry* 131, 220–224.

Kumar, T., Dweikat, I., Sato, S., Ge, Z., Nersesian, N., Chen, H., *et al.* (2012) Modulation of kernel storage proteins in grain sorghum (*Sorghum bicolor* (L.) Moench). *Plant Biotechnology Journal* 10, 533–544.

Kuwano, M., Mimura, T., Takaiwa, F. and Yoshida, K.T. (2009) Generation of stable 'low phytic acid' transgenic rice through antisense repression of the 1d-myo-inositol 3-phosphate synthase gene (RINO1) using the 18-kDa oleosin promoter. *Plant Biotechnology Journal* 7, 96–105.

Lee, M., Huang, T., Toro-Ramos, T., Fraga, M., Last, R.L. and Jander, G. (2008) Reduced activity of Arabidopsis thaliana HMT2, a methionine biosynthetic enzyme, increases seed methionine content. *The Plant Journal* 54, 310–320.

Li, G., Yang, S., Li, M., Qiao, Y. and Wang, J. (2009) Functional analysis of an *Aspergillus ficuum* phytase gene in *Saccharomyces cerevisiae* and its root-specific, secretory expression in transgenic soybean plants. *Biotechnology Letters* 31, 1297–1303.

Li, K., Wang, S.-Q., Wu, F.-Q., Li, S.-C., Deng, Q.-M., Wang, L.-X., *et al.* (2008) Analysis on T4 progeny of transgenic rice with lysine-rich protein gene (sb401) mediated by *Agrobacterium tumefaciens* transformation. *Chinese Journal of Rice Science* 22, 131–136.

Li, Z., Meyer, S., Essig, J.S., Liu, Y., Schapaugh, M.A., Muthukrishnan, S., *et al.* (2005) High-level expression of maize γ-zein protein in transgenic soybean (*Glycine max*). *Molecular Breeding* 16, 11–20.

Liu, Q., Singh, S.P. and Green, A.G. (2002) High-stearic and high-oleic cottonseed oils produced by hairpin RNA-mediated post-transcriptional gene silencing. *Plant Physiology* 129, 1732–1743.

Liu, Q.-Q., Li, Q.-F., Jiang, L., Zhang, D.-J., Wang, H.-M. and Gu, M.-H. (2006) Transgenic expression of the recombinant phytase in rice (*Oryza sativa*). *Rice Science* 13, 79–84.

Lott, J.N.A., Ockenden, I., Raboy, V. and Batten, G.D. (2000) Phytic acid and phosphorus in crop seeds and fruits: a global estimate. *Seed Science Research* 10, 11–33.

Lucas, D.M., Taylor, M.L., Hartnell, G.F., Nemeth, M.A., Glenn, K.C. and Davis, S.W. (2007) Broiler performance and carcass characteristics when fed diets containing lysine maize (LY038 or LY038 × MON 810), control, or conventional reference maize. *Poultry Science* 86, 2152–2161.

Ma, X.-F., Tudor, S., Butler, T., Ge, Y., Xi, Y., Bouton, J., *et al.* (2012) Transgenic expression of phytase and acid phosphatase genes in alfalfa (*Medicago sativa*) leads to improved phosphate uptake in natural soils. *Molecular Breeding* 30, 377–391.

McNaughton, J., Roberts, M., Smith, B., Rice, D., Hinds, M., Sanders, C., *et al.* (2008) Comparison of broiler performance when fed diets containing event DP-3Ø5423-1, non-transgenic near-isoline control, or commercial reference soybean meal, hulls, and oil. *Poultry Science* 87, 2549–2561.

Molvig, L., Tabe, L.M., Eggum, B.O., Moore, A.E., Craig, S., Spencer, D., *et al.* (1997) Enhanced methionine levels and increased nutritive value of seeds of transgenic lupins (*Lupinus angustifolius* L.) expressing a sunflower seed albumin gene. *Proceedings of the National Academy of Sciences* 94, 8393–8398.

Mueller, S.C., Undersander, D. and Putman, D.H. (2008) Alfalfa for industrial and other uses. In: Summers, C.G. and Putman, D.H. (eds) *Irrigated Alfalfa Management in Mediterranean and Desert Zones*. University of California, Davis, California, Chapter 19, pp. 1–6.

Naylor, R.L., Goldburg, R.J., Primavera, J.H., Kautsky, N., Beveridge, M.C.M., Clay, J., *et al.* (2000) Effect of aquaculture on world fish supplies. *Nature* 405, 1017–1024.

Naylor, R.L., Hardy, R.W., Bureau, D.P., Chiu, A., Elliott, M., Farrell, A.P., *et al.* (2009) Feeding aquaculture in an era of finite resources. *Proceedings of the National Academy of Sciences* 106, 15103–15110.

NOAA-USDA (2011) *The Future of Aquafeeds*. NOAA/USDA Alternative Feeds Initiative. NOAA, Washington, DC.

Nyannor, E.K.D., Williams, P., Bedford, M.R. and Adeola, O. (2007) Corn expressing an *Escherichia coli*-derived phytase gene: a proof-of-concept nutritional study in pigs. *Journal of Animal Science* 85, 1946–1952.

Nykiforuk, C., Shewmaker, C., Harry, I., Yurchenko, O., Zhang, M., Reed, C., *et al.* (2012) High level accumulation of gamma linolenic acid (C18:3Δ6.9,12 cis) in transgenic safflower (*Carthamus tinctorius*) seeds. *Transgenic Research* 21, 367–381.

Ohnoutkova, L., Kalaninova, J., Vaskova, J., Pauk, A., Marchetti, S. and Harwood, W. (2010) Expression of *Aspergillus niger* phytase gene in barley – Abstract of a poster presented at the

2010 *In Vitro* Biology Meeting and IAPB 12th World Congress. *In Vitro Cellular and Developmental Biology – Animal* 46, 147–148.

Oliver, A.L., Grant, R.J. and Pedersen, J. (2004) Comparison of brown midrib-6 and -18 forage sorghum with conventional sorghum and corn silage in diets of lactating dairy cows. *Journal of Dairy Science* 87, 637–644.

Parisi, C., Rodriguez-Cerezo, E. and Thangaraj, H. (2013) Analysing patent landscapes in plant biotechnology and new plant breeding techniques. *Transgenic Research* 22, 15–29.

Peng, J., Rapp, W., Tian, K., Wicker, J., Nadig, G., Vaduva, G., *et al.* (2004) Development of soybeans with improved amino acid composition. Paper presented at the Plant Biology Conference 2004, July 24–28, 2004. Lake Buena Vista, Florida.

Peng, R.-H., Yao, Q.-H., Xiong, A.-S., Cheng, Z.-M. and Li, Y. (2006) Codon-modifications and an endoplasmic reticulum-targeting sequence additively enhance expression of an *Aspergillus* phytase gene in transgenic canola. *Plant Cell Reports* 25, 124–132.

Piquemal, J., Chamayou, S., Nadaud, I., Beckert, M., Barrière, Y., Mila, I., *et al.* (2002) Down-regulation of caffeic acid O-methyltransferase in maize revisited using a transgenic approach. *Plant Physiology* 130, 1675–1685.

Ponstein, A.S., Bade, J.B., Verwoerd, T.C., Molendijk, L., Storms, J., Beudeker, R.F., *et al.* (2002) Stable expression of phytase (phyA) in canola (*Brassica napus*) seeds: towards a commercial product. *Molecular Breeding* 10, 31–144.

Raboy, V. (2001) Seeds for a better future: 'low phytate', grains help to overcome malnutrition and reduce pollution. *Trends in Plant Science* 6, 458–462.

Reddy, M.S.S., Chen, F., Shadle, G., Jackson, L., Aljoe, H. and Dixon, R.A. (2005) Targeted down-regulation of cytochrome P450 enzymes for forage quality improvement in alfalfa (*Medicago sativa* L.). *Proceedings of the National Academy of Sciences of the United States of America* 102, 16573–16578.

Reisen, P., McCaslin, M. and Fitzpatrick, S. (2009) *Roundup Ready Alfalfa Update and New Biotech Traits*. Forage Genetics International, Nampa, Idaho.

Ruiz-Lopez, N., Sayanova, O., Napier, J.A. and Haslam, R.P. (2012) Metabolic engineering of the omega-3 long chain polyunsaturated fatty acid biosynthetic pathway into transgenic plants. *Journal of Experimental Botany* 63, 2397–2410.

Sato, S., Xing, A., Ye, X., Schweiger, B., Kinney, A., Graef, G., *et al.* (2004) Production of γ-linolenic acid and stearidonic acid in seeds of marker-free transgenic soybean. *Crop Science* 44, 646–652.

Scott, M.P., Edwards, J.W., Bell, C.P., Schussler, J.R. and Smith, J.S. (2006) Grain composition and amino acid content in maize hybrids representing 80 years of commercial maize varieties. *Maydica* 51, 417–423.

Shewry, P., Powers, S., Field, J., Fido, R., Jones, H., Arnold, G., *et al.* (2006) Comparative field performance over 3 years and two sites of transgenic wheat lines expressing HMW subunit transgenes. *TAG Theoretical and Applied Genetics* 113, 128–136.

Shi, J., Wang, H., Schellin, K., Li, B., Faller, M., Stoop, J.M., *et al.* (2007) Embryo-specific silencing of a transporter reduces phytic acid content of maize and soybean seeds. *Nature Biotechnology* 25, 930–937.

Shukla, V.K., Doyon, Y., Miller, J.C., DeKelver, R.C., Moehle, E.A., Worden, S.E., *et al.* (2009) Precise genome modification in the crop species *Zea mays* using zinc-finger nucleases. *Nature* 459, 437–441.

Stein, A.J. and Rodríguez Cerezo, E. (2009) The global pipeline of new GM crops: implications of asynchronous approval for international trade. JRC Scientific and Technical Reports. Office for Official Publications of the European Communities, Luxembourg.

Sun, X.-B., Fang, R., Yu, G.-H., Zhang, P.-P., Xu, L. and Ma, H.-X. (2010) Transfer of high lysine gene *cflr* into wheat and analysis for protein and lysine content in transgenic wheat seeds. *Jiangsu Journal of Agricultural Sciences* 6, 1162–1169.

Swiatkiewicz, S. and Arczewska Wlosek, A. (2011) Prospects for the use of genetically modified crops with improved nutritional properties as feed materials in poultry nutrition. *World's Poultry Science Journal* 67, 631–642.

Takada, R. and Otsuka, M. (2007) Effects of feeding high tryptophan GM-rice on growth performance of chickens. *International Journal of Poultry Science* 6, 524–526.

Tamás, C., Kisgyörgy, B., Rakszegi, M., Wilkinson, M., Yang, M.-S., Láng, L., *et al.* (2009) Transgenic approach to improve wheat (*Triticum aestivum* L.) nutritional quality. *Plant Cell Reports* 28, 1085–1094.

Taylor, J. and Taylor, J.R.N. (2011) Protein biofortified sorghum: effect of processing into traditional African foods on their protein quality. *Journal of Agricultural and Food Chemistry* 59, 2386–2392.

Tocher, D. (2009) Issues surrounding fish as a source of omega-3 long-chain polyunsaturated fatty acids. *Lipid Technology* 21, 13–16.

Ufaz, S. and Galili, G. (2008) Improving the content of essential amino acids in crop plants: goals and opportunities. *Plant Physiology* 147, 954–961.

Ullah, A.H., Sethumadhavan, K., Mullaney, E.J., Ziegelhoffer, T. and Austin-Phillips, S. (2002) Cloned and expressed fungal *phyA* gene in alfalfa produces a stable phytase. *Biochemical and Biophysical Research Communications* 290, 1343–1348.

Undersander, D. (2010) Round-Up Ready, low lignin and other new traits in alfalfa's future. In: Laboski, C. and Boerboom, C. (eds) *Proceedings of the 2010 Wisconsin Crop Management Conference*. University of Wisconsin, Madison, Wisconsin, pp. 95–101.

Ursin, V.M. (2003) Modification of plant lipids for human health: development of functional land-based omega-3 fatty acids. *Journal of Nutrition* 133, 4271–4274.

Van Deynze, A., Bradford, K.J. and Van Eenennaam, A. (2004) *Crops Biotechnology: Feeds for Livestock*. ANR Publications 8145. University of California, California.

Vasal, S.K. (2000) The quality protein maize story. *Food and Nutrition Bulletin* 21, 445–450.

Wakasa, K., Hasegawa, H., Nemoto, H., Matsuda, F., Miyazawa, H., Tozawa, Y., *et al.* (2006) High-level tryptophan accumulation in seeds of transgenic rice and its limited effects on agronomic traits and seed metabolite profile. *Journal of Experimental Botany* 57, 3069–3078.

Wallis, J.G., Watts, J.L. and Browse, J. (2002) Polyunsaturated fatty acid synthesis: what will they think of next? *Trends in Biochemical Sciences* 27, 467–473.

Wambugu, F. (2007) *Developing Sustainable Food Security Through Agricultural Biotechnology: The ABS Project Model. Science Based Improvements of Rural/Subsistence Agriculture*. Academy of Science of South Africa (ASSAf), Pretoria, South Africa.

Wang, J., Chen, L., Liu, Q.Q., Sun, S.S., Sokolov, V. and Wang, Y.P. (2011) Transformation of LRP gene into *Brassica napus* mediated by *Agrobacterium tumefaciens* to enhance lysine content in seeds. *Genetika* 47, 1616–1621.

Weichert, N., Saalbach, I., Weichert, H., Kohl, S., Erban, A., Kopka, J., *et al.* (2010) Increasing sucrose uptake capacity of wheat grains stimulates storage protein synthesis. *Plant Physiology* 152, 698–710.

Wu, X.R., Chen, Z.H. and Folk, W.R. (2003) Enrichment of cereal protein lysine content by altered tRNAlys coding during protein synthesis. *Plant Biotechnology Journal* 1, 187–194.

Wu, X.R., Kenzior, A., Willmot, D., Scanlon, S., Chen, Z., Topin, A., *et al.* (2007) Altered expression of plant lysyl tRNA synthetase promotes tRNA misacylation and translational recoding of lysine. *The Plant Journal* 50, 627–636.

Yang, S., Li, G., Li, M. and Wang, J. (2011) Transgenic soybean with low phytate content constructed by *Agrobacterium* transformation and pollen-tube pathway. *Euphytica* 177, 375–382.

Yu, J., Peng, P., Zhang, X., Zhao, Q., Zhy, D., Sun, X., *et al.* (2004) Seed-specific expression of a lysine rich protein sb401 gene significantly increases both lysine and total protein content in maize seeds. *Molecular Breeding* 14, 1–7.

Zhang, K., Qian, Q., Huang, Z., Wang, Y., Li, M., Hong, L., *et al.* (2006) Gold hull and internode2 encodes a primarily multifunctional cinnamyl-alcohol dehydrogenase in rice. *Plant Physiology* 140, 972–983.

Zhao, Q. and Dixon, R.A. (2011) Transcriptional networks for lignin biosynthesis: more complex than we thought? *Trends in Plant Science* 16, 227–233.

Zhao, Z.-Y. (2007) The Africa Biofortified Sorghum Project – applying biotechnology to develop nutritionally improved sorghum for Africa biotechnology and sustainable agriculture 2006 and beyond. In: Xu, Z., Li, J., Xue, Y. and Yang, W. (eds) *Biotechnology and Sustainable Agriculture 2006 and Beyond*. Springer, Netherlands, pp. 273–277.

Zhao, Z., Glassman, K., Sewalt, V., Wang, N., Miller, M., Chang, S., *et al.* (2003) Nutritionally improved transgenic sorghum. In: Vasil, I. (ed.) *Plant Biotechnology 2002 and Beyond*. Kluwer Academic Publishers, the Netherlands, pp. 413–416.

13 Cultivation and Developments in the Field of GM Plants in Asia

Jie Wen* and Ranran Liu

Institute of Animal Sciences, Chinese Academy of Agricultural Sciences, P.R. China

13.1 The Current Status of GM Plants in Asia

According to an International Service for the Acquisition of Agro-biotech Applications (ISAAA) report (Clive, 2011), in 2011 Asia grew about 17.7 million hectares (Mha) of genetically modified (GM) crops, with a 3.54-fold increase in 5 years. In China and India, which together account for more than one-third of the world's population, over 14 million small farmers benefit from 14.5 Mha of GM crops, the majority being Bt cotton, which carries a gene that codes for the *Bacillus thuringiensis* (Bt) toxin (see Chapters 3 and 14). The quantity of GM crop products imported into the Asian region for processing into food and animal feed is substantial, and almost every country imports GM food.

13.1.1 China

Due to the importance of agriculture to China and the expected increasing role of GM plants in agriculture, they have received much attention and support. It is estimated that China is developing the largest plant biotechnology capacity outside of North America. According to the 'Long and Mid-term National Development Plan for Science and Technology (2006–2020)' of China, the programme will focus on crop research (rice, wheat, maize and cotton). The target is to develop new varieties of GM crops with traits such as resistance to insect disease,

stress resistance and high yields. China has been investing in a US\$3.5 billion research and development (R&D) initiative on GM plants (from 2008 to 2020). According to the Annual Report on Bioindustry in China, in 2009 over 50 types of GM plants had been developed; the top ten consisted of cotton, rice, maize, potato, tomato, wheat, rapeseed, tobacco, poplar and soybean. Seven types of GM plants developed by Chinese scientists have been approved for commercial production, including Bt cotton, Bt rice, storage-tolerant tomato, virus-resistant sweet pepper, dwarf morning glory, virus-resistant papaya and phytase maize. More than 60 versions of GM plant have been approved for field trials and release, including major crops – rice, maize and wheat, as well as cotton, potato, tomato, soybean, groundnut and rape (Li *et al.*, 2010; NDRC/CSB, 2010).

Despite large achievements, GM plants in China only make up a small share of the world biotechnology market in comparison with some developed countries. With regard to quantity, production scale of GM plants and technology, China still lags behind the USA.

13.1.2 India

The government of India recognized the importance of biotechnology and set up the National Biotechnology Board in 1982. The successful adoption of Bt cotton made agricultural biotechnology one of the fastest

**E-mail: wenj@iascaas.net.cn*

growing sections of the Indian biotech industry. Since its introduction in 2002, the Bt cotton area has expanded to over 90% of the total area in cotton and accounted for more than 95% of India's cotton production in 2010 (see Chapter 14). India is the second largest producer and exporter of cotton in the world. The government of India has approved six types of GM cotton and more than 300 hybrids for cultivation.

In addition to cotton, private seed companies and public research institutions are working on the development of various biotech crops mainly for traits like pest resistance, nutritional enhancement, drought tolerance and high yields. Several varieties of crops are undergoing development and field trials for regulatory approval – banana, castor, cotton, maize, rice, tomato, mustard, potato, sorghum and papaya.

On 14 October 2009, the Genetic Engineering Approval Committee (GEAC) recommended to the Ministry of Environment and Forest approval for the environmental release of Bt aubergine; however, this is still awaiting a final decision (USDA, 2011a).

13.1.3 Japan

Japan is the world's largest per capita importer of GM foods and feeds. It annually imports about 16 million tonnes (Mt) of maize and 4 Mt of soybeans, approximately three-quarters of which are from GM plants. Japan also imports billions of dollars worth of processed foods containing GM plant-derived oils, sugars, yeasts, etc. As of June 2011, over 95 GM varieties of seven crops had been approved for environmental release, including cultivation. As yet, there is no commercial production of biotech crops for food/feed, and the biotech blue rose released by Suntory in 2009 is the only GM plant commercially cultivated in Japan (USDA, 2011b).

13.1.4 Philippines

The Philippines was the first country in Asia to approve a biotech crop for food and feed,

and has developed a strong public institutional capacity for pioneering biotechnology-related R&D. Several biotech crops (rice, papaya, banana, sugarcane, potato and tomato) are in development, and field trials with government oversight have been conducted on rice and maize. It is noteworthy that Golden Rice, a GM rice that contains enhanced levels of beta carotene, bred (see Chapters 7 and 12) by the International Rice Research Institute (IRRI), is approaching the completion of its regulatory requirements in the Philippines and Bangladesh. It is expected that Golden Rice will first be released in the Philippines in 2013/2014 (Clive, 2011).

13.1.5 Korea

Korea is dependent on imported food (except rice) and feed grains, most of which come from the USA. A limited amount of food products are made from biotech ingredients, given consumer concerns about biotechnology, but the majority of feed is made from GM maize and soybean meal. Korea is developing a variety of GM crops, such as herbicide-tolerant rice and virus-resistant pepper (USDA, 2011c).

In addition, governments in Vietnam, Indonesia, Malaysia, the Philippines, Thailand and Singapore have given high priority to plant biotechnology research, and proof-of-concept research is ongoing on a variety of GM plants. Most of the remaining countries, Bangladesh, Pakistan and Nepal, have both upstream (functional genomics) and downstream (backcrossing biotech crop parents with local crops) biotech research activities (see also Chapters 3, 12 and 14).

13.2 The Current R&D Status of GM Plants in Asia

13.2.1 The current R&D status of GM plants in China

China has taken the lead in research on GM plants in Asia. Achievements in functional genomics, transgenic technology and the safety assessment of food/feed from GM plants in China are reviewed below and in Section 13.3 (see also Chapters 2 and 12).

Functional genomics

In China, studies on molecular technology, the construction of molecular linkage maps, gene mapping, cloning and identification of functional genes have achieved substantial advances. Molecular linkage maps of important plants, such as rice, wheat, maize, soybean and cotton, have been constructed. Quantitative trait loci (QTLs) underlying the production, quality and resistance traits of these plants have been mapped. More than 1000 genes and QTLs in plants have been mapped in the past two decades. In excess of 500 gene loci have been fine-mapped (these loci explained more than 10% of the phenotypic variance and the genetic distance from genetic markers of less than 2 cM) with regard to traits influencing production, disease and insect resistance, anti-adversity, quality and nutrient absorption efficiency (Chen *et al.*, 2006; NDRC/CSB, 2010; Qiu *et al.*, 2011).

The first draft map of the rice genome (*Oryza sativa* L. ssp. *indica*) was published in October 2001 by Chinese scientists and the fine map was finished in 2002. Subsequently, the genome sequence and analysis of rice (*Oryza sativa* L. subsp. *Japonica* var. Nipponbare) chromosome 4 has been finished by the China National Center for Gene Research. It is one of the two first-sequenced chromosomes in the world. At the same time, 80% of the genome sequence and analysis of rice (*Guangluai-4*) chromosome 4 and the sequence analysis in the centromere of chromosome 4 were also finished. Thereafter, the deep sequencing of other plants such as wheat, cotton, cucumber, tomato and watermelon has been finished or is in progress (Wan, 2011).

In China, many achievements have been made in the construction of mutant libraries in rice. Through the use and reconstruction of the enhancer trap system of GAL4/VP16-UAS, a 2,700,000 strain T-DNA insertion mutant library was obtained and the

bioinformatics database (Rice Mutant Database, RMD) has been developed. At the same time, full-length cDNA libraries of some crops, such as rice, wheat and maize, and research platforms of DNA microarrays for rice and maize were built. In addition, 239 expression profiles of stress traits including anti-adversity, disease resistance, low phosphorus and low nitrogen were developed; numerous key genes related to different stressors were identified (Chen *et al.*, 2006; NDRC/CSB, 2010; Qiu *et al.*, 2011).

Many important genes in different crops have been cloned and studies of their function and pathways performed. For example, using map-based cloning and the mutant library, genes underlying production, quality and plant types in crop traits, such as MONOCULM 1 (*MOC1*, a gene that is important in the control of rice tillering), *Ghd7* (an important regulator of heading date and yield potential in rice), *GS3* (a major gene for grain length and weight), *GIF1* (a gene controlling rice grain-filling and yield) have been isolated and cloned (Li *et al.*, 2003; Fan *et al.*, 2006; Wang *et al.*, 2008; Xue *et al.*, 2008); the genes underlying disease resistance, anti-adversity and insect resistance, such as *xa13* (a gene for bacterial blight resistance in rice), *Xa26* (a gene conferring resistance to *Xanthomonas oryzae* pv. *oryzae* in rice), *Bph14* (a gene conferring resistance to brown planthopper in rice), etc., have also been isolated and cloned (Sun *et al.*, 2004; Chu *et al.*, 2006; Du *et al.*, 2009). In addition, some genes in other crops have also been cloned; for example, genes underlying fibre cell elongation in cotton, *GhACT1* and *GmDET2*, and the development of soybean, *TFL*, etc. (NDRC/CSB, 2010; Tian *et al.*, 2010; see also Chapter 2).

Transgenic technology

Transgenic technology consists of gene cloning and genetic transformation.

GENE CLONING TECHNIQUE. In recent years, technology platforms for genome bio-informatics, proteomics, biochip and geno-typing have been developed and have laid the foundation for large-scale gene cloning in China. The main gene cloning techniques used include gene express sequence tag (EST)-based cloning technique, QTL mapping-based cloning, transposon tagging technique and differentially expressed gene-based cloning. Genome-wide association study (GWAS) and high-throughput sequencing began to be applied in this field. The expression library transformation method, which systematically can identify functional genes in crops, was first developed in China. These provide a methodological base for large-scale cloning of genes.

GENETIC TRANSFORMATION TECHNOLOGY. In China, genetic transformation technology for plants includes mainly gene gun-mediated, agrobacterium-mediated and pollen tube pathway transformation methods. Through integration and optimization, genetic transformation systems have built up for some crops, such as rice, cotton, wheat, maize, soybean and poplar. Due to the establishment of a large-scale agrobacterium-mediated genetic transformation system in rice, transformation efficiency has improved from 40% to 83%, transformation periods have shortened to 3–4 months and ~5000 genes can be transformed each year. Through the integration of the three transformation methods, a highly efficient genetic transformation platform has been developed for cotton, with 10,000 strains being transformed each year. The transformation efficiency of wheat with the agrobacterium-mediated method reached 2% and a transformation technique of the mature embryo in wheat has been established, which circumvents the seasonal influence in wheat transgenesis. Transformation with agrobacterium-mediated methods in the immature embryo and shoot tip of maize was developed and transformation efficiency can reach about 5%. In soybean, the cotyledonary node and hypocotyl of agrobacterium-mediated transformation methods were used and about 1% transformation efficiency was achieved.

With the global advancement of high-efficiency, multi-gene transformation techniques, new progress has also been achieved in China. Scientists have developed binary bacterial artificial chromosomes (BIBAC) and artificial chromosomes based on P1-derived artificial chromosome (TAC) vectors to transform multiple genes simultaneously. For manipulation of foreign gene expression, progress has been made on improving expression efficiency by using strong promoter, enhancer and matrix attachment regions, etc. Marker knockout technology, timed degradation of target genes and marker-free transgenic technology have been under continuous development.

Foreign gene transformation vectors were constructed using the Cre/lox deletion system, plant green tissue special promoter (rbcs) and the R/RS marker knockout system, which have been applied in transgenic tobacco and rice.

Through constructing the expression vector of double T-DNA insect resistance in monocotyledons, the first marker-free, double insect-resistant transgenic rice in the world has been developed. The updated transformation system has been constructed, which can transform multiple target genes in minimal constructs and prevent the selective marker and vector backbone from entering into receiver seedlings. These improved transgenic technologies will facilitate the production of GM plants in China (Wan, 2011).

13.2.2 The current R&D status of GM plants in Japan

Japan is conducting broad research on agricultural biotechnology. In Japan, the Ministry of Agriculture, Forestry and Fisheries (MAFF) is devoting significant resources towards research in genomics and biotech crop development. An example of this effort can be seen in Japan's contribution to rice genome sequencing, as well as genome analysis of other plants such as soybeans (Yu *et al.*, 2002; Schmutz *et al.*, 2010). Initial releases will most likely come from Japanese public sector research. Priority traits could include high yield, disease-resistant rice, drought-tolerant rice and wheat, nutritionally altered rice and heavy-metal-accumulating rice.

13.3 Safety Assessment of Food/ Feed from GM Plants in Asia

With increased plantings of GM plants and their application in food and feed production, more attention has been given to their safety/risk assessment. In terms of nutrition, the biosafety of GM plants is one of the most important topics (see also Chapter 3). Comparative approaches have been utilized, i.e. food and feed are compared with their non-GM counterparts in order to identify intended and unexpected differences, which subsequently are assessed with respect to their potential impact on safety for humans and animals, along with nutritional quality. Generally, with regard to food/feed from GM plants, scientists from China have studied the key nutritional components, acute toxicity, immune toxicity, allergenicity and reproductive toxicity of introduced proteins (Cry1Ab, Cry1Ac, cowpea trypsin inhibitor/CpTI, etc.) and food (maize, rice, etc.). Feeding safety has been assessed for imported transgenic maize, soybean, rapeseed, etc.

Rice is one of the main crops in the world, with 92% of the total area planted in Asia and 31% in China. With the maturing of GM rice technology, more effort has been put into safety assessment. In addition, as an important crop for both food and feed, the biosafety assessment of GM maize is also an important issue. This review will introduce the biosafety assessment of GM rice and maize in China. The assessment procedure consists of five aspects, including substantial equivalence of nutrition, nutritional assessment in animals, *in vivo* and *in vitro* toxicological studies, allergenicity assessment and horizontal transformation of the introduced gene.

13.3.1 Substantial equivalence (SE) of nutrition

Studies were conducted to assess whether the key nutrients (carbohydrates, protein, amino acids, key minerals and vitamins) in transgenic plant components used for feed or food had been changed. The results found

the same nutritional value as in their non-transgenic counterparts in Bt rice and Bt maize (Wang *et al.*, 2002b; Li *et al.*, 2004a). The total protein content and essential amino acids were increased in maize carrying the lysine-rich protein gene (Tang, 2008; see also Chapters 3 and 4).

13.3.2 Nutritional assessment in animals

Li *et al.* (2004b) and Zhao *et al.* (2005) evaluated the influence of GM rice containing the disease-resistance gene, *Xa21*, and the saline-tolerance gene, *codA*, on the physiological metabolism and genetic horizontal transformation in fed laboratory rats. No toxicity or other adverse effects were found in the transgenic rice group. Pigs are good models for human nutrition because of anatomical similarities to humans (e.g. body size, skin, cardiovascular system and urinary system) and because of functional similarities (immune system and gastrointestinal system). Chinese experimental minipigs were used to assess rice that was genetically modified with the *SCK* gene (a modified cowpea trypsin inhibitor gene). Body growth (weight, height, body length, thoracic circumference, etc.) and feed intake of animals were recorded after 62 days of feeding. It was concluded that the feeding value of GM rice and parental rice was similar and no detrimental or unexpected effects were observed in animals fed the GM rice (Yang *et al.*, 2005). The digestibility of protein and amino acids in GM rice were compared to parental rice in minipigs; except for the decreased digestibility of lysine, there were no significant differences in the apparent and true digestibility of the other 17 amino acids (Han *et al.*, 2004).

Rice is eaten by humans after cooking. The effect of raw or cooked GM rice flour on the development of silkworm larvae has been compared. In contrast to normal feed, the weight of silkworm larvae, the number of grown silkworms, the number of cocoons, the weight of cocoons and cocoon layers at different periods were significantly less or delayed in the group fed raw GM rice. There was no difference between the normal fed group and that fed with cooked GM rice, indicating that Bt transgenic rice lost its toxicity to silkworms when the Bt toxin protein was denatured by cooking (Wang *et al.*, 2002a; see also Chapters 5 and 6).

13.3.3 Toxicology studies

Animal feeding trials (90 days) in rodents have been recommended by the FAO/WHO for risk assessment (toxicity testing) of GM plants (FAO, 2001). According to the industry standard 'Safety assessment of genetically modified plant and derived products' issued by the Ministry of Agriculture in China, 90-day feeding trials are also required for the safety assessment of food/feed from GM plants. Wang *et al.* (2002b) conducted a 90-day feeding trial to evaluate the toxicology of transgenic rice flour with a synthetic *cry1Ab* gene and no toxic effect was detected. To assess the teratogenicity of GM rice, rats were fed for 90 days with transgenic rice that expressed the insecticidal proteins, CpTI and Xa21. They concluded that this transgenic rice had no maternal toxicity, embryotoxicity or teratogenicity (Li *et al.*, 2004b; Zhuo *et al.*, 2004). The subchronic toxicity of transgenic high-lysine maize was assessed by feeding rats for 90 days, and the results showed that the GM maize had no harmful effects on the growth and nutrition of the rats (Tang, 2008; see also Chapters 3 and 5).

13.3.4 Allergenicity assessment

With regards to introduced protein, the hygromycin B phosphotransferase gene (*hpt*) has been widely used as a selectable marker in the process of plant genetic engineering. Lu *et al.* (2007) conducted *in vitro* digestibility and animal studies to assess the safety of GM plants. The feeding HPT protein was digested by simulated gastric fluid within 40 s, and the protein did not induce detectable levels of serum-specific IgE antibodies or histamine in the test animals. They concluded that HPT had a low probability of inducing allergenic reactions (see Chapter 3).

13.3.5 Horizontal transformation of introduced genes

With advances in genetic transformation technology, horizontal transformation of introduced antibiotic genes will not be a threat to public health. On the basis of the regulations from the Ministry of Agriculture, GM plants with antibiotic gene markers will not be approved for production in China.

13.4 Regulations on Administration of GM Plant Safety in Asia

Most Asian countries have guidelines for research on genetically modified organisms (GMOs), and some countries have set up a series of regulations or administrative measures for the management of the biosafety of GM plants and their products (see also Chapter 3).

China has implemented a series of regulations related to GM crops since the 1990s. In 1993, the then State Science and Technology Commission issued the 'Administrative Measures on the Safety of Genetic Engineering'. An initial legal framework on GMO regulation was then established. In 2001, the State Council passed new 'Regulations on Administration of Agricultural Genetically Modified Organisms Safety' (http://www.biosafety.gov.cn/image20010518/5420.pdf), which aimed to enhance the biosafety management of GMOs during the activities of research, experimentation, production, processing, marketing, importation and exportation. What is noteworthy about this regulation is that, first, it is a regulation, not ministerial administrative measures, which means that it is more comprehensive in nature. Second, it was not issued by the Ministry of Agriculture but by the superior authority, the State Council. This change enhanced the legal effect of the act.

In order to implement this regulation, the Ministry of Agriculture subsequently issued the following, more detailed, ministerial acts: the 'Administrative Measures on the Safety of the Import of Agricultural GMOs', the 'Administrative Measures on the Labeling of Agricultural GMOs' and the 'Administrative Measures on the Safety Assessment of Agricultural GMOs' in July 2002. As specified by these regulations, agricultural GMOs are classified into Classes I, II, III and IV according to the extent of their risks to humans, animals, plants, microorganisms and the ecological environment; the biosafety assessment of agricultural GMOs shall go through five stages: namely, research in the laboratory, restricted field testing, enlarged field testing, production testing and application of a safety certificate of agricultural GMOs. The following 16 items of agri-GMOs are on the first list of labelling: soybean seeds, soybean, soybean powder, soybean oil, soybean meal; maize seeds, maize, maize oil, maize powder; rape seeds, rape oil, rape meal; cotton seeds; tomato seeds, fresh tomato and tomato paste. They shall be clearly labelled when they are sold within the territory of China. In addition, the 'Administrative Measures on GM Food Hygiene' was issued by The Ministry of Health in 2002 and the 'Administrative Measures on the Inspection and Quarantine of the Import and Export of GMO Products' was issued by The General Administration of Quality Supervision, Inspection and Quarantine in 2004. Taken together, one regulation combined with five administrative measures normalize the procedures of biosafety assessment, labelling, production, marketing and inspection and quarantine of imports and exports with regards to agricultural GMOs (Guo, 2011).

In Japan, the commercialization of biotech plant products requires food, feed and environmental approval. Four ministries are involved in the regulatory framework: MAFF, the Ministry of Health, Labor and Welfare (MHLW), the Ministry of Environment (MOE) and the Ministry of Education, Culture, Sports, Science and Technology (MEXT). These ministries are also involved in environmental protection and in regulating laboratory trials. The Food Safety Commission (FSC), an independent risk assessment body, performs food and feed safety risk assessment for the MHLW and MAFF. The regulatory framework for the biosafety of GM plants consists of the 'Basic Law on Food Safety', the 'Law Concerning the Safety and Quality Improvement of Feed (the Feed Safety Law)' and the 'Law Concerning Securing of Biological Diversity (Regulation of the Use of Genetically Modified Organisms)' (USDA, 2011b).

Other countries are in the process of establishing a legislative framework for the biosafety and commercial release of GM crops. India has established a Biosafety Committee and the GEAC to oversee the biosafety and applications of GM crops. In the Philippines, the National Committee on Biosafety mandates the guidelines and approvals. The approval permit stipulates that the performance of the GM crop and its effect on the environment as well as human and animal health are assessed. The Philippine government released guidelines, taking effect as of 1 July 2003 that regulated the importation and commercialization of GM crops.

13.5 The Future of GM Plants in Asia

There will likely be more than five billion people in Asia by 2025. Traditional farming practices and equipment, however, are reaching their limits of effectiveness in increasing agricultural productivity. The types of GM crops that may become available in the future could enhance both yields and the nutritional value of staple foods and eliminate chemicals that are harmful to the environment. Asia, therefore, has the potential to lead the world in applying biotechnology for these new classes of products, with the way paved by GM crops and food (see Chapter 1). Meanwhile, to promote healthy development of the agricultural biotechnology industry, several major challenges need to be overcome by:

1. Improving biotechnology R&D, such as identifying and patenting plant genes of great value and refining the biosafety assessment system.

2. Developing rigorous evidence on which to base biosafety policy.

3. Improving public education and increasing overall confidence in the risk assessment and management policies regarding GM plants used in food and feed.

References

Chen, H.D., Karplus, V.J., Ma, H. and Deng, X.W. (2006) Plant biology research comes of age in China. *The Plant Cell* 18, 2855–2864.

Chu, Z.H., Fu, B.Y., Yang, H., Xu, C.G., Li, Z.K., Sanchez, A., *et al.* (2006) Targeting xa13, a recessive gene for bacterial blight resistance in rice. *Theoretical and Applied Genetics* 112(3), 455–461.

Clive, J. (2011) *Global Status of Commercialized Biotech/GM Crops: ISAAA Brief No 43.* ISAAA, Ithaca, New York.

Du, B., Zhang, W.L., Liu, B.F., Hu, J., Wei, Z., Shi, Z.Y., *et al.* (2009) Identification and characterization of Bph14, a gene conferring resistance to brown planthopper in rice. *Proceedings of the National Academy of Sciences* 106(52), 22163–22168.

Fan, C.C., Xing, Y.Z., Mao, H.L., Lu, T.T., Han, B., Xu, C.G., *et al.* (2006) GS3, a major QTL for grain length and weight and minor QTL for grain width and thickness in rice, encodes a putative transmembrane protein. *Theoretical and Applied Genetics* 112, 1164–1171.

FAO (2001) Evaluation of allergenicity of genetically modified foods. Report of a joint FAO/WHO Expert Consultation on Allergenicity of Foods Derived from Biotechnology 22–25 January 2001 (http://www.who.int/foodsafety/publications/biotech/en/ec_jan2001.pdf, accessed 1 May 2013).

Guo, J.C. (2011) Development of new DNA-based methods for analysis of genetically modified plants and their derived products. PhD thesis, The Shanghai Jiao Tong University, Shanghai, PR China.

Han, J.H., Yang, Y.X., Men, J.H., Bian, L.H., Yang, X.G. and Zhu, Y.Z. (2004) The comparison of ileal digestibility of protein and amino acids in GM rice and parental rice. *Acta Nutrimenta Sinica* 26(5), 362–365.

Li, X.Y., Qian, Q., Fu, Z.M., Wang, Y.H., Xiong, G.S., Zeng, D.L., *et al.* (2003) Control of tillering in rice. *Nature* 422, 618–621.

Li, Y., Wang, J.K., Qiu, L.J., Ma, Y.Z., Li, X.H. and Wan, J.M. (2010) Crop molecular breeding in China: current status and perspectives. *Acta Agronomica Sinica* 36, 1425–1430.

Li, Y.H., Piao, J.H., Chen, X.P., Zhuo, Q., Mao, D.Q., Yang, L.C., *et al.* (2004a) Nutrition assessment of transgenic rice. *Journal of Hygiene Research* 33(3), 303–306.

Li, Y.H., Piao, J.H., Chen, X.P., Zhuo, Q., Mao, D.Q., Yang, L.C., *et al.* (2004b) Study on the teratogenicity effects of genetically modified rice with Xa21 on rats. *Journal of Hygiene Research* 33(6), 710–712.

Lu, Y., Xu, W., Kang, A., Luo, Y., Guo, F., Yang, R., *et al.* (2007) Prokaryotic expression and allergenicity assessment of hygromycin B phosphotransferase protein derived from genetically modified plants. *Journal of Food Science* 72(7), 228–232.

National Development and Reform Commission and Chinese Society of Biotechnology [NDRC/CSB] (2010) Annual report on bioindustry in China/2009. Chemical Industry Press, Beijing.

Qiu, L.J., Guo, Y., Li, Y., Wang, X.B., Zhou, G.A., Liu, Z.X., *et al.* (2011) Novel gene discovery of crops in China: status, challenging, and perspective. *Acta Agronomica Sinica* 37(1), 1–17.

Schmutz, J., Cannon, S.B., Schlueter, J., Ma, J.X., Mitros, T., Nelson, W., *et al.* (2010) Genome sequence of the palaeopolyploid soybean. *Nature* 463, 178–183.

Sun, X.L., Cao, Y.L., Yang, Z.F., Xu, C.G., Li, X.H., Wang, S.P., *et al.* (2004) Xa26, a gene conferring resistance to *Xanthomonas oryzae pv. oryzae* in rice, encodes an LRR receptor kinase-like protein. *The Plant Journal* 37(4), 517–527.

Tang, M.Z. (2008) Study of the food safety assessment and the model to evaluate allergenicity of transgenic maize inserted with lysine-rich protein gene. PhD thesis, The Chinese Academy of Agricultural Sciences, Beijing.

Tian, Z., Wang, X., Lee, R., Li, Y., Specht, J., Nelson, R., *et al.* (2010) Artificial selection for determinate growth habit in soybean. *Proceedings of the National Academy of Sciences* 107, 8563–8568.

USDA (2011a) India Agricultural Biotechnology Annual 2011 (http://gain.fas.usda.gov/Recent GAIN Publications/Agricultural Biotechnology Annual_New Delhi_India_9-15-2011.pdf, accessed 1 May 2013).

USDA (2011b) Japan Biotechnology Annual Report 2011 (http://gain.fas.usda.gov/Recent%20 GAIN%20Publications/Agricultural%20Bio technology%20Annual_Tokyo_Japan_9-19-2011.pdf, accessed 1 May 2013).

USDA (2011c) Korea – Republic of – Agricultural Biotechnology Annual Biotechnology Annual

Report 2011 (http://gain.fas.usda.gov/Recent%20GAIN%20Publications/Agricultural%20Biotechnology%20Annual_Seoul_Korea%20-%20Republic%20of_9-23-2011.pdf, accessed 1 May 2013).

Wan, J.M. (2011) Research and development status and future strategy of transgenic plants in China. *Chinese Bulletin of Life Sciences* 23(2), 157–167.

Wang, E.T., Wang, J.J., Zhu, X.D., Hao, W., Wang, L.Y., Li, Q., *et al.* (2008) Control of rice grain-filling and yield by a gene with a potential signature of domestication. *Nature Genetics* 40, 1370–1374.

Wang, Z.H., Wang, Y., Cui, H.R., Shu, Q.Y. and Xia, Y.W. (2002a) A preliminary study on toxicological evaluation of transgenic rice flour with a synthetic Cry1Ab gene from *Bacillus thuringiensis*. *Journal of the Science of Food and Agriculture* 82, 738–744.

Wang, Z.H., Shu, Q.Y., Cui, H.R., Xu, M.K., Xie, X.B. and Xia, Y.W. (2002b) The effect of Bt transgenic rice flour on the development of silkworm larvae and the sub-micro-structure of its midgut. *Scientia Agricultura Sinica* 35(6), 714–718.

Xue, W.Y., Xing, Y.Z., Weng, X.Y., Zhao, Y., Tang, W.J., Wang, L., *et al.* (2008) Natural variation in Ghd7 is an important regulator of heading date and yield potential in rice. *Nature Genetics* 40, 761–767.

Yang, X.X., Chen, S.R., Han, J.H., Yang, J.M. and Yang, X.G. (2005) The nutritional evaluation of genetically modified rice-comparative feeding study of minipigs. *Acta Nutrimenta Sinica* 27(1), 38–45.

Yu, J., Hu, S.N., Wang, J., Wong, G.S., Li, S.G., Liu, B., *et al.* (2002) A draft sequence of the rice genome (*Oryza sativa L. ssp. indica*). *Science* 296, 79–92.

Zhao, Z.H., Yang, L.T., Ai, X.J., Zhang, D.B. and Zou, S.X. (2005) Influence of genetically modified rice containing *codA* gene on physiological metabolism and genetic horizontal transformation in fed rats. *Journal of Agricultural Biotechnology* 13(3), 341–346.

Zhuo, Q., Chen, X.P., Piao, J.H., Han, C. and Yang, X.G. (2004) Study on the teratogenicity effects of genetically modified rice which expressed cowpea trypsin inhibitor on rats. *Journal of Hygiene Research* 33(1), 74–77.

14 Socio-economic Aspects of Growing GM Crops

Matin Qaim*

Georg-August-University of Goettingen, Department of Agricultural Economics and Rural Development, Goettingen, Germany

14.1 Introduction

The global area under genetically modified (GM) crops grew from 1.7 million hectares (Mha) in 1996 to 170 Mha in 2012 (see Fig. 1.2). Today, around 17 million farmers worldwide grow GM crops in 28 countries, including 20 developing countries (James, 2012). So far, most of the commercial applications involve herbicide tolerance and insect resistance, but other GM traits are in the research pipeline and might be commercialized in the short- to medium-term future.

The rapid global spread of GM crops has been accompanied by an intense public debate. Supporters see great potential in this technology to raise agricultural productivity and reduce seasonal variations in food supply due to biotic and abiotic stresses. Against the background of increasing demand for agricultural products and natural resource scarcities, productivity increases are necessary for achieving long-term food security (see Chapter 1). Second-generation GM crops, such as crops with higher micronutrient contents (biofortified crops), could also help reduce specific nutritional deficiencies among the poor (see Chapters 7, 10 and 12). Furthermore, the technology could contribute to rural income increases, which is particularly relevant for poverty reduction in developing countries. Finally, supporters argue that reductions in the use of chemical pesticides through GM crops could alleviate environmental and health problems associated with intensive agricultural production systems.

In contrast, biotechnology opponents emphasize the environmental and health risks associated with GM crops. Moreover, doubts have been raised with respect to the socio-economic implications in developing countries. Some consider high-tech applications as inappropriate for smallholder farmers and disruptive to traditional cultivation systems. Also, it is feared that the dominance of multinational companies in biotechnology and the international proliferation of intellectual property rights (IPRs) would lead to exploitation of poor agricultural producers. In this view, GM crops are rather counterproductive for food security and development.

While emotional public controversies continue, there is a growing body of literature providing empirical evidence on the impacts of GM crops in different countries. This chapter reviews recent socio-economic studies, focusing on peer-reviewed academic papers. Claims and studies by narrow interest groups are not included, as they are not objective and usually build on information that is not representative. We review studies on the farm-level impacts of herbicide-tolerant and insect-resistant crops. Moreover, we summarize some macro-level research looking at global impacts. Finally, we discuss the potential effects of future GM crop applications,

*E-mail: mqaim@uni-goettingen.de

including crops with other improved agronomic traits and nutritional traits.

14.2 Impacts of Herbicide-tolerant Crops

Herbicide-tolerant (HT) crops are tolerant to certain broad-spectrum herbicides like glyphosate or glufosinate, which are more effective, less toxic and usually cheaper than selective herbicides. HT technology so far is used mostly in soybean, maize, cotton and rapeseed, and to a lesser extent in sugarbeet and a few other crops. The dominant crop is HT soybean, which was grown on 81 Mha in 2012, mostly in the USA, Brazil, and Argentina, but also in a number of other countries. Likewise, HT maize is cultivated primarily in North and South America, with smaller areas in South Africa and the Philippines. In maize, HT is often stacked with insect-resistance genes. The same is true for HT cotton in the USA. HT rapeseed is grown predominantly in Canada and the USA (James, 2012; Fig. 1.2).

14.2.1 Agronomic and economic effects

HT-adopting farmers benefit in terms of lower herbicide expenditures. Total herbicide quantities applied were reduced in some situations but not in others. In Argentina, herbicide quantities were increased significantly (Table 14.1). This was due largely to the fact that herbicide sprays were substituted for tillage. In Argentina, the share of soybean farmers using no-till has almost doubled to 80% since the introduction of HT technology. Also, in the USA and Canada, no-till practices expanded through HT adoption (Fernandez-Cornejo and Caswell, 2006). In terms of yield, there is no significant difference between HT and conventional crops in most cases. Only in a few examples, where certain weeds were difficult to control with selective herbicides, did the adoption of HT and the switch to broad-spectrum herbicides result in better weed control and higher crop yields. Examples are HT soybeans in Romania and HT maize in Argentina (Brookes and Barfoot, 2012).

Overall, HT technology reduces the cost of production through lower expenditures on herbicides, labour, machinery and fuel. Yet, the innovating companies charge a technology fee on seeds, which varies between crops and countries. Several early studies of HT soybeans in the USA showed that the fee was of a similar magnitude or sometimes higher than the average cost reduction, so that profit effects were small or negative (Naseem and Pray, 2004). Comparable results were also obtained for HT cotton and HT rapeseed in the USA and Canada. The main reason for farmers in such situations to still use HT technology was easier weed control and the saving of

Table 14.1. Average effects of HT soybeans in Argentina. (From Qaim and Traxler, 2005.)

	Conventional soybeans	HT soybeans	Change (%)
Herbicide expenditure (US$/ha)	33.64	19.10	−43.2
Herbicide quantity (l/ha)	2.68	5.57	107.8
Of which:			
In toxicity classes I–III (l/ha)	1.10	0.07	−93.6
In toxicity class IV (l/ha)	1.58	5.50	248.1
Share of farmers using no-till practices	0.42	0.80	90.5
Number of tillage passes per plot	1.66	0.69	−58.4
Labour time (h/ha)	3.92	3.30	−15.8
Machinery time (h/ha)	2.52	2.02	−19.8
Fuel (l/ha)	53.03	43.70	−17.6
Cost of production (US$/ha)	212.99	192.29	−9.7
Soybean yield (t/ha)	3.02	3.01	−0.3
Profit (US$/ha)	271.66	294.65	8.5

management time. Fernandez-Cornejo *et al.* (2005) showed that the saved management time for US soybean farmers translated into higher off-farm incomes. Moreover, farmers are heterogeneous; that is, many adopters have benefitted in spite of zero or negative mean profit effects. The average farm-level profits seem to have increased over time, due partly to seed price adjustments and farmer learning effects.

In South America, the average profit effects of HT crops, especially HT soybeans, are larger. While the agronomic advantages are similar to those in North America, the fee charged on seeds is lower. The reason for this is that HT soybean technology is not patented in most South American countries. Many soybean farmers in Brazil and Argentina use farm-saved GM seeds. Qaim and Traxler (2005) showed for Argentina that the average profit gain through HT soybean adoption was in a magnitude of US$23/ha (see Table 14.1). The technology is so attractive for farmers that HT is now used on almost 100% of the Argentine soybean area. In Paraguay and Uruguay, adoption rates of HT soybeans are similarly high; in Brazil, this technology was adopted on 88% of the national soybean area in 2012 (James, 2012).

While farmers in these middle-income countries benefit significantly from HT soybeans, most soybean growers operate relatively large-scale and fully mechanized farms. So far, HT crops have not been widely adopted in the small farm sector of developing countries. Smallholders often weed manually, so that HT crops are inappropriate, unless labour shortages or weeds that are difficult to control justify conversion to chemical practices. In some regions of Asia and Africa, smallholder farmers have switched to the so-called system of rice intensification (SRI), where rice is grown with intermittent irrigation under aerobic conditions. This saves water and is associated with less greenhouse gas emissions, but problems with weeds tend to increase. Under such conditions, HT rice might be an interesting alternative.

14.2.2 Environmental effects

Adoption of HT crops does not lead to reductions in herbicide quantities in most cases, but selective herbicides, which are often relatively toxic to the environment, are substituted by less toxic broad-spectrum herbicides (see Table 14.1). Glyphosate, for instance, has little residual activity and is decomposed rapidly to organic components by microorganisms in the soil. According to the international classification of pesticides, it belongs to toxicity class IV, the lowest class for 'practically non-toxic' pesticides. Also, the reduction in tillage operations and the expansion of no-till practices through HT technology adoption brings about environmental benefits in terms of a reduction in soil erosion, fuel use and greenhouse gas emissions (Brookes and Barfoot, 2012).

On the other hand, weed species might develop resistance to glyphosate and other broad-spectrum herbicides, which would require increasing amounts of pesticides to be applied. Glyphosate resistance in certain weed species has already been reported in some locations. Furthermore, the high profitability of HT soybeans has led many farmers in Argentina and Brazil to convert bush and grass land into soybean land and cultivate the same crop year after year. Although the soybean area in these countries has been growing over the past 20 years, growth has accelerated since the introduction of HT technology. Area conversion and soybean monocultures might contribute to biodiversity loss and other environmental problems. These are not technology-inherent risks, as they would occur in any situation where the profitability of one particular crop increases considerably. But appropriate policies and regulations are required to avoid negative environmental effects.

14.3 Impacts of Insect-resistant Crops

Insect-resistant GM crops grown commercially so far involve different genes from the soil bacterium *Bacillus thuringiensis* (Bt) that make the plant resistant to certain lepidopteran and coleopteran pest species. The most widely used examples are Bt maize and Bt cotton. In 2012, Bt maize was grown on 47 Mha (see Fig. 1.2) in more than 15 different countries. The biggest Bt maize areas are found in the USA, Argentina, South Africa, Canada and the Philippines. Bt cotton was grown on 23 Mha in 2012, mostly in India, China, Pakistan and the USA, but also in a number of other countries (James, 2012).

14.3.1 Agronomic and economic effects

If insect pests are controlled effectively through chemical pesticides, the main effect of switching to Bt crops will be a reduction in insecticide applications. However, there are also situations where insect pests are not controlled effectively by chemical means, due to the unavailability of suitable insecticides or other technical, financial or institutional constraints. In those situations, Bt adoption can help reduce crop damage and thus increase effective yields (Qaim and Zilberman, 2003). Table 14.2 confirms that both insecticide-reducing and yield-increasing effects of Bt crops can be observed internationally.

In cotton, high amounts of chemical insecticides are normally used to control the bollworm complex, which is the main Bt target pest. Accordingly, Bt cotton adoption allows significant insecticide reductions, ranging from 20% to 80% on average. Yield effects are also significant, especially in developing countries. In some countries, such as Argentina, conventional cotton farmers underuse chemical insecticides, so that insect pests are not controlled effectively (Qaim and Janvry, 2005). In India, China and Pakistan, chemical input use is much higher, but the insecticides are not always very effective, due to low quality, resistance in pest populations and incorrect timing of sprays (Huang *et al.*, 2003; Qaim *et al.*, 2006).

For Bt maize, similar effects can be observed, albeit generally at a lower magnitude. Except for Spain, where the percentage reduction in insecticide use is large, the more important result of Bt maize is an increase in effective yields. In tropical and subtropical areas, mean yield effects are higher, because there is more pest pressure. The average Bt maize yield gain of 11% in South Africa refers to large commercial farms. These farms have been growing yellow Bt maize hybrids for several years. Gouse *et al.* (2006) also analysed on-farm trials that were carried out with smallholder

Table 14.2. Average effects of Bt cotton and Bt maize. (From Qaim, 2009; Kouser and Qaim, 2013.)

Country	Insecticide reduction (%)	Increase in yield (%)	Increase in profit (US$/ha)
		Bt cotton	
Argentina	47	33	23
Australia	48	0	66
China	65	24	470
India	41	37	135
Mexico	77	9	295
Pakistan	21	28	504
South Africa	33	22	91
USA	36	10	58
		Bt maize	
Argentina	0	9	20
Philippines	5	34	53
South Africa	10	11	42
Spain	63	6	70
USA	8	5	12

farmers and white Bt maize hybrids in South Africa; they found average yield gains of 32% on Bt plots. In the Philippines, average yield advantages of Bt maize are 34%. These patterns suggest that smallholder farmers face bigger constraints in controlling insect damage in their conventional crops.

The profit effects of Bt technologies are also shown in Table 14.2. Bt seeds are more expensive than conventional seeds, because they are sold mostly by private companies that charge a special technology fee. The fee is correlated positively with the strengths of IPR protection in a country. In all countries, Bt-adopting farmers benefit financially; that is, the economic advantages associated with insecticide savings and higher effective yields more than outweigh the technology fee charged on GM seeds. The absolute gains differ remarkably between countries and crops. On average, the extra profits are higher for Bt cotton than Bt maize. They are also higher in developing than developed countries. Apart from agroecological and socio-economic differences, GM seed costs are often lower in developing countries, due to weaker IPRs, seed reproduction by farmers and subsidies or other types of government price interventions (Basu and Qaim, 2007; Krishna and Qaim, 2008).

14.3.2 Social effects

The majority of the world's poor are smallholder farmers or agricultural labourers. Therefore, GM crops may also have important implications for poverty and income distribution in developing countries. Bt crops are generally suitable for the small farm sector. Especially in China, India and South Africa, Bt cotton is often grown by farms with less than 5 ha of land. In South Africa, many smallholders grow Bt white maize as their staple food. Several studies show that the Bt advantages for small-scale farmers are of a similar magnitude as for larger-scale producers, in some cases even higher (Pray *et al.*, 2001; Morse *et al.*, 2004; Qaim, 2009).[1]

Subramanian and Qaim (2010) have analysed the broader socio-economic outcomes of Bt cotton in India, including the effects on rural employment and household incomes. Building on a village-modelling approach, they show that Bt technology is employment generating, especially for hired female agricultural labourers. This is due to significantly higher yields being harvested. But employment is also generated in other local rural sectors, like trade and services, which are linked to cotton production. The impacts on rural household incomes, including the farm and non-farm community, are shown in Fig. 14.1. Each additional hectare of Bt cotton produces 82% higher aggregate incomes than conventional cotton, implying a remarkable gain in overall economic welfare through technology adoption. All types of households – including those below the poverty line – benefit considerably more from Bt cotton than from conventional cotton. These findings demonstrate that Bt crops can contribute to poverty reduction and rural development.

Recent long-term studies for India suggest that these technological benefits have been stable or even increasing over time (Krishna and Qaim, 2012). Kathage and Qaim (2012) show that farm households adopting Bt cotton have increased their living standards significantly, as measured by higher food and non-food consumption values. Because of higher incomes, Bt cotton-adopting households can not only afford more calories but also better dietary quality. Qaim and Kouser (2013) confirm that the introduction of Bt technology in India has contributed to a reduction of food insecurity by 15–20% among cotton-growing households.

These results cannot be generalized, because impacts do not depend on the technology only but also on the context (Glover, 2010; Stone, 2011; Kathage and Qaim, 2012). A conducive institutional environment is important to promote wide and equitable access to new seed technologies. Well-functioning input and output markets, including efficient micro-credit schemes, will spur the process of innovation adoption. Unfortunately, such conditions first need to be established in the poorest countries of Africa and Asia, so that the GM crop impacts

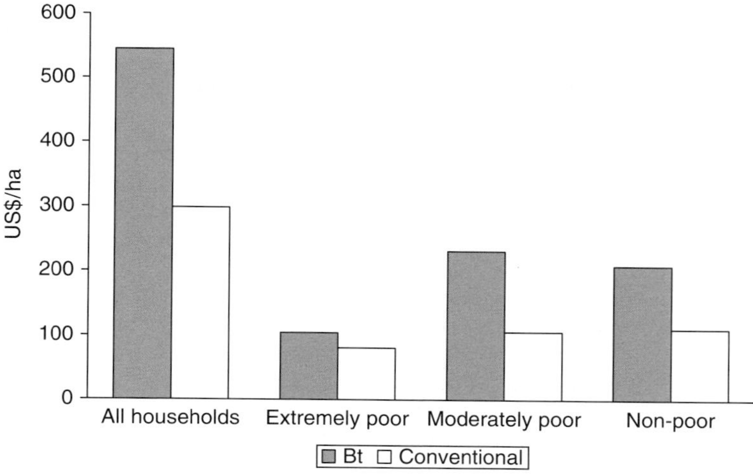

Fig. 14.1. Household income effects of Bt cotton compared to conventional cotton in India. (From Qaim et al., 2009).

observed so far in China, India and other middle-income countries cannot simply be extrapolated. Like any agricultural technology, GM crops are not a substitute for but a complement to much needed institutional change in rural areas of developing countries.

14.3.3 Environmental and health effects

Bt crops also have environmental and health implications. In the public debate, potential environmental risks, such as undesirable gene flow or impacts on non-target organisms, are often in the fore. Also, food safety concerns are being raised. Bradford *et al.* (2005) and Wolfenbarger *et al.* (2008) have reviewed such risks, concluding that most of them are not connected to the technique of genetic modification but would be present for any conventionally produced crops with the same heritable traits. While potential risks need to be further analysed and managed, Bt crops can also bring about substantial environmental and health benefits.

The main environmental benefits are related to reductions in chemical insecticides, so far especially in cotton. Worldwide, cotton is the crop that receives the largest quantities of chemical insecticide sprays, so the percentage reductions in insecticides discussed above also translate into huge reductions in absolute quantities. Brookes and Barfoot (2012) estimated that between 1996 and 2010 Bt cotton was responsible for a global saving of 170 million kg of pesticide active ingredients, reducing the environmental impact of total cotton pesticides by 26%. In their study on Bt cotton in Pakistan, Kouser and Qaim (2013) tried to quantify and monetize some of the environmental advantages. They showed that benefits resulting from less damage to beneficial insects and lower pesticide contamination in soils and groundwater added significantly to the financial gains resulting from Bt cotton adoption.

In the first years of Bt crop deployment, it was predicted that insect populations would soon develop Bt resistance, which would undermine the technology's effectiveness and lead to declining insecticide reductions over time. However, until now, Bt resistance development has not been observed under practical field conditions, which might be due partly to successful resistance management strategies, such as the planting of non-Bt refuges (Tabashnik *et al.*, 2008). But even in countries where no such strategies are implemented, Bt resistance has not been reported on a significant scale.

There are also other factors that can lead to changes in Bt effects over time. In China, for instance, insecticide applications somewhat increased again after several years of Bt cotton use, in spite of the absence of Bt resistance. Wang *et al.* (2008) attributed this to secondary pests, which might have become more important through the Bt-induced reduction in broad-spectrum insecticides. Secondary pests are mirids, mealybugs and other sucking pests that are not controlled by Bt. Using long-term field trial data from China, Lu *et al.* (2010) also found that secondary pest populations increased in Bt cotton. Krishna and Qaim (2012) analysed pesticide use patterns in India over a period of 7 years. They found that farmers with Bt cotton increased their sprays against sucking pests. Nevertheless, pesticide reductions through Bt increased over time, because the rise in sprays against secondary pests was more than offset by the decline in sprays against bollworms. Krishna and Qaim (2012) found that conventional cotton growers in India could reduce their sprays as well, because the widespread adoption of Bt cotton led to area-wide suppression of bollworm populations. Similar effects were reported for Bt cotton in China and Bt maize in the USA (Wu *et al.*, 2008; Hutchison *et al.*, 2010).

Bt crops are also associated with health benefits. Direct health advantages for farmers occur due to less insecticide exposure during spraying operations. Often, the health hazards for farmers applying pesticides are greater in developing than developed countries, because environmental and health regulations are laxer, pesticides are mostly applied manually and farmers are less educated and less informed about negative side effects. Pray *et al.* (2001) and Huang *et al.* (2003) showed for China that the frequency of pesticide poisonings was significantly lower among Bt cotton adopters than non-adopters. Hossain *et al.* (2004) used econometric models to establish that this observation was related causally to Bt technology. Bennett *et al.* (2003) and Kouser and Qaim (2011) obtained similar results for Bt cotton in South Africa and India.

For consumers, Bt crops can bring about health benefits through lower pesticide residues in food and water. Furthermore, in a variety of field studies, Bt maize has been shown to contain significantly lower levels of certain mycotoxins, which can cause cancer and other diseases in humans (Wu, 2006). Especially in maize, insect damage is one factor that contributes significantly to mycotoxin contamination (see Table 6.1). In the USA and other developed countries, maize is inspected carefully so that lower mycotoxin levels primarily might reduce the costs of testing and grading. But in many developing countries, strict mycotoxin inspections are uncommon. In such situations, Bt technology could contribute to lowering the actual health burden (Wu, 2006; Parrott, 2010).

14.4 Macro-level Effects of GM Crops

The studies discussed so far build on micro-level data collected through farm surveys and field observations. But GM crops are now grown on 170 Mha worldwide, so impacts are also observable at the macro level. Sexton and Zilberman (2012) tried to evaluate these macro-level effects. Based on several years of data, they estimated cross-country regressions, where the production quantities of different agricultural crops in a country were explained by land area and area grown with GM crops. In all regressions, the GM crop area has large and significant positive effects, implying that GM technology adoption has increased country-level agricultural output. For GM soybean, the average production-increasing effect in technology-adopting countries was 13%, for GM rapeseed it was 25% and for GM maize and cotton it was 46% and 65%, respectively (Sexton and Zilberman, 2012). Not all of these increases are net yield gains of GM technology. Technology-adopting farmers may also have increased their fertilizer applications. In some cases, better weed control with GM allows farmers to grow a second crop per year, as is observed partly for HT soybeans in South America. But GM technology has triggered these effects, so the technology already contributes to considerable global production increases.

During the past 10 years, global food prices have shown an increasing trend, because growth in demand has outpaced growth in supply. Especially during the food crisis in 2008, when prices rose sharply over a short period of time, the number of undernourished people in developing countries increased by over 100 million (FAO, 2009). Poor people often spend a substantial portion of their income on food. Hence, there is little buffer to make up for rising prices. Sexton and Zilberman (2012) used their cross-country regression results and a global multi-market equilibrium model to predict how food prices would have looked in 2008 without GM crops. As can be seen in Fig. 14.2, without GM crop adoption, prices for important commodities would have been 30–40% higher. This is plausible given that 81% of all soybeans worldwide, 35% of all maize and 30% of all rapeseed are already genetically modified (James, 2012). Figure 14.2 also shows the price effects for wheat, even though GM wheat is not yet commercialized anywhere in the world. This effect is due to market spillovers. As wheat competes with maize and other crops in production and consumption, price developments are correlated across markets.

The global demand for food, feed and bioenergy is likely to double by 2050 (Godfray *et al.*, 2010). But land, water, fuel and other resources needed for agricultural production are becoming increasingly scarce (see Chapter 1). Moreover, climate change may affect food production negatively (Whitford *et al.*, 2010; World Bank, 2010). Against this background, many predict that food prices will rise further in the future, unless new technologies are being developed and implemented that can boost productivity in a sustainable way. GM crops could play an important role in this connection, provided that the technological potentials are harnessed for many crops and traits other than those that have been commercialized so far.

14.5 Potential Impacts of Future GM Crops

14.5.1 Crops with improved agronomic traits

As discussed above, most of the GM crops commercialized so far involve herbicide tolerance or Bt insect resistance. So far, Bt is used mainly in maize and cotton. Yet, there are also other Bt crops that are soon likely to be commercialized (Romeis *et al.*, 2008). Especially, Bt rice and Bt aubergine have been field tested extensively in China and India. Data from these trials are in line with results for Bt maize and Bt cotton (see Chapter 13). That is, insecticide-reducing

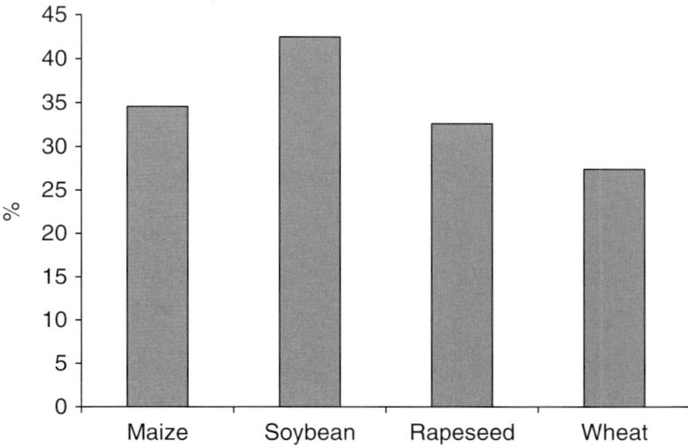

Fig. 14.2. World market prices would be much higher without GM crop adoption. From own presentation based on Sexton and Zilberman, 2012).

and yield-increasing effects have been reported (Huang *et al.*, 2005; Krishna and Qaim, 2008).

In their *ex ante* study of Bt aubergine in India, Krishna and Qaim (2008) projected that the technology, which controls the aubergine fruit and shoot borer, would reduce chemical insecticide use by up to 50% and increase yields by 40% on average. This will not only improve farmers' profits, but will also lower market prices and thus improve consumer access to vegetables. This could have positive nutrition effects among the poor. Moreover, Bt aubergine will be less contaminated with pesticide residues, which have become a real health problem in some parts of India. In spite of the expected positive economic, environmental and health effects, Bt aubergine – as the first GM food crop to be commercialized in India – has aroused controversial public debate. After a careful review of the biosafety and food safety data, the Genetic Engineering Approval Committee, which is the responsible authority in India (see Chapters 3 and 13), declared Bt aubergine to be safe and approved this technology in October 2009 (Kumar, 2009). However, after a series of public hearings, which were influenced heavily by anti-biotech campaigns, the Minister of Environment and Forests suspended the commercialization of Bt aubergine for an indefinite period. This example demonstrates how much the regulatory procedures, which should be science based, are influenced by subjective views of certain lobbying groups.

Also, for other pest-resistant GM traits such as fungal, virus, nematode or bacterial resistance, which are being developed in different crops, pesticide-reducing and yield-increasing effects can be expected. As already observed for Bt technologies, positive yield effects will be more pronounced in developing countries, where pest pressure is often higher and farmers face more severe constraints in controlling pest damage (Table 14.3). Especially in the non-commercial and semi-commercial crop sectors, where technical and economic constraints impede a more widespread use of chemicals, pest-related crop losses are often 50% and higher (Oerke, 2006). Based on conditions of pest pressure and current crop protection, the biggest yield gains are expected in South and South-east Asia and sub-Saharan Africa.

The effects of GM crops with tolerance to abiotic stresses will also be situation specific. A drought-tolerant GM variety can lead to substantially higher yields than conventional varieties under water stress, whereas the effect may be small when sufficient water is available. Especially in the semi-arid tropics, many small-scale farmers are operating under drought-prone conditions, so that the benefits of drought tolerance could be sizeable. In a study referring to eight low-income countries in Asia and sub-Saharan Africa, Kostandini *et al.* (2009) calculate that the average yield gains of GM drought tolerance traits may be 18% in maize, 25% in wheat and 10% in rice.

While the development of drought-tolerant varieties is a major priority both in public and private sector crop improvement programmes, biotech researchers are also working on tolerance to other abiotic stress factors such as heat, salinity, flood and

Table 14.3. Expected yield effects of pest-resistant GM crops in different regions. (From Qaim and Zilberman, 2003.)

Region	Pest pressure	Availability of chemical alternatives	Adoption of chemical alternatives	Yield effect of GM crops
Developed countries	Low to medium	High	High	Low
Latin America (commercial)	Medium	Medium	High	Low to medium
China	Medium	Medium	High	Low to medium
Latin America (non-commercial)	Medium	Low to medium	Low	Medium to high
South and South-east Asia	High	Low to medium	Low to medium	High
Sub-Saharan Africa	High	Low	Low	High

coldness (see also Chapter 1). Climate change (Reynolds, 2010) is associated with more frequent weather extremes, so that more tolerant GM crops can help reduce the risks of crop failures and food crises. Furthermore, research is under way to develop crops with higher nutrient efficiency, especially with respect to nitrogen and phosphorus. Nutrient-efficient crops will reduce chemical fertilizer use and the associated environmental problems in intensive agricultural production systems, while they will contribute to yield gains in regions where fertilizers are currently underused, as in large parts of sub-Saharan Africa. Some of these traits are genetically complex, so that commercialization may not be expected in the short run. But in the medium and long run, the contribution to food security could be sizeable (Qaim, 2011; see also Chapter 1).

14.5.2 Crops with improved nutritional traits

Nutritionally enhanced GM crops that researchers are working on include oilseeds with improved fatty acid profiles or staple foods with enhanced contents of essential amino acids, minerals and vitamins (see Chapters 7 and 12). Enhancing food crops with higher nutrient contents through conventional breeding or GM approaches is also called biofortification (Qaim *et al.*, 2007). A well-known example of a GM biofortified crop is Golden Rice, which contains significant amounts of provitamin A. Golden Rice could become available in some Asian countries in 2013 or 2014 (Beyer, 2010; see Chapters 12 and 13).

Biofortified crops do not involve direct productivity and income effects for farmers or consumers, so the benefits need to be evaluated differently. Especially in developing countries, micronutrient deficiencies are widespread. Children and women in poverty households are particularly affected. Adverse health outcomes include impaired physical and mental development, higher incidence of infectious diseases and premature deaths. If biofortified staple crops were widely grown and consumed, micronutrient deficiencies could be reduced, entailing important health advantages and economic benefits. Qaim *et al.* (2007) suggested a framework for evaluating the potential benefits: since micronutrient malnutrition caused significant health costs, which could be reduced through biofortification, they quantified the health costs with and without biofortified crops and interpreted the difference – that is, the health cost saved – as the technological benefit.

In their *ex ante* analysis of the impact of Golden Rice, Stein *et al.* (2008) used representative household data from India to show that this technology could reduce the health costs of vitamin A deficiency by up to 60%. They also calculated a high cost-effectiveness of Golden Rice, which compared favourably with other micronutrient interventions such as food supplementation or industrial fortification.

Significant economic and health benefits can also be expected for other biofortified crops, like iron and zinc dense staple foods or crops containing higher amounts of essential amino acids (Qaim *et al.*, 2007; De Steur *et al.*, 2012; see Chapters 7 and 12). The high potential cost-effectiveness of biofortification is due to the fact that the approach is self-targeting to the poor, with biofortified seeds spreading through existing formal and informal distribution channels. However, possible issues of consumer acceptance have to be considered. And, especially when no price premium is paid in the output market, suitable strategies to convince farmers to adopt biofortified crops are needed. A combination of nutritional traits with interesting agronomic traits might be a practicable avenue.

14.6 Conclusions

GM crops have been used commercially for over 15 years. So far, mostly HT and Bt crops have been employed. Available impact studies show that these crops are beneficial to farmers and consumers. While HT crops lead to cost savings in weed control and tillage operations, Bt crops entail significant chemical pesticide reductions and higher effective yields. The average economic benefits for adopting farmers are sizeable. Moreover, Bt crops bring about environmental and health advantages. They are also well suited for small-scale farmers when they are embedded in a conducive institutional environment. In India, for instance, Bt cotton contributes to more employment, higher household incomes and poverty reduction. In many cases, farmers in developing countries benefit more from GM crop adoption than farmers in developed countries, due to weaker IPR protection and differences in agroecological conditions.

GM technologies that are still in the research pipeline (see Chapter 12) include crops that are tolerant to abiotic stresses and crops that contain higher amounts of nutrients. The benefits of such applications could eventually be bigger than the ones already observed. Against the background of a dwindling natural resource base and growing demand for agricultural products, GM crops could contribute significantly to global food security and poverty reduction. New technologies will have to play the main role for the necessary production increases in the future. So far, multinational companies dominate GM crop developments, mostly focusing on crops with large international markets. More public research and public–private partnerships will be necessary to ensure that technologies that are particularly relevant for poor farmers and consumers in developing countries are also made available.

In spite of the large potential of GM crops, the technology lacks public acceptance, especially in Europe (see Chapter 15). Concerns about new risks and the lobbying efforts of anti-biotech groups have led to complex and costly biosafety, food safety and labelling regulations, which slow down innovation rates and lead to a bias against small countries, minor crops, and small firms and public research organizations. Overregulation has become a real threat for the further development and use of GM crops. The costs in terms of foregone benefits might be large, especially for developing countries. This is not to say that zero regulation would be desirable, but the trade-offs associated with regulation have to be considered. In the wider public, the risks of GM crops seem to be overrated, while the benefits are underrated.

Note

[1] Especially for India, there are still reports by biotech critics that Bt cotton ruins smallholder farmers. However, such reports do not build on representative data. Gruère and Sengupta (2011) showed that the occasional claim of a link between Bt cotton adoption and farmer suicides could not be substantiated.

References

Basu, A.K. and Qaim, M. (2007) On the adoption of genetically modified seeds in developing countries and the optimal types of government intervention. *American Journal of Agricultural Economics* 89, 784–804.

Bennett, R., Morse, S. and Ismael, Y. (2003) Bt cotton, pesticides, labour and health: a case study of smallholder farmers in the Makhathini Flats, Republic of South Africa. *Outlook on Agriculture* 32, 123–128.

Beyer, P. (2010) Golden rice and 'golden' crops for human nutrition. *New Biotechnology* 27, 478–481.

Bradford, K.J., Van Deynze, A., Gutterson, N., Parrott, W. and Strauss, S.H. (2005) Regulating transgenic crops sensibly: lessons from plant breeding, biotechnology and genomics. *Nature Biotechnology* 23, 439–444.

Brookes, G. and Barfoot, P. (2012) *GM Crops: Global Socioeconomic and Environmental Impacts 1996–2010.* PG Economics, Dorchester, UK.

De Steur, H., Gellynck, X., Van Der Straeten, D., Lambert, W., Blancquaert, D. and Qaim, M. (2012) Potential impact and cost-effectiveness of multi-biofortified rice in China. *New Biotechnology* 29, 432–442.

FAO (2009) *The State of Food Insecurity in the World: Economic Crises – Impacts and Lessons Learned.* FAO and WFP, Rome.

Fernandez-Cornejo, J. and Caswell, M. (2006) The first decade of genetically engineered crops in the United States. Economic Information Bulletin 11, United States Department of Agriculture, Washington, DC.

Fernandez-Cornejo, J., Hendricks, C. and Mishra, A. (2005) Technology adoption and off-farm household income: the case of herbicide-tolerant soybeans. *Journal of Agricultural and Applied Economics* 37, 549–563.

Glover, D. (2010) Is Bt cotton a pro-poor technology? A review and critique of the empirical record. *Journal of Agrarian Change* 10, 482–509.

Godfray, H.C.J., Beddington, J.R., Crite, I.R., Haddad, L., Lawrence, D., Muir, J.F., *et al.* (2010) Food security: the challenge of feeding 9 billion people. *Science* 327, 812–818.

Gouse, M., Pray, C., Schimmelpfennig, D. and Kirsten, J. (2006) Three seasons of subsistence insect-resistant maize in South Africa: have smallholders benefited? *AgBioForum* 9, 15–22.

Gruère, G. and Sengupta, D. (2011) Bt cotton and farmer suicides in India: an evidence-based assessment. *Journal of Development Studies* 47, 316–337.

Hossain, F., Pray, C.E., Lu, Y., Huang, J. and Hu, R. (2004) Genetically modified cotton and farmers' health in China. *International Journal of Occupational and Environmental Health* 10, 296–303.

Huang, J., Hu, R., Pray, C., Qiao, F. and Rozelle, S. (2003) Biotechnology as an alternative to chemical pesticides: a case study of Bt cotton in China. *Agricultural Economics* 29, 55–68.

Huang, J., Hu, R., Rozelle, S. and Pray, C. (2005) Insect-resistant GM rice in farmers' fields: assessing productivity and health effects in China. *Science* 308, 688–690.

Hutchison, W.D., Burkness, E.C., Mitchell, P.D., Moon, R.D., Leslie, T.W., Fleischer, S.J., *et al.* (2010) Areawide suppression of European corn borer with Bt maize reaps savings to non-Bt maize growers. *Science* 330, 222–225.

James, C. (2012) *Global Status of Commercialized Biotech/GM Crops: 2011*. ISAAA Briefs No 44. International Service for the Acquisition of Agri-Biotech Applications, Ithaca, New York.

Kathage, J. and Qaim, M. (2012) Economic impacts and impact dynamics of Bt (*Bacillus thuringiensis*) cotton in India. *Proceedings of the National Academy of Sciences USA* 109, 11652–11656.

Kostandini, G., Mills, B.F., Omamo, S.W. and Wood, S. (2009) *Ex ante* analysis of the benefits of transgenic drought tolerance research on cereal crops in low-income countries. *Agricultural Economics* 40, 477–492.

Kouser, S. and Qaim, M. (2011) Impact of Bt cotton on pesticide poisoning in smallholder agriculture: a panel data analysis. *Ecological Economics* 70, 2105–2113.

Kouser, S. and Qaim, M. (2013) Valuing financial, health, and environmental benefits of Bt cotton in Pakistan. *Agricultural Economics* 44(3), 323–335.

Krishna, V.V. and Qaim, M. (2008) Potential impacts of Bt eggplant on economic surplus and farmers' health in India. *Agricultural Economics* 38, 167–180.

Krishna, V.V. and Qaim, M. (2012) Bt cotton and sustainability of pesticide reductions in India. *Agricultural Systems* 107, 47–55.

Kumar, P.A. (2009) Bt brinjal: a pioneering push. *Biotech News* 4, 108–111.

Lu, Y., Wu, K., Jiang, Y., Xia, B., Li, P., Feng, H., *et al.* (2010) Mirid bug outbreaks in multiple crops correlated with wide-scale adoption of Bt cotton in China. *Science* 328, 1151–1154.

Morse, S., Bennett, R. and Ismael, Y. (2004) Why Bt cotton pays for small-scale producers in South Africa. *Nature Biotechnology* 22, 379–380.

Naseem, A. and Pray, C. (2004) Economic impact analysis of genetically modified crops. In: Christou, P. and Klee, H. (eds) *Handbook of Plant Biotechnology*. John Wiley and Sons, Chichester, UK, pp. 959–991.

Oerke, E.-C. (2006) Crop losses to pests. *Journal of Agricultural Science* 144, 31–43.

Parrott, W. (2010) Genetically modified myths and realities. *New Biotechnology* 27, 545–551.

Pray, C.E., Ma, D., Huang, J. and Qiao, F. (2001) Impact of Bt cotton in China. *World Development* 29, 813–825.

Qaim, M. (2009) The economics of genetically modified crops. *Annual Review of Resource Economics* 1, 665–693.

Qaim, M. (2011) Genetically modified crops and global food security. In: Carter, C., Moschini, G. and Sheldon, I. (eds) *Genetically Modified Food and Global Welfare*. Emerald Publishing, Bingley, UK, pp. 29–54.

Qaim, M. and Janvry, A. de (2005) Bt cotton and pesticide use in Argentina: economic and environmental effects. *Environment and Development Economics* 10, 179–200.

Qaim, M. and Kouser, S. (2013) Genetically modified crops and food security. *PLOS ONE* 8(6), e64879.

Qaim, M. and Traxler, G. (2005) Roundup Ready soybeans in Argentina: farm level and aggregate welfare effects. *Agricultural Economics* 32, 73–86.

Qaim, M. and Zilberman, D. (2003) Yield effects of genetically modified crops in developing countries. *Science* 299, 900–902.

Qaim, M., Subramanian, A., Naik, G. and Zilberman, D. (2006) Adoption of Bt cotton and impact variability: insights from India. *Review of Agricultural Economics* 28, 48–58.

Qaim, M., Stein, A.J. and Meenakshi, J.V. (2007) Economics of biofortification. *Agricultural Economics* 37 (Suppl. 1), 119–133.

Qaim, M., Subramanian, A. and Sadashivappa, P. (2009) Commercialized GM crops and yield. *Nature Biotechnology* 27, 803–804.

Reynolds, M.P. (2010) *Climate Change and Crop Production*. CAB International, Wallingford, UK, 292 pp.

Romeis, J., Shelton, A.S. and Kennedy, G.G. (eds) (2008) *Integration of Insect-Resistant Genetically Modified Crops within IPM Programs*. Springer, New York.

Sexton, S. and Zilberman, D. (2012) Land for food and fuel production: the role of agricultural biotechnology. In: Zivin, J.S.G. and Perloff, J.M. (eds) *The Intended and Unintended Effects of US Agricultural and Biotechnology Policies*. University of Chicago Press, Chicago, Illinois, pp. 269–288.

Stein, A.J., Sachdev, H.P.S. and Qaim, M. (2008) Genetic engineering for the poor: Golden Rice and public health in India. *World Development* 36, 144–158.

Stone, G.D. (2011) Field versus farm in Warangal: Bt cotton, higher yields, and larger questions. *World Development* 39, 387–398.

Subramanian, A. and Qaim, M. (2010). The impact of Bt cotton on poor households in rural India. *Journal of Development Studies* 46, 295–311.

Tabashnik, B.E., Gassmann, A.J., Crowder, D.W. and Carrière, Y. (2008) Insect resistance to Bt crops: evidence versus theory. *Nature Biotechnology* 26, 199–202.

Wang, S., Just, D.R. and Pinstrup-Andersen, P. (2008) Bt cotton and secondary pests. *International Journal of Biotechnology* 10, 113–121.

Whitford, R., Gilbert, M. and Langridge, P. (2010) Biotechnology in agriculture. In: Reynolds, M.P. (ed.) *Climate Change and Crop Production*. CAB International, Wallingford, UK, pp. 219–244.

Wolfenbarger, L.L.R., Naranjo, S.E., Lundgren, J.G., Bitzer, R.J. and Watrud, L.S. (2008) Bt crop effects on functional guilds of non-target arthropods: a meta-analysis. *PLOS One* 3, e2118.

World Bank (2010) *World Development Report: Development and Climate Change*. World Bank, Washington, DC.

Wu, F. (2006) Bt corn's reduction of mycotoxins: regulatory decisions and public opinion. In: Just, R.E., Alston, J.M. and Zilberman, D. (eds) *Regulating Agricultural Biotechnology: Economics and Policy*. Springer, New York, pp. 179–200.

Wu, K.M., Lu, Y.-H., Feng, H.-Q., Jiang, Y.-Y. and Zhao, J.-Z. (2008) Suppression of cotton bollworm in multiple crops in China in areas with Bt toxin-containing cotton. *Science* 321, 1676–1678.

15 Public Acceptance of GM Plants

Joachim Scholderer[1]* and Wim Verbeke[2]

[1]Aarhus University, School of Business and Social Sciences, Department of Business Administration, Aarhus, Denmark; [2]Ghent University, Faculty of Bio-Science Engineering, Department of Agricultural Economics, Ghent, Belgium

15.1 Introduction

There were times when genetically modified (GM) plants were a much-contested issue. In the early days of genetic engineering – the 1970s and early 1980s – the acceptability of the new technology was debated mainly among scientists themselves (e.g. Berg, 1974) and mainly in terms of laboratory safety. Most countries issued best-practice guidelines at the time to ensure that potential biohazards were contained safely in the laboratory. The situation changed in the mid-1980s, when the first GM organisms were tested outside the laboratory. Non-governmental organizations entered the debate and gradually changed its nature, reframing it in terms of environmental risk, bioethics and the precautionary principle. In the European Union (EU), the debate became more heated throughout the 1990s, largely proportional to the commercial success of pest-resistant and herbicide-tolerant crops, and culminated in a 5-year moratorium on the approval of new transgenic cultivars that lasted from 1999 to 2004 (for details, see Scholderer, 2005).

Since then, the regulatory framework in Europe has been completely overhauled and the value-laden debate of the 1990s has been replaced by a rather more technical discussion, focusing on good agricultural practices that can ensure the coexistence of GM with conventional and organic crop production systems (Scholderer and Verbeke, 2012). Although certain NGOs have made every effort to keep the value debate alive and export it to other regions of the world, this has never really succeeded (Aerni and Bernauer, 2005). On the one hand, this can be considered good news for everybody who plans to research, or invest in, GM crop production systems. On the other hand, the acreage covered by transgenic crops continues to be disproportionally low in Europe (James, 2011), the region in the world where the debate was led most destructively. We believe that much can be learned from this. This chapter sheds light on the mechanics of what we, in the title, so innocuously call 'public acceptance'.

15.2 What Is Public Acceptance?

Public acceptance is a multi-layered concept: who should be considered the public, and what particular types of action or inaction should be taken as indicators of acceptance? In public relations, the line of work that is concerned professionally with the public acceptance of commercial and political issues, it is customary to refer to 'publics' in the plural rather than to 'public' in the singular. A public is understood as a group of persons or organizations that have an interest or stake in an issue (hence the modern expression 'stakeholder'), and these interests or stakes are assumed to vary between different publics. In the context of

*E-mail: sch@asb.dk

GM plants, we can distinguish at least five different publics.

The first is the general public in their role as potential buyers and consumers of products that contain ingredients based on GM plants. The second is the same general public, but in a very different role: as voters, or opinion poll participants, who may endorse or reject particular agricultural and food policies on GM plants. The third public is the farmers; that is, the potential customers of seed companies who market GM cultivars. The fourth public is the customers of these farmers: food companies who might decide to use GM ingredients in their products, plus their customers, the retail chains that may or may not decide to buy the resulting products and sell them on to consumers. Finally, there is a host of pressure groups, lobby organizations and political bodies who seek to influence agricultural and food policies directly (through their lobbying activities) and indirectly (via the media and their assumed influence on the opinions held by the general public) to further the interests of their respective patrons. It is important not to confuse these five publics; arguably, much of the confusion in the debate about GM plants was caused by an astonishing lack of ability among political decision makers to distinguish them and weight their influence in an appropriate manner.

15.3 The Buying Behaviour of Consumers

In many ways, the buying behaviour of consumers can be regarded as the most straightforward indicator of public acceptance. In order to make a decision to buy or not to buy a product containing GM ingredients, consumers have to be able to ascertain whether or not such ingredients are contained in the product. This requires a labelling policy: a clear indication in the ingredients list, usually on the back of the package and in rather small print, which ingredients have been derived from GM plant material. Consumer surveys throughout the world speak a clear language: a large majority of consumers would prefer such a labelling policy, including consumers in countries where such labelling has long been mandatory (such as the member states of the EU) and also in countries where such labelling is not mandatory (for example, the USA).

Empirical evidence collected in the field – that is, observations of the behaviour of consumers in actual retail settings, as opposed to the laboratory – suggests that labelling has negligible effects on the choices made by consumers. Econometric studies that compared the retail sales in different food categories before and after the introduction of mandatory labelling regimes (for example, Marks *et al.*, 2004, in the Netherlands; Lin *et al.*, 2008, in China) found no or only weak effects. Experimental selling studies in which products clearly labelled as GM were sold to consumers, for example at farmers' markets or roadside stalls, found no substantial effects either (Mather *et al.*, 2005, in New Zealand; Knight *et al.*, 2007, in Belgium, France, Germany, Sweden and the UK; Aerni *et al.*, 2011, in Switzerland).

The main result of these field studies was that consumers applied their usual decision criteria in the same manner to products containing GM ingredients as they did to conventional products from the same category. In other words, the attractiveness of a product to a consumer does not seem to depend much on the processes by which some of its ingredients have been produced – after all, consumers buy products, not technologies. Instead, the attractiveness of a product depends on its quality, its price and the appropriateness of its package size. The 'it depends on the product' character of these findings is reflected by the extremely heterogeneous results of laboratory experiments in which small groups of participants are asked to make hypothetical choices between GM products and their conventional counterparts. In a meta-analysis of 51 primary studies, Dannenberg (2009) found effect sizes that varied from an average willingness to pay a 240% price premium for a GM product in one study to a 784% premium for a conventional counterpart in another study (the median willingness to pay was an 18% premium for products without GM ingredients). In light of these results, it makes little sense to say that consumers accept or reject GM foods per se – it always depends on the particular product and its properties and not so much on the technologies that have been used in its production.

15.4 Political Attitudes Held by Citizens

Although people do not seem to distinguish much between products with and without GM ingredients in their role as consumers, this does not mean that they do not have any opinions about the process of genetically modifying a living organism and whether the use of the underlying technologies should be promoted by public policy. In their role as citizens, many people do indeed have strong opinions about this. The European Commission has monitored the attitudes of EU citizens since 1991 in a special Euro-barometer series. Figure 15.1 shows the development over time of the index 'optimism about biotechnology' (Gaskell *et al.*, 2010). The index compares the estimated proportion of citizens who expect that bio-technology will have mainly positive future impacts with the proportion of citizens who expect mainly negative impacts. Positive values of the index indicate that the majority is optimistic, whereas negative values indicate that the majority is sceptical. The trend lines showed that, at the beginning of the 2010s, the majority of EU citizens were as optimistic again about the future impact of biotechnology as they had been in the early 1990s. However, it is also apparent that the heated debate of the mid-1990s had a considerable negative impact: it took almost a decade until optimism had reached the same levels again as before the debate.

The index 'optimism about biotechnology' is a rather general measure of the future expectations of European citizens. Besides this index, all Eurobarometer surveys have measured attitudes towards particular groups of applications, including pest-resistant crops ('taking genes from plant species and transferring them into crop plants to make them more resistant to insect pests'), processed GM foods with altered properties ('using modern biotechnology in the production of foods, for example to give them a higher protein content, to be able to keep them longer, or to change the taste'; see Chapters 6, 7 and 12), and transgenic and cisgenic apples ('some European researchers think there are new ways of controlling common diseases in apples – things like scab and mildew. There are two new ways of doing this. Both mean that the apples could be grown with limited use of pesticides, and so pesticide residues on the apples would be minimal. The first way is to artificially introduce a resistance gene from another species such as a bacterium or animal into an apple tree to make it resistant to mildew and scab. The second way is to artificially introduce a gene that exists naturally in wild/crab apples which provides resistance to mildew and scab'; see Gaskell, *et al.*, 2003, 2006 and 2010). The average attitudes of European citizens towards these example applications have oscillated around the neutral point of the response scale over the past 10 years. When asked specifically about the usefulness and moral acceptability of these applications, Europeans tended to evaluate pest-resistant crops (including transgenic and cisgenic apples) slightly above the neutral point and processed foods with altered properties slightly below the neutral point. When asked specifically about risks, Europeans tended to evaluate all applications slightly above the neutral point (i.e. as slightly risky). Taken together, European citizens seem to have rather neutral attitudes towards GM crops, at least on average.

On an individual level, there is con-siderable variation: certain gene technology applications can polarize people's opinions. In light of the fact that people are rarely confronted with products that are labelled as containing GM ingredients (either because there are hardly any products on the market, such as in Europe, or because there is no mandatory labelling regime, as in the USA) and that gene technology is not much of an issue in the popular press these days,

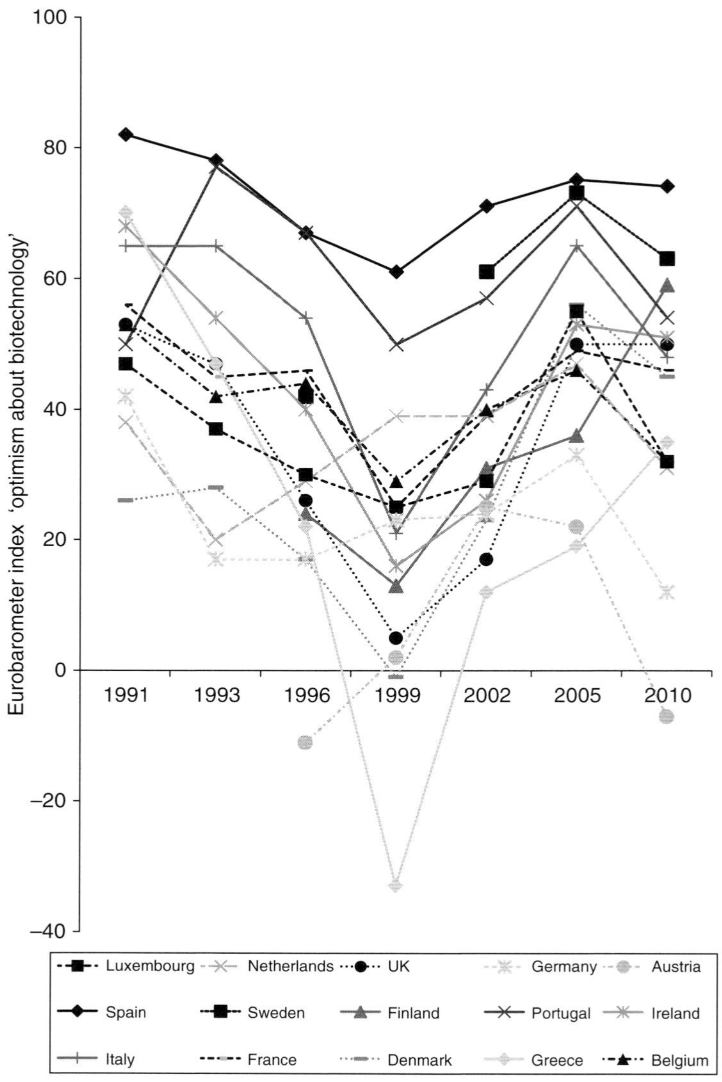

Fig. 15.1. Development of the Eurobarometer index 'optimism about biotechnology' from 1991 to 2010 (based on data reported by Gaskell *et al.*, 2010).

social scientists have often wondered where people's attitudes towards GM crops and foods actually come from. Detailed analyses of people's attitude structures suggest that attitudes towards gene technology are, in a way, 'constructed'. Even in 2010, 16% of the Eurobarometer respondents openly admitted that they had never heard about GM foods before (Gaskell *et al.*, 2010). Other respondents may have heard about the

concept before but may never have reflected in depth about the pros and cons of gene technology and its applications. It appears that, in such a situation, people try to set the core concept of gene technology – the planned modification of the genome of a living organism – into a relation to their general socio-political attitudes and values and evaluate the concept based on its perceived match or mismatch with those general attitudes and values (for a detailed review, see Scholderer and Verbeke, 2012).

Empirically, the strongest correlations can be found with attitudes towards the environment, attitudes towards technological progress, trust in the actors and institutions that commercialize new technologies and regulate their risks, and general tendencies to reject the unknown (Borre, 1990; Sparks *et al.*, 1995; Siegrist, 1998, 2000; Bredahl, 2001; Søndergaard *et al.*, 2007). Considering the nature of the general attitudes and values that are associated strongly with attitudes towards gene technology, it can be argued that attitudes towards gene technology are *moral* judgements, not evaluations of the risks and benefits of the technology or its applications. This interpretation is consistent with another, often replicated finding: attitudes towards gene technology are highly resistant to change and cannot be influenced by typical mass communication techniques. People may even interpret such communications as attempts to undermine the legitimacy of their value orientations, which can lead to serious boomerang effects (Frewer *et al.*, 2003; Scholderer and Frewer, 2003). Even though this may be seen as politically incorrect, the best strategy for many seed companies and biotechnology associations is therefore to try not to communicate with the general public at all. The general public may not love gene technology, but they do not love pesticides or mutagenesis either (apparently even less; see Hagemann and Scholderer, 2007). And despite all the scepticism, everybody eats.

15.5 Adoption Decisions Made by Farmers

Although farmers are the actual 'users' of transgenic crops, their voice has largely been ignored in the public debate on gene technology (Guehlstorf, 2008). There are relatively few empirical studies on the attitudes of farmers, most are relatively recent, most are based on small, non-representative samples of farmers and almost all of them are about the attitudes of farmers towards the use of transgenic crops in horticulture (e.g. Chong, 2005; Kondoh and Jussaume, 2006; Heller, 2007; Hall, 2008; Kaup, 2008; Mauro and McLachlan, 2008; Areal *et al.*, 2011; Mal *et al.*, 2012; Skevas *et al.*, 2012). Due to the heterogeneity of these studies – with crops ranging from aubergine to cotton to maize and regions ranging from Northern India to Washington State to Portugal – it is difficult to draw general conclusions from them. A consistent finding is that the farmers who participated in these studies tended to evaluate transgenic crops quite pragmatically. Higher yields and reduced spending on pest and weed control are typically seen as the most tangible benefits of transgenic varieties. Coexistence measures and the bureaucratic burdens associated with them tend to make farmers hesitant. However, this hesitation appears to become less pronounced as experience with the successful implementation of coexistence measures accumulates in a region (for a detailed review, see Scholderer and Verbeke, 2012). With the obvious exception of organic farmers and farmers in countries where certain transgenic varieties are banned, there seem to be no strong tendencies for or against GM crops in general.

15.6 The Role of Food Manufacturers and Retailers

The situation becomes a little more complicated at the levels of the value chain that immediately precede the final consumer markets. In theory, every food manufacturer is free to source transgenic ingredients and use them in their products, as long as the ingredients are approved for marketing in the countries where the products will be sold. In reality, it is not quite so straightforward. After all, food manufacturers do not sell their products to consumers. They sell them to retailers, and therefore the retailers decide whether a product that contains GM ingredients will ever make it on to the shelves of a supermarket. It is important to understand this: retailers are the gatekeepers to consumer markets and can exercise almost unlimited power over the rest of the value chain. From the point of view of a retailer, a good product is one that has high profit margins and is frequently bought. Neither of these characteristics favour products with transgenic ingredients: conventional substitutes are available for every transgenic ingredient that has to be labelled (in countries where labelling regulations exist), and typically also at a price that would not really influence the retailer's margin from the end product. In other words, retailers have no positive incentive to sell products with transgenic ingredients that have to be labelled. And in the absence of a positive incentive, the presence of a business risk – however small it may be – will already be an exclusion criterion.

And that business risk exists. Retailers in Europe were made painfully aware of this in September 1998 (for historical details, see Scholderer, 2005): on the same day the EU regulation on novel foods and food ingredients entered into force in Germany, Nestlé launched their US product, Butterfinger ®, as a test case, clearly labelled as containing GM ingredients. The environmental pressure group, Greenpeace, responded immediately with a major campaign, picketing stores in several German cities where the product was sold and orchestrating global media coverage. After just 1 week, most major European retailers had declared that they would not sell any GM foods until their safety had been proven and until consumers actively demanded that they should be sold. Although Greenpeace continued this campaign for many years (and boldly documented it; see Holbach and Keenan, 2005), the market for labelled GM foods has effectively been dead in Europe since September 1998. The situation continues unabated: no retailer in their right mind will risk a Greenpeace blockade of their stores.

15.7 The World of Pressure Groups and Lobbying

The events surrounding the Butterfinger ® launch in Germany in September 1998 are an excellent case to demonstrate the twisted realities that coexist, side by side, in the world of lobbies and pressure groups. Although the immediate cause of the disappearance of GM foods from European supermarket shelves was a threat campaign, levelled by a highly organized and hierarchically structured pressure group (Greenpeace) against the European retail sector, every interest group created their own historical narratives and myths in the aftermath of the events, explaining the disappearance of GM foods in arbitrary ways that were consistent with their policies and appeared to legitimize them. The most common of these narratives is that 'consumer rejection' (not the store blockade arranged by Greenpeace) prompted European retailers to take the products off the shelves.

This is interesting insofar as, already at that time, the available empirical evidence indicated that consumer choices in real shopping situations were largely unaffected by the presence or absence of GM labels on consumer products. And curiously enough, it seems that even the pro-GM lobbies began to believe this narrative. Arguably, the policy that would have served their interests best at the time would have been the widespread introduction of very stringent labelling regimes: if virtually all consumer products in all retail stores carried 'may contain genetically modified ingredients' labels, anti-GM campaigners would have been unable to focus their resources on particular products or retail stores, and their campaigns would most likely have dissipated. Instead, pro-GM lobbies across the globe fought desperately against the introduction of any form of labelling regime (in many countries even successfully), alienating potential cooperation partners and inadvertently creating an image of gene technology as something clandestine, something that needed to be hidden.

The political sphere really did not do any better. In hindsight, it appears that at some point in the second half of the 1990s, political and administrative elites began to confuse the general public with the *images* of the general public that the various lobbies and pressure groups created for them. As a result, gene technology became a priority on the political agenda. In some regions of the world, the regulatory systems were completely overhauled to regain the trust of the population (which was assumed to have been lost). In other countries, PR campaigns were launched to bring biotechnology closer to the public, assuming that the public rejected biotechnology and that this, in turn, would be a great risk for the future of the economy. In yet other countries, agriculture ministers adopted radical positions, even banning the cultivation of certain transgenic crop varieties (probably assuming that this would swing a great number of votes). Was it really worth it?

15.8 Conclusions

A dispassionate look at the results of the numerous opinion surveys and consumer studies conducted in the last two decades suggests that, in the eyes of most 'publics' – consumers, citizens, food manufacturers, retailers – gene technology has never really been an issue of particularly high importance. On average, their attitudes tend to be neutral. And even at times when controversies between stakeholders escalated, for example in the EU in the second half of the 1990s, only relatively minor changes could be observed in the attitudes of the general public. A 'crisis of confidence' it certainly was not. We believe that this is an important lesson to remember: although heated arguments may be exchanged between stakeholders, and the points of disagreement may seem all-important at the time, they are still likely to be ignored by almost everybody who is *not* concerned professionally with GM crops. We would like to close with an appeal to reason: things should always be kept in perspective.

References

Aerni, P. and Bernauer, T. (2005) Stakeholder attitudes toward GMOs in the Philippines, Mexico, and South Africa: the issue of public trust. *World Development* 34, 557–575.

Aerni, P., Scholderer, J. and Ermen, D. (2011) How would Swiss consumers decide if they had freedom of choice? Evidence from a field study with organic, conventional and GM corn bread. *Food Policy* 36, 830–838.

Areal, F.J., Riesgo, L. and Rodriguez-Cerezo, E. (2011) Attitudes of European farmers towards GM crop adoption. *Plant Biotechnology Journal* 9, 945–957.

Berg, P. (1974) Potential biohazards of recombinant DNA molecules. *Science* 185, 303.

Borre, O. (1990) Public opinion on gene technology in Denmark 1987 to 1989. *Biotech Forum Europe* 7, 471–477.

Bredahl, L. (2001) Determinants of consumer attitudes and purchase intentions with regard to genetically modified foods: results of a cross-national survey. *Journal of Consumer Policy* 24, 23–61.

Chong, M. (2005) Perception of the risks and benefits of Bt eggplant by Indian farmers. *Journal of Risk Research* 8, 617–634.

Dannenberg, A. (2009) The dispersion and development of consumer preferences for genetically modified food – a meta-analysis. *Ecological Economics* 68, 2182–2192.

Frewer, L.J., Scholderer, J. and Bredahl, L. (2003) Communicating about the risk and benefits of genetically modified foods: the mediating role of trust. *Risk Analysis* 23, 1117–1133.

Gaskell, G., Allum, N. and Stares, S. (2003) *Europeans and Biotechnology in 2002. Eurobarometer 58.0.* Office for Official Publications of the European Communities Luxembourg (http://ec.europa.eu/public_opinion/archives/ebs/ebs_177_en.pdf, accessed 9 April 2013).

Gaskell, G., Allansdottir, A. and Allum, N. (2006) *Europeans and Biotechnology in 2005: Patterns and Trends.* Office for Official Publications of the European Communities, Luxembourg (http://ec.europa.eu/research/press/2006/pdf/pr1906_eb_64_3_final_report-may2006_en.pdf, accessed 9 April 2013).

Gaskell, G., Stares, S. and Allansdottir, A. (2010) *Europeans and Biotechnology in 2010: Winds of Change?* European Commission, Directorate General for Research, Brussels (http://ec.europa.eu/public_opinion/archives/ebs/ebs_341_winds_en.pdf, accessed 9 April 2013).

Guehlstorf, N.P. (2008) Understanding the scope of farmer perceptions of risk: considering farmer opinions on the use of genetically modified (GM) crops as stakeholder voice in policy. *Journal of Agricultural and Environmental Ethics* 21, 541–558.

Hagemann, K. and Scholderer, J. (2007) Consumer versus expert hazard identification: a mental models study of mutation-bred rice. *Journal of Risk Research* 10, 449–464.

Hall, C. (2008) Identifying farmer attitudes towards genetically modified (GM) crops in Scotland: Are they pro- or anti-GM? *Geoforum* 39, 204–212.

Heller, C. (2007) Techne versus technoscience: divergent (and ambiguous) notions of food 'quality' in the French debate over GM crops. *American Anthropologist* 109, 603–615.

Holbach, M. and Keenan, L. (2005). *No Market for GM Labelled Food in Europe*. Greenpeace International, Amsterdam (http://www.green peace.org/eu-unit/Global/eu-unit/reports-brief ings/2009/3/no-market-for-gm-labelled-food. pdf, accessed 9 April 2013).

James, C. (2011) *Global Status of Commercialized Biotech/GM Crops: 2011*. ISAAA Brief No 43. ISAAA, Ithaca, New York.

Kaup, B.Z. (2008) The reflexive producer: the influence of farmer knowledge upon the use of Bt corn. *Rural Sociology* 73, 62–81.

Knight, J.G., Mather, D.W., Holdsworth, D.K. and Ermen, D.F. (2007) Acceptance of GM food – an experiment in six countries. *Nature Biotechnology* 25, 507–508.

Kondoh, K. and Jussaume, R.A. (2006) Contextualizing farmers' attitudes towards genetically modified crops. *Agricultural and Human Values* 23, 341–352.

Lin, W., Tuan, F., Dai, Y., Zhong, F. and Chen, X. (2008) Does biotech labelling affect consumers' purchasing decisions? A case study of vegetable oils in Nanjing, China. *AgBioForum* 11, 123–133 (http://www.agbioforum.org/v11n2/ v11n2a06-lin.pdf, accessed 9 April 2013).

Mal, P., Anik, A.R., Bauer, S. and Schmitz, P.M. (2012) Bt cotton adoption: a double-hurdle approach for North Indian farmers. *AgBioForum* 15, 294–302 (http://www.agbioforum.org/v15n3/ v15n3a05-mal.pdf, accessed 9 April 2013).

Marks, L.A., Kalaitzandonakes, N. and Vickner, S. (2004) Consumer purchasing behavior toward GM foods in Europe. In: Evenson, R. and Santaniello, V. (eds) *Consumer Acceptance of Biotech Foods*. CAB International, Wallingford, UK, pp. 23–39.

Mather, D., Knight, J. and Holdsworth, D. (2005) Pricing differentials for organic, ordinary and genetically modified food. *Journal of Product and Brand Management* 14, 387–392.

Mauro, I.J. and McLachlan, S.M. (2008) Farmer knowledge and risk analysis: postrelease evaluation of herbicide-tolerant canola in western Canada. *Risk Analysis* 28, 463–476.

Scholderer, J. (2005) The GM foods debate in Europe: history, regulatory solutions, and consumer response research. *Journal of Public Affairs* 5, 1–12.

Scholderer, J. and Frewer, L.J. (2003) The biotechnology communication paradox: experimental evidence and the need for a new strategy. *Journal of Consumer Policy* 26, 125–157.

Scholderer, J. and Verbeke, W. (2012) *Genetically Modified Crop Production: Social Sciences, Agricultural Economics, and Costs and Benefits of Coexistence*. VDF, Zurich, Switzerland (http:// www.vdf.ethz.ch/service/3494/3495_genetically- modified-crop-production_oa.pdf, accessed 9 April 2013).

Siegrist, M. (1998) Belief in gene technology: the influence of environmental attitudes and gender. *Personality and Individual Differences* 24, 861–866.

Siegrist, M. (2000) The influence of trust and perceptions of risks and benefits on the acceptance of gene technology. *Risk Analysis* 20, 195–204.

Skevas, T., Kikulwe, E.M., Papadopoulou, H., Skevas, I. and Wesseler, J. (2012) Do European farmers reject genetically modified maize? Farmer preferences for genetically modified maize in Greece. *AgBioForum* 15, 242–256 (http://www.agbioforum.org/v15n3/v15n3a02- skevas.pdf, accessed 9 April 2013).

Søndergaard, H.A., Grunert, K.G. and Scholderer, J. (2007) Consumer attitudes towards novel enzyme technologies in food processing. In: Rastall, B. (ed.) *Novel Enzyme Technology for Food Applications*. Woodhead, Cambridge, UK, pp. 85–97.

Sparks, P., Shepherd, R. and Frewer, L.J. (1995) Assessing and structuring attitudes toward the use of gene technology in food production: the role of perceived ethical obligation. *Basic and Applied Social Psychology* 16, 267–285.

Index